No. 1650
$19.95

CMOS/TTL A USER'S GUIDE WITH PROJECTS

BY JOSEPH J. CARR

TAB BOOKS Inc.
BLUE RIDGE SUMMIT, PA. 17214

FIRST EDITION

FIRST PRINTING

Printed in the United States of America

Library of Congress Cataloging in Publication Data

Carr, Joseph J.
CMOS/TTL: a user's guide with projects.

Includes index.
1. Digital electronics—Amateurs' manuals.
2. Metal oxide semiconductors, Complementary—Amateurs' manuals. 3. Transistor-transistor logic circuits—Amateurs' manuals. I. Title.
TK9965.C32 1984 621.3819'5835 83-24349
ISBN 0-8306-0650-5
ISBN 0-8306-1650-0 (pbk.)

Contents

Other TAB books by the Author

901 *CET License Handbook—2nd Edition*
1070 *Digital Interfacing With An Analog World*
1182 *Complete Handbook of Radio Receivers (The)*
1194 *How To Troubleshoot & Repair Amateur Radio Equipment*
1224 *Complete Handbook of Radio Transmitters (The)*
1230 *Complete Handbook of Amplifiers, Oscillators and Multivibrators (The)*
1250 *Digital Electronics Troubleshooting*
1271 *Microcomputer Interfacing Handbook: A/D and D/A*
1273 *Amateur Radio Novice Class License Study Guide—3rd Edition*
1290 *IC Timer Handbook . . . with 100 projects & experiments*
1327 *Amateur Radio Advanced Class License Study Guide—3rd Edition*
1351 *Amateur Radio General Class License Study Guide—3rd Edition*
1396 *Microprocessor Interfacing*
1436 *104 Weekend Electronics Projects*
1482 *Commercial FCC License Handbook—3rd Edition*
1550 *Linear IC/OP Amp Handbook—2nd Edition*
1636 *The TAB Handbook of Radio Communications*
1643 *8-Bit and 16-Bit Microprocessor Cookbook*

Introduction

It wasn't many years ago that digital electronics was almost unknown to hobbyists and amateurs, and even many professionals. In fact, most people outside of aerospace, military, and the computer industries never heard of digital electronics—only some vague notions about "pulse circuits" and the like. Digital electronics either didn't exist for the individual, or it was an arcane mystery shared only by initiates in the above-mentioned corners of the electronics industry.

Although one would be way out of line to claim that the old-fashioned "analog" electronics industry is down and out (it is not!), the truth is that all electronics people, hobbyists, servicers, and amateurs must be at least aware of the basic tenets of digital electronics. It is no longer safe to merely ignore the world of digital electronics. Even consumer electronics technicians, who were insulated from "digital" for so many years, must become familiar with digital because they find digital electronics circuits in their work. Computerized color-TV tuning and stereo-FM tuning are fast becoming the rule rather than the exception; relatively low-cost units have these features that were once reserved for the top-of-the-line.

Many people correctly believe that digital electronics, at least in its basics, is simpler conceptually than most areas of analog electronics. The devices used in digital electronics recognize only two states, i.e., "on" and "off." This makes the digital device very similar (at least in concept) to simple on-off switches and electromechanical relays. Anyone who can understand relay and switch logic circuits can also understand digital electronic circuits; it's that simple.

It can probably be stated without fear of contradiction that digital electronics took a great leap forward with the introduction of integrated circuit technology in the early 1960s. Previously, digital circuits had been made from discrete transistors, diodes, and resistors. Everyone who was in the field in those early times will remember seeing piles of 2N404 transistors and 1N60 germanium diodes in digital equipment. A popular digital frequency counter, one of the first made with real digits rather than columns of lamps, used a pile of these transistors/diodes and some mighty strange display decoder devices. Since high-voltage *Nixie®* tubes were used, ordinary transistors could not safely be employed. The early frequency counters used a combination of photoresistors and NE-2 neon lamps to decode the binary coded decimal (BCD) data from the counters into 1-of-10 code for the *Nixie®* tubes.

The first IC digital devices were of the resistor-transistor-logic (RTL) family, and were followed shortly thereafter with the diode-transistor-logic (DTL) devices. Things really took off, with the invention of transistor-transistor-logic (TTL), and considerable improvements resulted when complementary-metal-oxide-silicon devices made their debut. Today, the RTL and DTL are obsolete for all new design, and everything is done in TTL and CMOS. There are reasons why each might be popular, so both are still needed in abundance. It is a fair bet that any modern digital electronic product will contain at least one of these devices, and may include many others.

In this book, we will study the different devices available in each family (but not an exhaustive list—that would fill several books), their application, precautions in their use, selection from the two families, and the advantages and disadvantages of each. Chapters 3 and 4 are a catalog of different members of the TTL and CMOS families, respectively, and are intended to present useful information in an easily understood form—somewhat easier than the data given in manufacturer's data books. The approach is prose in place of "truth tables."

Special appreciation is extended to Texas Instruments, who provided the TTL pinouts for Chapter 3, and to RCA Solid-State who provided the CMOS pinouts for Chapter 4.

Chapter 1

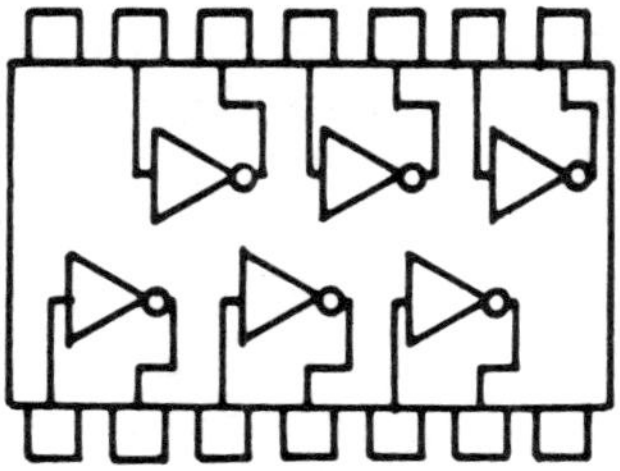

Digital Electronics

What makes a circuit "digital?" A digital circuit is one in which the devices represent quantities in discrete integer forms. In digital electronics, the specific form is the binary, or base-2, number system. Although the binary system is needed in some areas of digital electronics, we can do without it for the purposes of this book. If you are of a mind to learn the system, there are a number of books in the TAB BOOKS Inc. catalog that deal with this topic, including several written by me. You may obtain a free copy of the TAB BOOKS catalog by writing to: TAB BOOKS Inc., Blue Ridge Summit, PA 17214. In this book, we will content ourselves with discussing binary logic states.

Digital circuits, being binary in nature, can respond to only *two* different input states, and have only two different output states. These states are called "1" and "0" after the digits of the base-2 numbers system, "true" and "false" (in older texts), "logical-1" and "logical-0", or, "HIGH" and "LOW." In this book, we will tend to use the "HIGH/LOW" designations because that is really descriptive of what is happening in the circuit, and this is a practical, applications-oriented book. The "1-0" designation will be used in some cases, notably truth tables where the presentation makes graphic sense, and in the chapter on binary arithmetic circuits where the use is utterly appropriate.

Digital logic IC devices are grouped according to "families." A logic family is a series of IC devices that are designed to be interconnected with each other without concern for interfacing, and which use similar technology in their construction. TTL and CMOS are the logic families that we are considering in this book, although the other major families will at least be introduced in the next chapter.

It is required of any device in the family that its input and output properties be standardized so that it can be easily interconnected with other devices of the same series, without the need for external components to facilitate interfacing. For example, the TTL input is a *current source* that produces 1.8 milliamperes, while the TTL output is a *current sink* that will accept at least 1.8 milliamperes or some *integer multiple* of 1.8 milliamperes. The operating voltages, as well as input/output potentials, on the devices within the family are standardized. In TTL, for example, the dc potentials are ground and +5 volts dc (which must be regulated). In any given logic family, the input and output voltages and currents are fixed by convention (that is, agreement). The TTL family uses a standard method of determining how many TTL devices any given output will drive: *fan-in* and *fan-out*. These ratings are the only thing one needs to know about interfacing as long as only TTL devices are used in the circuit. In the next chapter, we will also discuss methods of interfacing nonsimilar logic families.

Fan-in is the load presented by one standard TTL input (i.e., 1.8 mA@5 V). Fan-in is dimensionless, and a fan-in of one (1) is as stated above, i.e., one standard TTL input. Thus, the fan-in represents the load that must be driven by the TTL outputs.

Fan-out refers to the drive capability of the TTL device, and is a dimensionless integer (i.e., 1, 2, 3 . . .) that refers to the number of standard TTL inputs that the device will successfully drive. Thus, a fan-out of ten will successfully drive up to ten standard TTL inputs. For the most part, unless otherwise stated all TTL outputs have a fan-out of ten. Some devices are available, however, which have a fan-out of 30, 50, or even 100. These devices are often used to drive microcomputer busses and in other situations where a given output line will have to drive a large number of loads.

POSITIVE VERSUS NEGATIVE LOGIC

Positive and negative logic is a concept that tends to confuse the newcomer to digital electronics. This situation is needlessly exacerbated by some textbook writers who fail to let the student know that the terms refer only to the designation of voltage levels as "1" and "0" (i.e., HIGH and LOW). The situation is shown using TTL terminology in Fig. 1-1. In Fig. 1-1A we see the *positive logic*

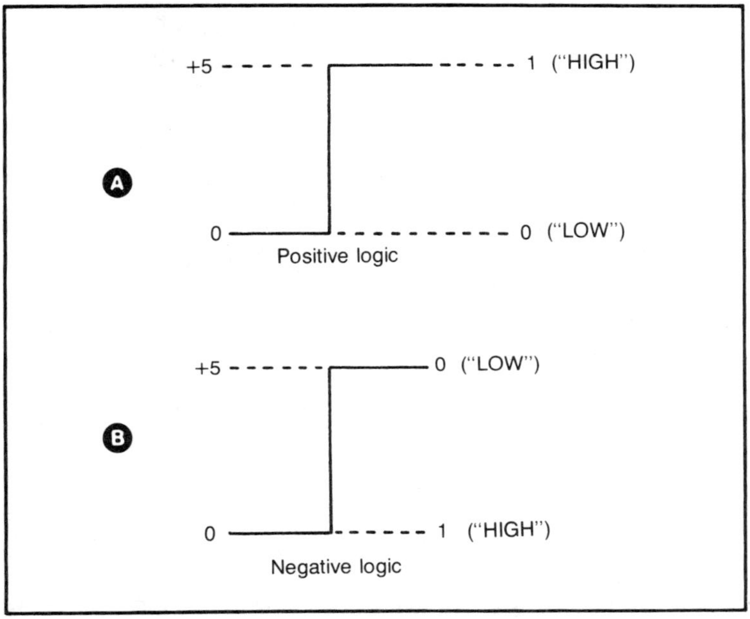

Fig. 1-1. (A) Positive-logic levels, (B) negative-logic levels.

case. Here, the HIGH or "1" state is represented by the positive voltage (nominally +5 volts), while the LOW or "0" state is represented by 0-volts. Positive logic is the most common used, and all device names reflect the positive logic case unless otherwise specified.

Negative logic is exactly the opposite of positive logic (isn't that convenient!). In the *negative logic* case, the positive voltage represents the binary digital zero (0), or LOW condition. The zero voltage represents HIGH or "1."

The problem of positive versus negative logic sometimes pops up in the matter of device names. There are certain devices which will take on different names depending upon which is intended, positive or negative. AND gates and OR gates, for example, will sometimes be listed in archaic texts as AND/OR or OR/AND, which merely reflects the fact that a positive logic AND gate is the same circuit as a negative logic OR gate; hence, AND/OR means "positive AND or negative OR." Fortunately, this method of denoting the devices seems to be going the way of the five-cent ice cream cone and all device names used in this text will reflect only positive logic names unless I tell you otherwise.

Note that CMOS devices might use some V+ voltage for

HIGH, and either 0-volts or some negative (V−) voltage for LOW. In those devices, positive logic equates (V+) = HIGH = 1, and (V−) = LOW = 0. There are, however, a *few* negative logic devices in the CMOS product line.

LOGIC LEVELS

Transistor-transistor-logic (TTL) works from a +5 volt dc power supply that must be regulated. The range of nominal dc power supply potentials is +4.5 to +5.2 volts, although one must be aware that potentials less than +4.75 volts dc are often the cause of erratic operation in some complex function devices. Although manufacturers tend to list the full range in some specification sheets, the user should be cautioned to maintain the higher potential for devices such as flip-flops and monostable multivibrators.

The logic levels used by TTL (or any family) reflect the dc operating potentials. These potentials must be maintained, or else problems result—the device will either (A) fail to work at all, (B) will work unpredictably, or (C) burn-out.

Figure 1-2 shows the input logic levels for TTL devices. The LOW zone is nominally zero volts, but may be anything within 0 to 0.8 volts (800 millivolts) and still be LOW. Any TTL device must be able to respond to any potential within this range as if it were officially a LOW condition.

The HIGH state will be recognized by any potential between +2.4 volts and +5.2 volts. Any device that purports to be TTL should be able to recognize potentials within these limits as an official HIGH condition.

An undefined zone exists between +0.8 volts (the upper end of the LOW zone), and +2.4 volts (the lower limit of the HIGH zone). If you apply potentials within this range to a TTL input the device will not know what to do with it, and will therefore operate in an unpredictable manner. This ambiguous situation must be avoided at all costs. The problem usually surfaces when one tries to drive too many TTL inputs for the rated fan-out of a device, or when there is an insufficient path to ground for a TTL device so that its output drops to some potential higher than 0.8 volts when it is supposedly LOW.

The smoke zone exists in the negative region. If a negative potential is applied to the TTL input, the device may very well die a rapid and untimely death especially if the source is not current-limited. The groan zone might also be a smoke zone, but not nearly as well defined as the negative potentials regions. If voltages greater than +5.2 volts are applied, the TTL device will overheat, and may burn up. One doesn't have to get much above +5.2 volts before the groan zone also becomes a smoke zone . . . so be careful In fact, you will find that the TTL reliability is poor when the voltages are over +5.0 volts. I recommend that one attempt to keep the levels between +4.75 volts and +5.0 volts for logic levels, and +4.9 volts and 5.05 volts for power supplies.

CMOS devices are able to use a much larger range of operating potentials. The CMOS device will operate from ± 4.5 volts to ± 15 volts (with some selected devices going to ± 18 volts dc). Furthermore, the two voltages, VDD and VSS, need not be equal. It is common practice to use a positive voltage and negative voltage, but this need not be always true. We could, for example, use a negative voltage and ground, or a positive voltage and ground. In fact, many CMOS devices (especially when used in conjunction with TTL devices) operate from TTL voltage levels, namely +5 volts dc and ground. Another alternative is to use unequal positive and negative voltages, e.g., +12 volts and −7 volts dc.

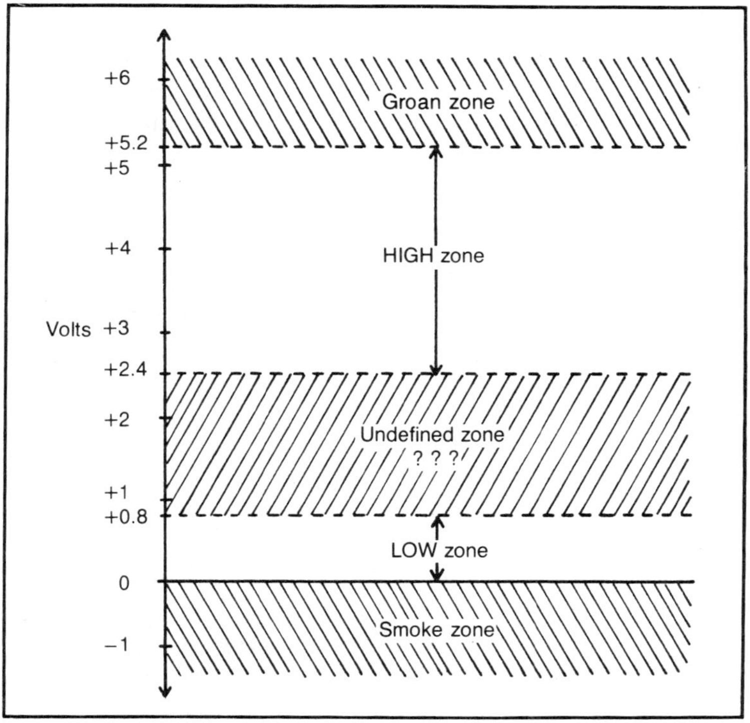

Fig. 1-2. TTL logic levels.

In TTL logic devices, the output transition occurs when the input goes above +2.4 volts, or below +0.8 volts. In CMOS devices, the output transition occurs when the input signal voltage is midway between the two power supply voltages, i.e., at a point equal to ((V+) + (V−)/2). For example, suppose we use ± 12 volts dc on the power supplies. The output transition occurs when the input signal voltage is ((+12) + (−12)) /2 = (+12 − 12)/2 = 0 volts. Similarly, when the CMOS device is operated from TTL supply levels: ((+5) + (0)) /2 = (+5)/2 = +2.5 volts.

Some CMOS devices, (like the TTL devices, it is the complex function devices that are affected) will fail to operate when the voltages are less than ± 7 volts dc. This problem is seen in some cases where the device is operated from TTL power supply potentials, and the symptom is erratic or unpredictable operation.

SPEED VERSUS POWER

The principle factors governing the operating speed (i.e., maximum operating frequency) of any digital IC device are the internal resistances and capacitances. If resistances are increased in an attempt to reduce power consumption, then the RC time constant increases. The longer time constant thus created makes the maximum operating speed slower. As a general rule, therefore, higher operating speed automatically means greater power dissipation. TTL devices will usually operate to 18 or 20 MHz with some selected types sprinting along as high as 50 or 80 MHz. These devices also require a substantial amount of current, typically 20 to 25 milliamperes per IC package. A circuit with 30 IC devices, therefore, could easily require (30 × 25 mA) which is 750 milliamperes. As a result, the TTL user is advised to have a bench power supply of at least 1 ampere (i.e., 1000 milliamperes), and more if possible. CMOS devices generally operate at slower maximum speeds than typical TTL devices: 4 to 5 MHz is common, with some going to 10 MHz. CMOS devices, however, are very low current devices.

CMOS devices typically draw microamperes of current under static conditions, and will draw current at a specified rate depending upon toggle frequency when operated in dynamic situations. The amount of current depends upon the supply voltages and the operating speed. At a supply voltage of 10 volts, for example, the CMOS device will draw a little less than 1 μA/kHz.

DESIGNING AND PROTOTYPING CIRCUITS

One of the most reliable signs of the neophyte designer is the habit of jumping into a project feet-first without the least little bit of planning! *That* is a very expensive and time-consuming practice, and should be avoided where possible . . . and it is avoidable!

The design of an electronic project is like the design of an addition to a home: it is a logical process that must be completed *before* you start nailing sticks together. In fact, if the home construction analogy holds true, we must remember that most jurisdictions require the plans to be approved before you begin work! Of course, at your workbench you are the approval authority, so will have to discipline yourself in this matter. A few hours spent at the desk pays rich dividends in dozens of hours saved on the workbench!

The very word *design* connotes intellectual activities such as planning, thought, intent, and it is a goal-directed process. Anyone with a little knowledge, the ability to research books and articles, and who takes a little time to plan activities, will be able to design and build hobbyist and amateur radio projects of substantial complexity. Although there are many projects that must be left to the design engineer, you need not be intimidated. To quote the author of one of the better op-amp books: ". . . the contriving of contrivances is a game for all."[1]

The design of a complex electronic project may require a lot of experience (which tells you what works and what doesn't), some knowledge (as found in this book), and no small amount of cleverness (which only *you* can provide). You will have to learn good design and laboratory practices in order to be efficient. You will also have to develop the habit of consulting data books and spec sheets in order to learn the maximum ratings of devices—or you will likely turn a lot of electronic devices into burnt-out piles of carbon (Hey! A physics first—a silicon to carbon converter!). If you follow the usual course, it is likely that your error will be discovered approximately 1.2 milliseconds before the power supply converts your wonderful pile of electronics into a pile of rubble.

Don't worry a lot if your earliest efforts seem a little futile—they are, after all, not totally wasted if you are the type of person who learns from past mistakes (that's called *negative feedback* in electronical terminology!). If you are the impatient type who des-

---

1. Smith, J.: *Operational Circuit Design* (John Wiley and Sons, New York)

pairs when not actively engaged in soldering and wiring thingees and suchees, then *stifle yourself!* The impatience to get on with the process of producing great hardware should be suppressed with "great vigah," as it will lead to disaster if insufficient preplanning is the case. The manual part of producing a project from scratch is the least of the effort, and is the last step in the process. If you are under pressure to produce something, for example in your job, then the temptation may be even greater.

Despite primordial urges to the contrary, tell anyone who will hurry you in the process to go suck a lollipop—or something. There is a little ploy that one can use to put off such turkeys; it has worked especially well in medical school and other university situations where the "customers" usually don't have the slightest idea what it is that engineers and technicians do for a living. Get some circuitry on a breadboard the first day—even the first hour if possible. Sprinkle the circuit with blinking LEDs, a speaker that goes *"BZZZZZZ"* or *"BEEP-BEEP"* or something; seven-segment readouts that are constantly counting 0-1-2-3-4-5-6-7-8-9-0-1-2 . . . are also a good touch. About every second day or so make some obvious *change* to the lash-up. For example, add an oscilloscope to the test instruments connected to the pile of rubbish (unstable Lissajous patterns are a neat idea, and they can be varied from time to time to impress the customer who, if the normal course is followed, will make a morning and afternoon inspection to see how things are coming along—all unofficially of course, and to the tune of "I don't mean to rush you." Always place a sign over the breadboard that reads: EXPERIMENT IN PROGRESS—DO NOT DISTURB. If you are in a scientific facility (as the medical school mentioned above) that sign will engender either a respect approaching worship, or will be unceremoniously ignored! The main idea behind the above deception is to buy you time to go off somewhere with a pile of books and paper and leisurely solve the design problem the correct way—something that some "customers" will never understand.

The First Step

Oddly enough, the first step in the design process is often completely overlooked by the novice designer: *define the problem*! Novice circuit designers are like the beginning physic student who habitually solved a more difficult problem than the professor assigned—mostly from ignorance than from a desire to excel. Although this advice may seem like either a case of overworked cynicism or perplexed callousness, it is actually based on observation. A lot of people start "designing" before they fully understand what the device is supposed to do! Defining the problem might involve all or some of the following: literature search, thinking, interviewing the eventual user, and talking with others who may have solved similar problems (and solved them successfully!).

Be sure that the problem remains the same from one end of the project to the other. It is very often the case that the device will take on additional capability as the eventual user thinks over his or her position. These accretions can sometimes be accommodated, but often times it is difficult or impossible. We can sometimes smuggle new capability into a design, but it is many times the case that the particular design approach selected early in the game will not accommodate new capability, and that you would have selected another way had you known what else the customer wanted. Of course, you may be able to anticipate other uses and make the design above to conform to new expectations. One sure way to keep the problems down is to avoid the customer until a design prototype is working. Then you will be in a position to say "Gee, Harry, if you had told me you wanted such-and-such last month when we started this project it would be easy and cheap to incorporate this new idea. But now it is expensive and time-consuming. Sorry." I once worked for a guy who came to the medical center from an aerospace contractor, and he claimed their automatic response to new NASA demands was the "two-n-two" formula: i.e., two months delay and two million dollars cost.

Do not underestimate the value of a literature search. Far too many designers—and that includes professionals in industry—seem to have a disdain for literature searches. Perhaps it is the well-known disdain for scholarship that graduate engineers are sometimes infected with, or perhaps it derives from N.I.H. (not invented here) syndrome. Do not be too proud to use perfectly good solutions from "prior art." Remember, please, that it is okay to admit that someone else once had a good idea, and that your job is to solve the problem. It is thoroughly unimportant how many clever state-of-the-art designs you turn out, nor is it terribly important how original are your ideas. The most elegant design is the simplest that will do the job properly.

The next step in the process is to formulate at least one approach to solving the problem. Try and figure out several approaches to the problem, and then select what seems like the best solution and develop it further. The word "contingencies" is heard a lot in well-managed design laboratories. Some wise souls will tell you that your first approach to solving a problem is usually the best (i.e., first impressions and all that rot): that's hogwash! In most cases, the first approach to solving the problem is usually among the worst that you will discover. Since this wisdom is true most of the time, it is wise to develop the habit of thinking about contingencies first.

One good method is to lay out a solution in block diagram form. Label the blocks with the circuit functions that it will perform—but don't fill in the blocks with specific circuits yet! Once you have solved the problem, then select the technology or logic family that will be needed. Obviously, if speed and +5-volt operation is paramount, then one would select TTL devices. If, on the other hand, portable operation and low current drain are absolutely essential, then use CMOS. The trade-offs are usually fairly straightforward, and will not present any great problem.

Now you may start drawing circuitry. One last thing to do before actually attempting to build the prototype of the device is to make a schematic drawing of the *complete electronic circuit,* and any special, critical mechanical components that will have to be manufactured. Keep the original inviolate, and actually work from copies made on a photocopier. Make all proposed and actual changes on the copy, and then transfer them to the master if it proves out. If you are planning to build a project from a textbook or magazine article, then copy it *in-toto* (by hand or on a copier) before starting.

The person who makes an electronic instrument for someone else but fails to properly document it is a contemptible, demonic, slug-like creature of wholly regrettable parentage! The most ill-mannered, contemptible thing you can do is not document your work, thereby making it difficult for whoever follows you on the project either to modify or repair the darn thing! Document as you work. You can obtain either an engineer's computation notebook, or a scientist's laboratory notebook at engineering/drafting supplies stores and most college bookstores. Once the project is completed, go back to the lab notebook and make a final drawing of the correct circuit. Make your work conform to a certain military specification: MIL-TD-41 (*M*ake *i*t *l*ike *t*he *d*rawing *f*or *o*nce!).

The main use of the lab notebook full of drawings is to help in the development of the final prototype of the instrument. Keep detailed records of key voltages and waveforms, signals and other measurements. Change the master drawing to reflect all changes in the circuitry. The lab notebook is viewed as critical in commercial laboratories because it will help establish a chain of evidence that will prove the company's claim to a patent (if the device is patentable), or, will help in reducing liability if one accidentally infringes another's patent but can prove independent development (you will still have to pay, however, but perhaps not as much as if you simply stole the idea!).

Also included in the final documentation for the project is alignment, calibration, and adjustment procedures (as appropriate). This step should be included even for projects where *you* are the only person who will ever work on the circuit . . . memory does grow cold, you know. The calibration or alignment might be self-evident when the circuit is still hot in your grubby little hands, but will be an arcane mystery to others . . . or to you a year from now!

Breadboarding

Another sign of the inept or novice designer is the habit of committing even relatively complex circuit designs to final form without first breadboarding the circuit. Every new design is not "finished" until it has been tested properly and found not lacking. Every idea that you conceive must be considered tentative and hypothetical until it has been proven valid. It often appears in our minds that an idea will work, but when the circuit is connected there is a nasty surprise waiting for us. This is why laboratory breadboards are fast-selling items in the professional engineering world. The kinds of problems that often show up include fallacious concepts (ideas that are scientifically wrong), and failure to recognize the effects of things like layout, lead-length, power-supply bypassing and the like. In digital circuits, there might also be incorrect logic. The *proper* way to proceed is to check the circuit out on a breadboard before setting it into concrete.

There are a number of useful breadboards (see Fig. 1-3) on the market. Several companies sell socket panels that can be made into a homebrew breadboard, or, you can use a breadboard such as the Heathkit or the AP Products, Inc. *Powerace 102*. This breadboard is designed especially for TTL digital projects (or CMOS operated from TTL voltages). It has a 5-volt dc (regulated) 1-ampere power supply, and several design aids in addition to the socket panels.

The final project will be committed to construction on a wire-wrap board, solder board, or a printed-circuit board. If you take pride in your work, it is likely that you will want to house your project in as nice a cabinet as possible. But cabinets are expensive,

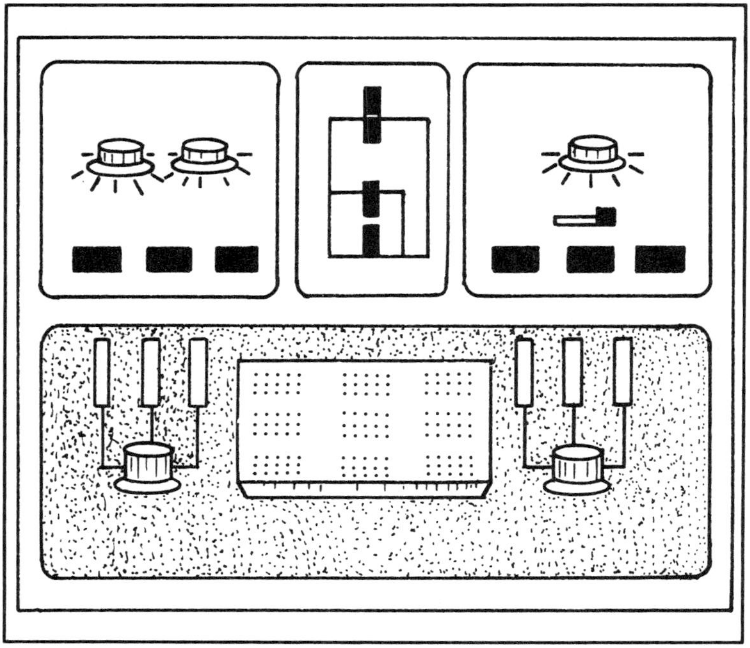

Fig. 1-3. Typical breadboarding system.

especially those that are a pleasure to use; so please, please, please breadboard the project first, prior to building it into a cabinet that is possibly inappropriate. You may find out the hard way that the cabinet is not useful, or, that you have erroneously drilled too many holes in that expensive front panel. Similarly, you may have selected a $40 cabinet that is appropriate for 4.5 × 6 inch *Vectorboard*® only to find out that the finished project will require 4.5 × 9 inch material that will not fit inside! The solution to *that* problem might be: (A) throw away all of the effort used to produce the incorrect board. (B) use an add-on "kluge board," or (C) cry a lot.

Before proceeding further let's review the steps necessary to produce a viable electronic project:

1. Study the problem that must be solved until you understand it completely.

2. Formulate several possible approaches to solving the problem.

3. Make block diagram drawings illustrating the best approaches, or those that seem most promising.

4. Make drawings, keep records, and document the process *as you work*.

5. Prototype the design on a laboratory breadboard such as the AP Products *Powerace 102* and correct any deficiencies found.

6. Test the circuit under realistic conditions and correct any problems.

7. Apply your best craftsmanship to making the final version of the project.

Wireboards

Most amateur and hobbyist builders of electronic circuits are going to use a wireboard instead of the more difficult to make printed circuit board. For many, the best solution will be to wire-wrap the connections. Figure 1-4 shows a typical wirewrap socket and the associated *Vector*® wirewrap tool. This process is simple to use, and is most reasonably priced for amateurs (other wirewrap instruments, intended for industrial use, are costly yet do not yield results that appear substantially superior to the results of this tool).

Figure 1-5 shows a *Vectorboard*® product that is designed for amateur and hobbyist use, and is relatively low cost. A pattern suitable for digital projects is printed on one side, while the other is blank. Other *Vector*® products are available with and without card-edge connector, and with or without foil patterns (also, various patterns are available). This board can be either wirewrapped or soldered, or both.

DESIGN AND TROUBLESHOOTING AIDS

There are several devices that are useful when trying to

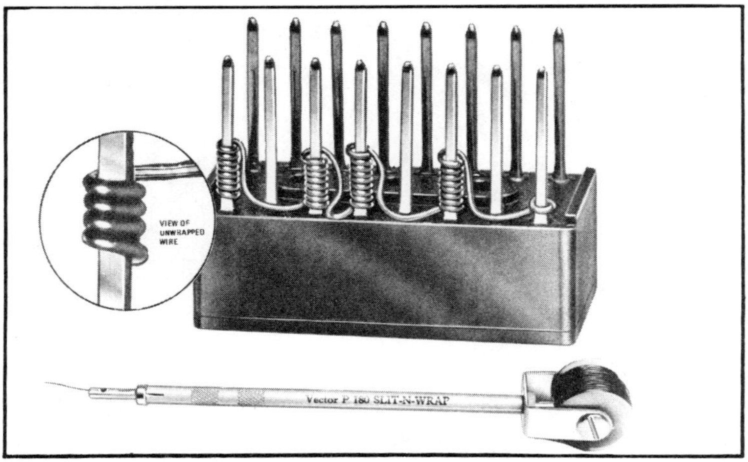

Fig. 1-4. Vector P180 slit-n-wrap tool.

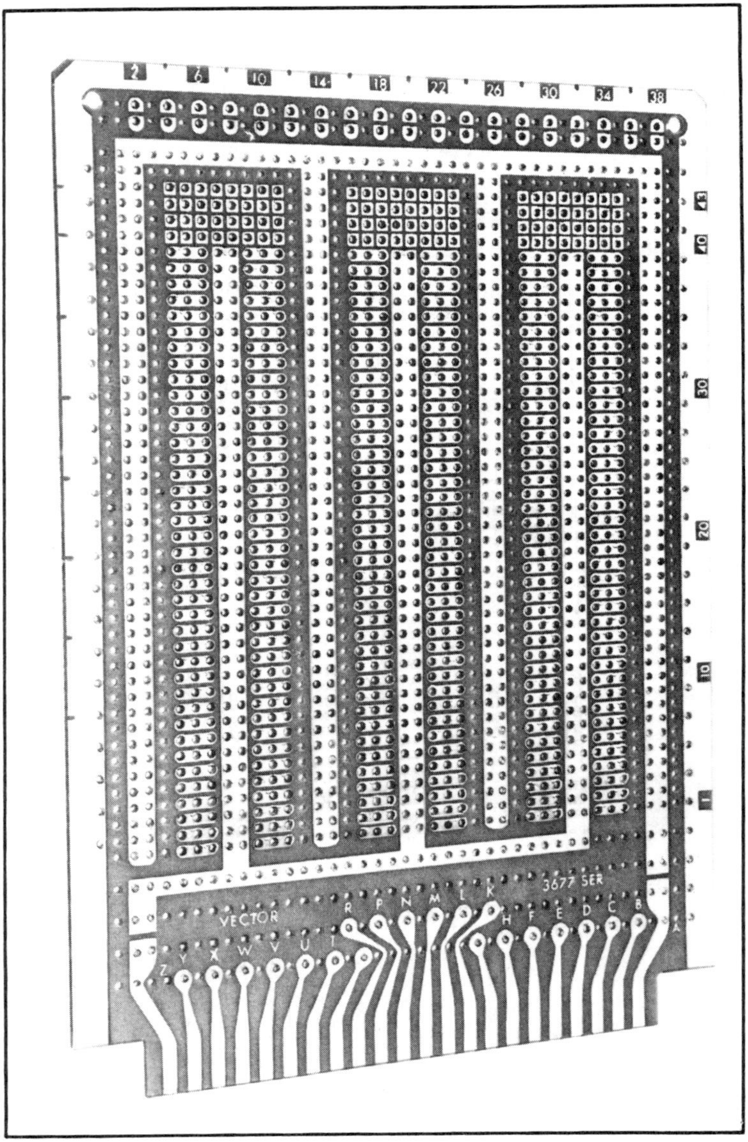

Fig. 1-5. Vector Electronics DIP board.

troubleshoot or debug digital circuits. A power supply monitor, for example, will allow you to find out whether or not the supply is working. Some equipment manufacturers will even include these circuits on each and every printed circuit board in the instrument so that you can tell at a glance whether or not the power supply is working. The power-supply monitor consists of a *light-emitting diode* (LED) connected in series with a resistor that will limit the current to 10 to 15 milliamperes. For TTL circuits, the resistor value will be 330 ohms, ½-watt. The anode end of the LED is connected to the +5 volt dc line, while the cathode end is connected to the ground terminal; it doesn't matter whether the resistor is in the cathode or anode circuits.

Logic-Level Probe. Another useful device is a logic level probe, which will use a circuit similar to those in Fig. 1-6, and may be built inside of a pen housing or similar probe. A CMOS version is shown in Fig. 1-6A (but will also work on TTL devices). In this

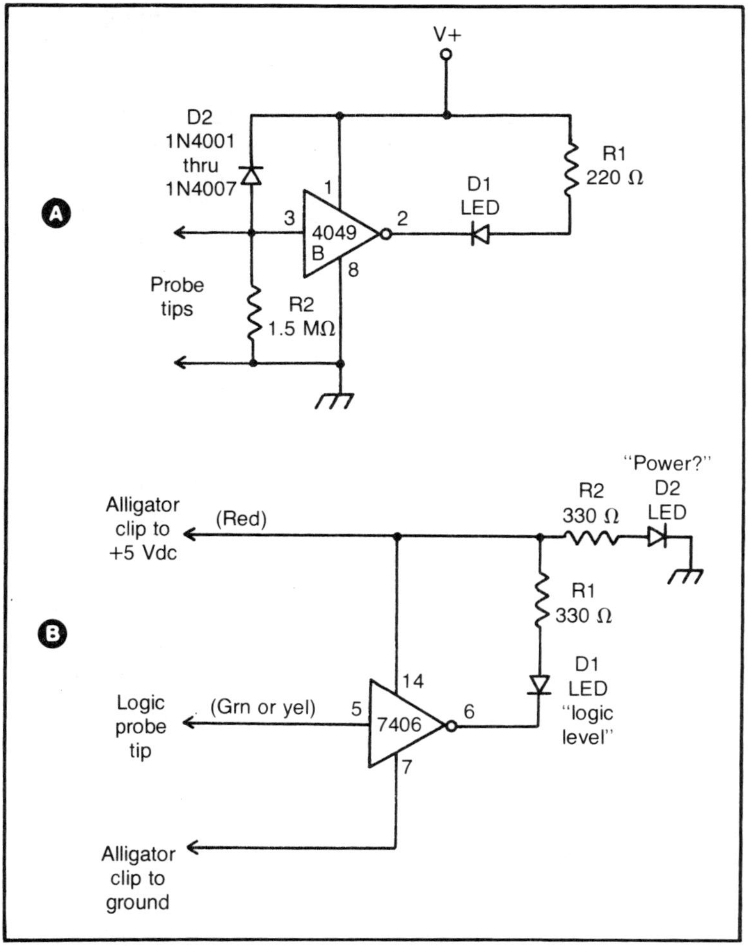

Fig. 1-6. (A) Logic probe (CMOS), (B) logic probe (TTL).

circuit, a light-emitting diode (LED) is connected such that it will turn on any time the output of the 4049B CMOS inverter is LOW. Therefore, when the input is HIGH, then the LED will turn on; when the input is LOW, then the LED is off.

A TTL version is shown in Fig. 1-6B. This logic level probe circuit uses a 7406 open-collector TTL inverter. Again, the LED is wired so that it will turn on when the output of the inverter is LOW, so indicates the presence of a HIGH condition on the input probe line.

Most published circuits such as Fig. 1-6B leave the user with an ambiguity that is difficult to live with in real troublshooting situations: how do you distinguish between an input-LOW condition and the situation where there is no dc power applied to the circuit? We solve this problem nicely in Fig. 1-6B by including a "Power?" monitor, LED D2. This diode will produce light whenever +5 volts dc is applied to the power line. Since this power line is an alligator clip, we will take power from the dc power supply of the equipment being tested, therefore making the test of power-on valid.

There are so-called logic clips available that will have fourteen- or sixteen-pin spring-operated clips that can be placed over a DIP IC package. The device contains several circuits such as this (up to 32 are known) and will tell the user the logic level on each pin, including +5-volts dc and ground. Such devices are immensely useful for logic troubleshooting, and are relatively low cost.

Logic-Level Generators. A circuit must sometimes be provided that will produce a logic level to the digital circuit in order to test it. A static level generator is shown in Fig. 1-7. This circuit

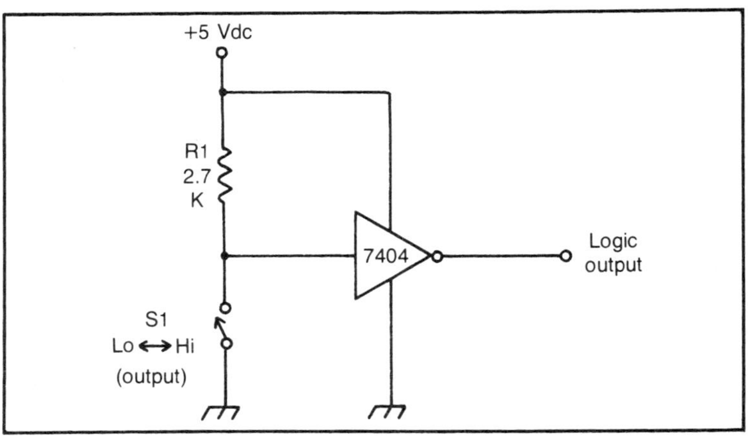

Fig. 1-7. Level generator.

uses a single inverter stage and a switch that determines whether the output level will be HIGH or LOW. A pull-up resistor (R1) is used to keep the input of the 7404 inverter HIGH any time that switch S1 is open. This condition corresponds to an output condition of LOW. When the switch is closed, the input of the 7404 is LOW, so the output will be HIGH.

A problem with the above circuit is that it is static, i.e., that the output will always be at the level that is set by switch S1. We sometimes want the line to remain at some level all of the time except when commanded to go to the alternate state by the operator.

The need is met by using a monostable multivibrator (more about them later). This circuit has the property of having only one

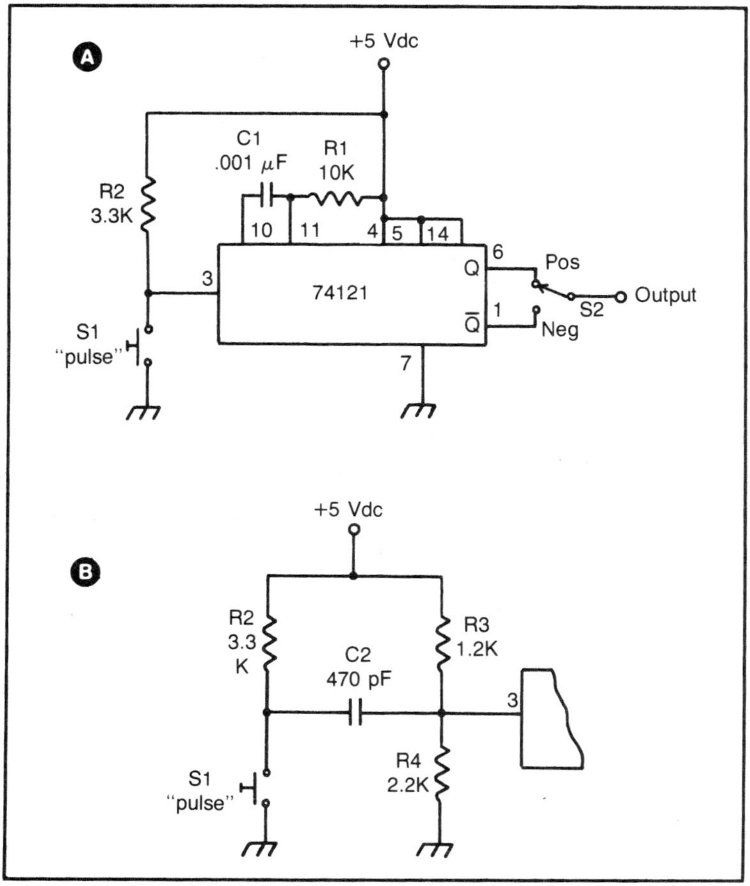

Fig. 1-8. (A) 1-μS pulser, (B) alternate circuit for input.

stable state. When the device is triggered, the output will go to its unstable state for a predetermined period of time, and then reverts back to the stable state. The circuit of Fig. 1-8A will produce a single output pulse every time the switch (S1) is pressed. The period of the pulse is set by resistor R1 and capacitor C1. The period is set approximately 0.7R1C1, so this circuit will produce an output pulse of approximately 0.7 microseconds. Longer or shorter duration pulses can be accommodated by changing the values of either R1, C1 or both. An alternative input triggering circuit is shown in Fig. 1-8B, and might be needed in cases where the output pulse is very short.

Pulse catcher circuits are shown in Figs. 1-9 and 1-10. These circuits are designed to remain dormant until a pulse is received on the input, and then produce a lighted LED for an output to let the operator know that a pulse has come along in the circuit.

The first version, Fig. 1-9, uses a pair of 7400 two-input NAND gates to form an R-S flip-flop. When switch S1 is momentarily closed, the output goes to the condition of Q being LOW and $\overline{Q}$ (i.e., "not-Q") being HIGH. This makes LED D1 glow, and D2 off. Light-emitting diode D1, then, indicates the *reset* state. When a pulse is received that brings the pin no. 4 input of the 7400 LOW momentarily, the output states change such that Q is HIGH and $\overline{Q}$ is

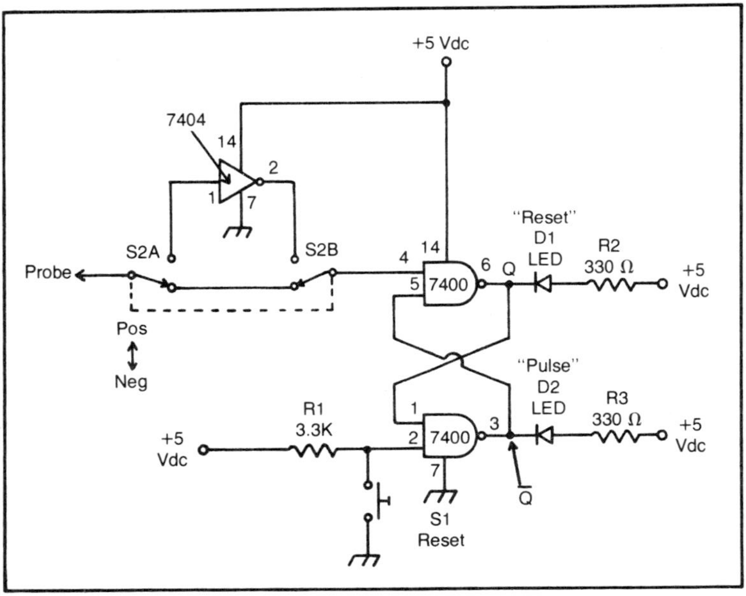

Fig. 1-9. Pulse catcher.

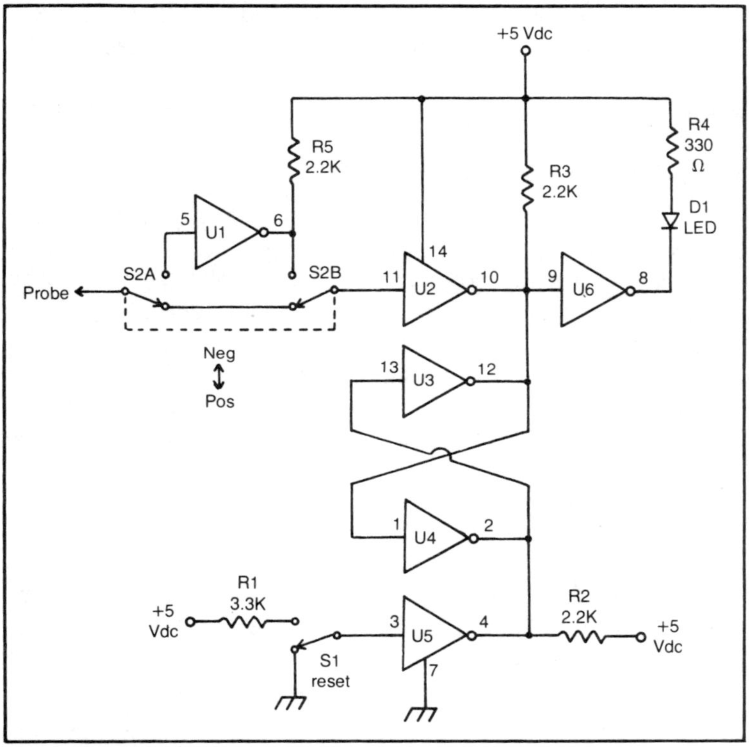

Fig. 1-10. Pulse catcher.

LOW; this condition turns off D1 and turns on D2, indicating a pulse was received.

We can accommodate either positive-going or negative-going pulses by including an inverter and switch in the input circuit of 7400. When switch S2 is in the down position, the probe is connected directly to the input of the 7400, so only a negative-going pulse will turn on the pulse LED. But, when the switch is in the up position, the input pulse will be inverted before being applied to the 7400 input. This means that the pulse LED will respond only to positive-going pulses.

The inverter can be any of the TTL inverters (see Chapter 3), or, it may be made from another 7400 two-input NAND gate. The 7400 device is actually a *quad* two-input NAND gate array, so it has two additional NAND gate sections besides those used in this circuit. We can use one of them as an inverter. A TTL inverter can be made from a two-input NAND gate by shorting the two inputs together to form a common input.

The version shown in Fig. 1-10 is based on a NOR-logic R-S flip-flop made from individual sections of an open-collector TTL hex-inverter circuit. It is clumsy, and second-rate compared with Fig. 1-9, but it is useful if you have only an inverter chip available.

DIGITAL CIRCUIT PROBLEMS

Many of the problems people have with TTL (especially) and CMOS circuits are of their own making. Probably the biggest problem is the matter of correct power-supply bypassing. The leading and trailing edges of certain TTL pulses are very sharp, meaning that they have frequency components into the megahertz (or even VHF) region, and clocking speeds may be up to 20 MHz. As a result, we have to take into account the stray capacitances and inductances of the power-supply lines. Since the devices also draw large currents, we must be able to reduce series resistances in the power-supply conductors.

Both the resistances and inductances tend to decrease markedly when the conductors are made of broad, flat conductors, as on a printed circuit board. As a result, it is important to not use ordinary hook-up wire or wirewrap wire except in relatively low-frequency, low-risetime circuits.

A solution to some of these problems is to use adequate power-supply bypassing capacitors. In general, the rule is place one 0.001 microfarad at each TTL package, and make it as close as possible to the package body. Also, place a capacitor of 1 to 100 microfarads (depending upon current load, the higher the current the more capacitance) at the point where the power-supply lines connect to the wiring board (C1 in Fig. 1-11). Also, some designers like to place another 1 to 100 μF unit either mid-way along the board or at the termination end of the line.

The purpose of these capacitors is to serve as local reservoirs

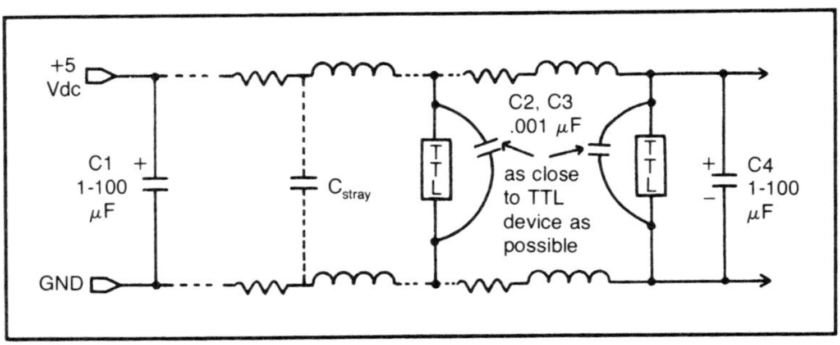

Fig. 1-11. TTL bypassing.

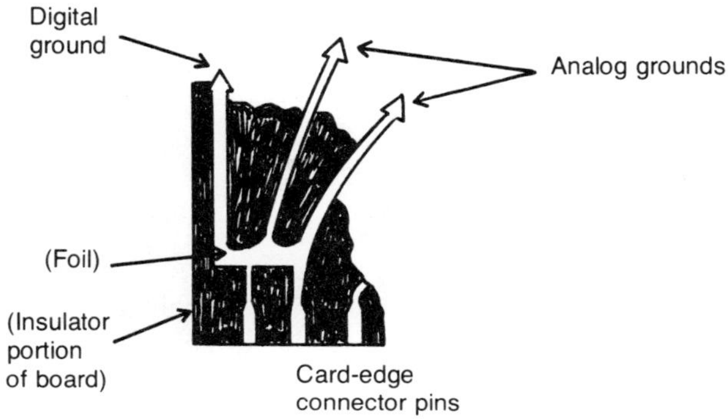

Fig. 1-12. Grounding on printed wiring board with mixed circuitry.

of current that can dump their charge into the circuit when the power-supply voltage drops out briefly. This drop-out occurs because of the RLC time constant of the power-supply lines. When an abrupt current demand is present, the supply may take a few nanoseconds to catch up with the demand. During this time, the chip can draw charge from the nearest capacitor.

The high current levels, and abrupt changes, present in TTL digital circuits makes grounding of considerable importance, especially where analog, rf or video circuits are also present on the same printed-circuit board. Figure 1-12 shows the proper method for grounding these circuits. Run the wiring lands for the digital and various types of analog circuits separately, and connect them together only at one point, preferably at the point where the power-supply ground enters the board (shown here at the card-edge connector). This tactic will greatly reduce the interference that can be caused between circuits when common grounds are run all over the board, or where the board is grounded at multiple points.

Chapter 2

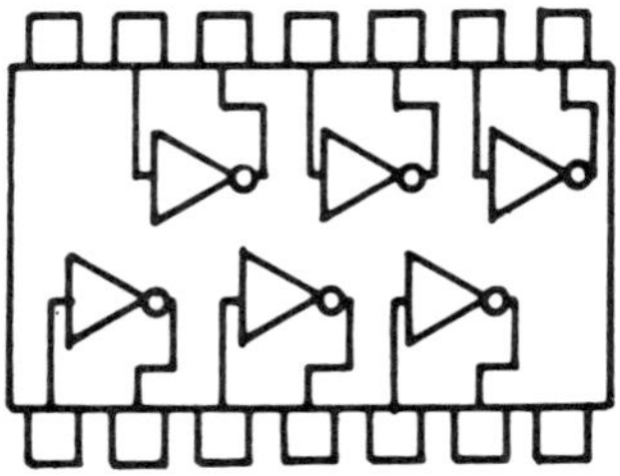

IC Logic Families

An IC logic family is a set of digital logic devices that can be connected together to form circuits without regard for interfacing problems such as voltage-level and impedance matching. The input and output voltages and currents are standardized in any given logic family, so one can interface them almost with impunity. For the most part, the different elements within any given family use the same fabrication technology: e.g., bipolar transistors for TTL and MOSFET transistors for CMOS. There are a number of different IC logic families still extant, but only TTL and CMOS are currently very popular. The obsolete families are: RTL, DTL, HTL (also called HNIL), and ECL. These families will be described briefly, but the main thrust of this book is TTL and CMOS devices.

RTL (RESISTOR-TRANSISTOR-LOGIC)

The resistor-transistor-logic (RTL) is obsolete, but was quite popular in the early 1960s. It was probably the first commercially available IC logic family. Figure 2-1 shows a typical RTL inverter circuit. An inverter is a digital circuit that produces an output opposite the input. In other words, a HIGH input produces a LOW output, and a LOW input produces a HIGH output.

RTL logic devices operate from +3.6 volt dc power supplies, so the two logic levels are 0-volts for LOW and +3.6-volts for HIGH. In Fig. 2-1, if the input is made LOW, then transistor Q1 is turned-off and no current flows in resistor R2—the output is therefore equal to the power supply voltage, or +3.6 Volts. Hence the

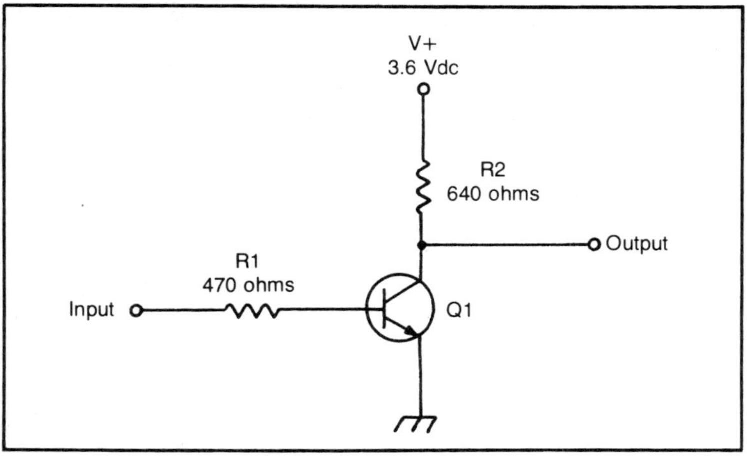

Fig. 2-1. RTL Inverter.

output is HIGH. Similarly, when the input is HIGH (i.e., made +3.6 volts dc), transistor Q1 is saturated, so the collector terminal (i.e., output) will be at ground potential (or very nearly so). Hence, a HIGH input will make a LOW output.

RTL devices usually carry type numbers in the uL900 range (mostly in 8- and 10-pin metal cans) and MC700 range (Motorola devices in 14-pin DIPs). It is interesting to note that, provided that no potential exists on the printed circuit board greater than +4 volts dc, no combination of opens or shorts will produce a blown RTL chip! This same cannot be said for certain other devices, such as TTL which very easily blow out if mishandled.

DTL (DIODE TRANSISTOR LOGIC) DEVICES

The next popular IC digital logic family to become commercially available was the diode transistor logic (DTL) line of devices. These ICs operated at speeds greater than RTL devices, and used +5 to +6 volt dc power supplies. In some older equipment, you may find DTL and TTL devices mixed, although this was regarded as generally poor practice.

Figure 2-2 shows a typical DTL inverter circuit; two NPN transistors are used in cascade with Q1 operating as an emitter-follower feeding the base of Q2. When the DTL input is HIGH, then diode D1 is reverse biased. In that condition, R1 will forward bias transistor Q1, which in turn forward biases D2 and Q2. Voltage levels in most digital circuits are selected to saturate the transistors, so when Q2 is turned on, it is turned on to full saturation. This condition means that the output of the inverter, which is the collector of Q2, goes nearly to ground, but the $V_{ce(sat)}$ voltage of the transistor—on the order of a few tenths of a volt, at most.

When the input is LOW, the cathode of D1 is grounded. Since D1 is now forward biased, the base of Q1 is essentially grounded. Under this condition Q1, D1, and Q2 are reverse biased. With Q2 cut off, then, the output voltage rises to $V_{CC(+)}$. Most DTL devices carried part numbers in the MC800 and MC900 ranges (Motorola designation).

TTL DEVICES

Probably the most widely used digital IC logic family is the transistor-transistor-logic (TTL) family. When most people speak of "digital ICs", it is the TTL family of devices to which they refer. Most TTL devices carry type numbers in the 7400 range. Those devices with 5400-series designations are military equivalents to the 7400-series device (i.e., a 5447 is a 7447 in uniform). The principal difference between the 5400 and 7400 devices in the operating temperature range (−55 °C to +125 °C for 5400 devices, and 0 °C to +70 °C for 7400).

Figure 2-3A shows the circuit for a typical TTL inverter. Like the DTL device, the TTL input acts as a *current source,* while the output acts as a *current sink*. The typical TTL input will *source* 1.8

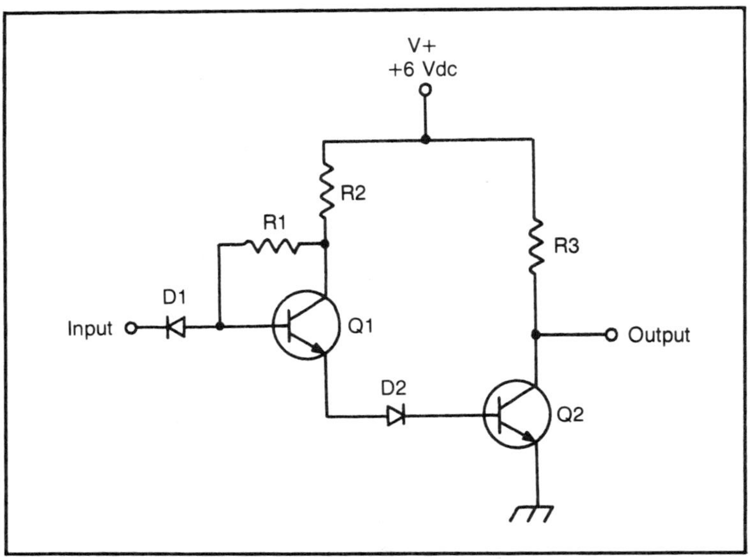

Fig. 2-2. DTL Inverter.

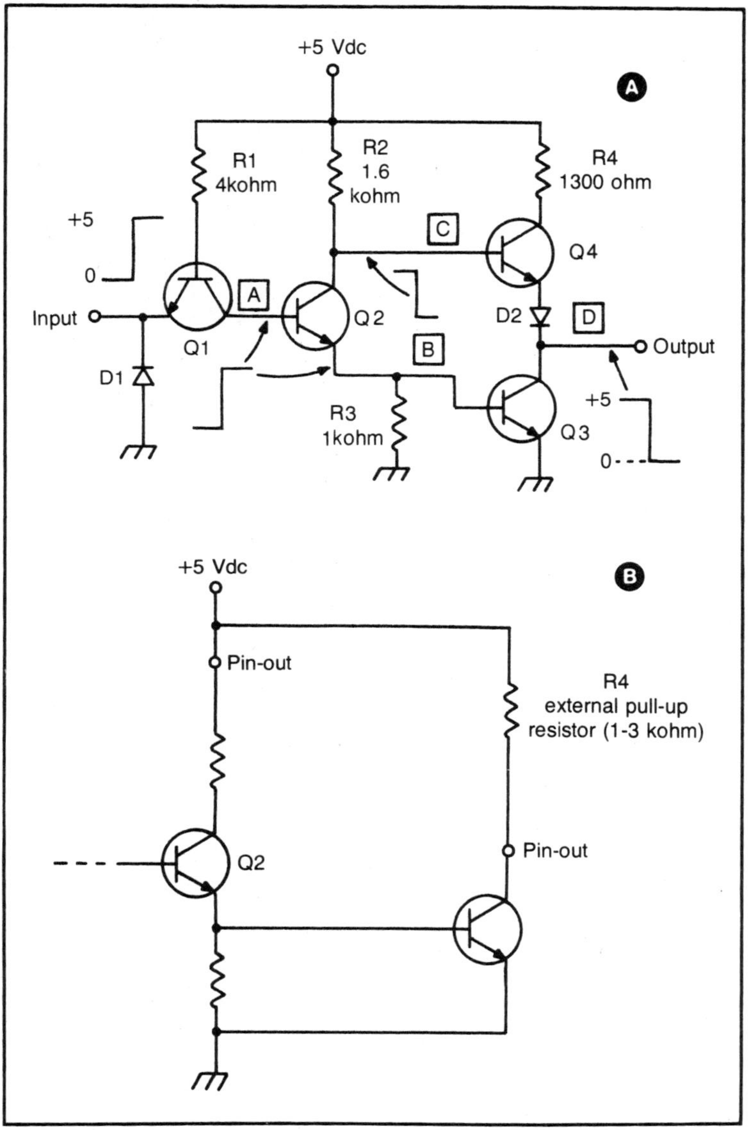

Fig. 2-3. (A) Regular TTL inverter, (B) open-collector TTL circuit.

mA, and is considered LOW if the voltage is 0 to 0.8 volts, or HIGH if 2.4 to 5.0 volts are applied. Performance at values of input potentials between 0.8 and 2.4 volts is not defined, so operation of the devices is unpredictable.

When the TTL input is HIGH, Q1 is cut off, so point *A* goes

HIGH. This condition turns on Q2 forcing point *B* HIGH and *C* LOW. We find, then, Q3 is turned on and Q4 is off. This forces the output LOW. Again, the transistors are operated either totally cut-off or totally saturated-on.

If the input is LOW, then exactly the opposite situation occurs: Q1 is turned on (forcing point *A* LOW), Q3 is off, and Q4 is turned on (i.e., it is connected to VCC(+)).

TTL devices must have a regulated dc power supply of +4.75 to +5.25 volts. In fact, there are some circuits or combinations of devices that require a more limited range of voltages nearer to +5 volts dc. Voltages greater than +5.25 volts often results in a high failure rate of TTL devices.

Some TTL ICs are described as being *open-collector* devices. These devices are essentially the same as regular TTL ICs, except that the output circuit is modified: i.e., Q4 and D2 are missing. An example of an open-collector circuit is shown in Fig. 2-3B. These devices require an external 1 to 2 kohm resistor between the output terminal and the 5 volt dc power supply line.

CMOS DEVICES

Complementary-metal-oxide, silicon (CMOS) IC devices are MOSFET transistors instead of PNP or NPN bipolar transistors that are used in other logic families. CMOS inputs, therefore, are very high impedance. Figure 2-4A shows a typical CMOS inverter circuit. Note that this family is called complementary because the output circuit consists of a complementary pair of MOSFET transistors; i.e., an n-channel and a p-channel connected in series.

CMOS devices can use a monopolar power supply, like TTL DTL, or may use a bipolar power supply after the fashion of operational amplifiers. CMOS power supplies the V+ can have any potential between +4 volts and +15 volts, while the V− may be −4 volts to −15 volts. In monopolar cases the V+ can also be +4 to +15 volts, while the "V−" is actually zero volts.

CMOS outputs are not directly TTL compatible, although some specific ICs in the CMOS line are designed to have a TTL output stage (i.e., the 4049 and 4050 devices). These TTL-compatible devices are often used to directly interface CMOS and TTL devices.

Figures 2-4B and 2-4C show the equivalent circuits for a CMOS inverter in both possible input conditions (i.e., input HIGH and input LOW). Recall that a p-channel MOSFET turns on when

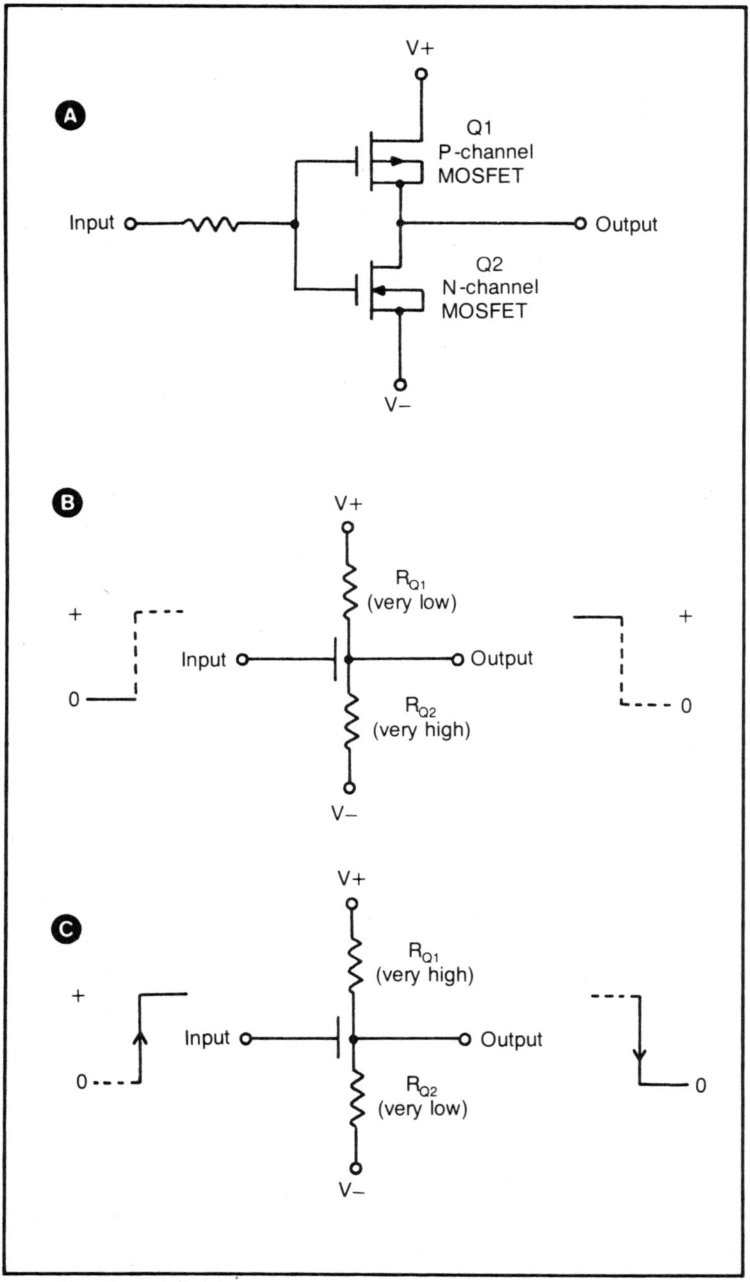

Fig. 2-4. (A) CMOS inverter, (B) equivalent circuit for LOW input, (C) equivalent circuit for HIGH input.

the gate is LOW, while the n-channel device turns on when the gate is HIGH.

Figure 2-4B shows the situation where the input is LOW. Transistor Q1 will have a very low (i.e., 200 ohms) channel resistance. In this case, the output is equivalent to a 200 ohm resistor in the V+ power supply line.

In Figure 2-4C we see the situation where the input is HIGH. Transistor Q2 now have a very high channel resistance, and Q1 has a very low channel resistance (again, about 200 ohms). In this case, the output looks like a 200 ohm resistance to ground, so the output is LOW.

The HIGH/LOW or LOW/HIGH output transition in a CMOS device occurs at a point where the input voltage is midway between the V+ and V− voltages. If V− and V+ are not equal, then the transition occurs at a potential of ½((V+) − (V−)). If, on the other hand, V− and V+ are equal, then the transition occurs at zero volts. If the V− potential is zero volts, then the transition occurs at ½(V+).

The CMOS output stage always looks like a high value and a low value resistor in series across the power supply (reexamine Figs. 2-4B and 2-4C), so negligible amounts of current are drawn from the power supply. A low resistance load exists only when the input voltage is at the transition point. The overall average current drain, therefore, is very small.

But CMOS devices do have a problem: they contain MOSFETs, so they are sensitive to static electricity. The A-series CMOS devices (i.e., 4001A) have this problem, but it is less severe in B-series (i.e., 4001B) devices. The B-series have built-in diode gate-protection to bypass high static potentials around the sensitive gate structure. Nonetheless, they should be handled with care.

HTL DEVICES

Noise pulses can be seen by logic circuits as valid input signals. This problem is especially bothersome in high speed TTL devices that are normally able to pass the high frequency, short duration, pulses that characterize noise. The solution in noisy environments is to use a digital IC logic family that requires a high input voltage to trigger. CMOS operated at high V− and V+ values meet this requirement, but the older bipolar *high threshold logic* (HTL) may also be used (Fig. 2-5).

HTL (also sometimes called *high noise immunity logic*, or HNIL), uses V+ values of 12 or 15 volts, depending upon the series.

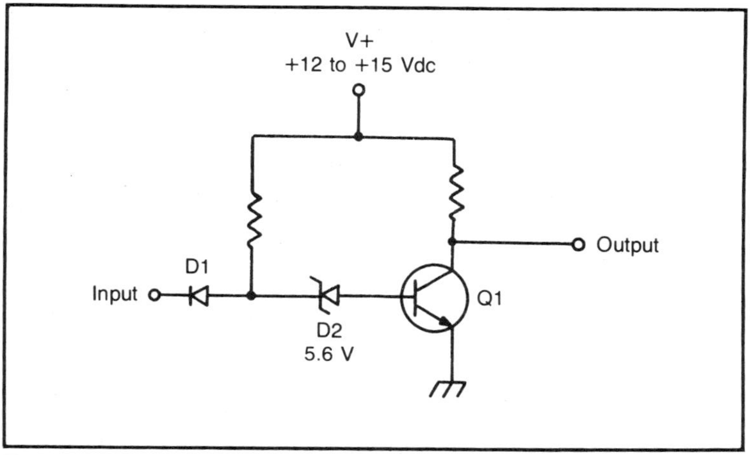

Fig. 2-5. HNIL or HTL logic inverter.

As a result, the logic levels are also high, so it requires a larger amplitude noise pulse to cause trouble.

EMITTER-COUPLED LOGIC

Up until now we have been talking about *saturated* logic families, i.e., the transistors in the ICs are either all the way on or all the way off (cutoff or saturated). Emitter-coupled logic, or ECL, is called ac logic because the transistors (bipolar PNP and NPN devices) are operated in the nonsaturated mode. As a consequence, ECL devices are capable of very fast operation. Most commonplace ECL devices operate to over 80 or 120 MHz, while some high-priced special-purpose ECL devices will operate to over 1 GHz (that's 1000 MHz!) The usual VHF/UHF prescaler for a digital frequency counter is an ECL frequency divider stage that divides the 500 MHz input signal down to 50 MHz, or less.

Note that it is necessary to use VHF/UHF circuit design and layout methods when working with ECL devices. The very high frequencies used are, after all, in the VHF and UHF frequency ranges. An example of this problem is that 50 ohm input and output impedance terminations are often necessary.

LOGIC FAMILY INTERFACING TECHNIQUES

When devices of the same logic family are interconnected one need not be concerned with the interconnection circuit—it is merely a conductor. That, in fact, is one of the factors that makes a logic *family*. When we try to cross-connect devices, however, we find that the various logic families are not necessarily compatible with each other. We cannot, for example, connect together all members of the CMOS and TTL families without some regard for external circuitry. In this section we will discuss some of the methods used to interface elements of *different* logic families, with special emphasis on TTL and CMOS interconnections.

There are several schemes shown in Fig. 2-6. Before proceeding to the illustrated cases, we can state that certain combinations are more or less universally accepted. For example, if a DTL device is operated from the +5 volt dc power supply required for TTL devices, then the DTL device can usually be treated as if it were a TTL device. Also, almost any CMOS device will be able to drive a single LS-series (low-power Schottky) or L-series (low power) TTL input. Similarly, the 4001A and 4002A devices (but not

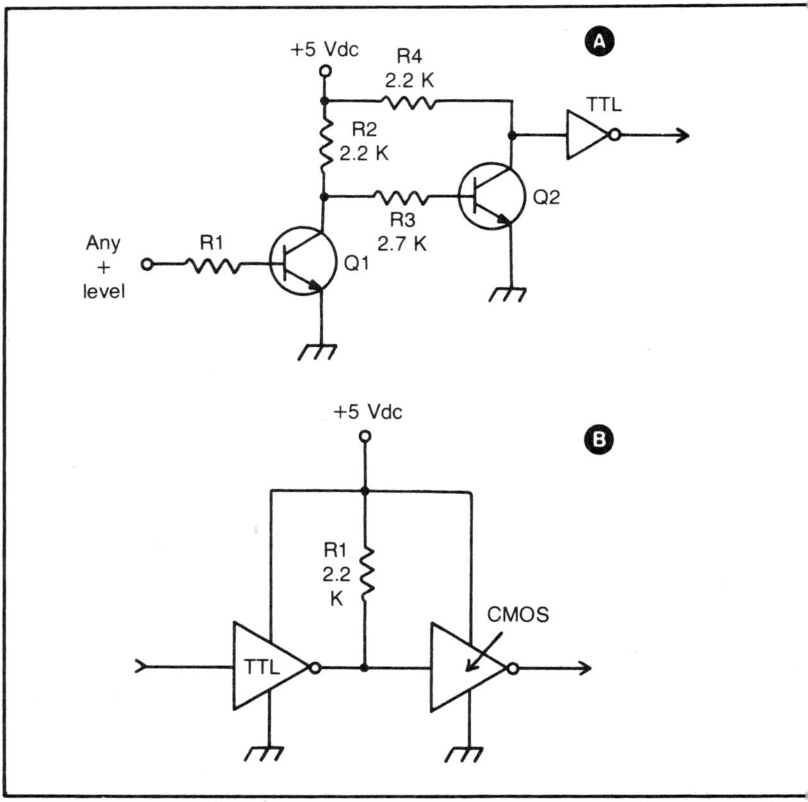

Fig. 2-6. (A) Logic-level translator for interfacing, (B) TTL-to-CMOS interfacing, (C) 4049/4050 devices will drive two TTL loads, (D) interfacing TTL to CMOS with greater than +5 Vdc supplies.

the B-versions) can drive a single regular TTL input. The 4049 and 4050 devices are a special case which will be discussed in a moment.

Figure 2-6A shows a circuit that could be used to interface almost any form of digital logic with TTL devices, even older circuits that are not IC. Any positive logic level applied to the input end of resistor R1 will cause transistor Q1 to saturate. The collector voltage of Q1 will then drop to zero, causing transistor Q2 to be unbiased, therefore cutoff. Under this condition, a HIGH (at +5 volts) will be applied to the TTL device's input. Alternatively, when the input end of resistor R1 is grounded, then transistor Q1 is cutoff, so its collector voltage rises to +5 volts dc. This potential will cause transistor Q2 to be forward biased into the saturation region. Under that condition, the collector voltage of Q2 will be zero, or

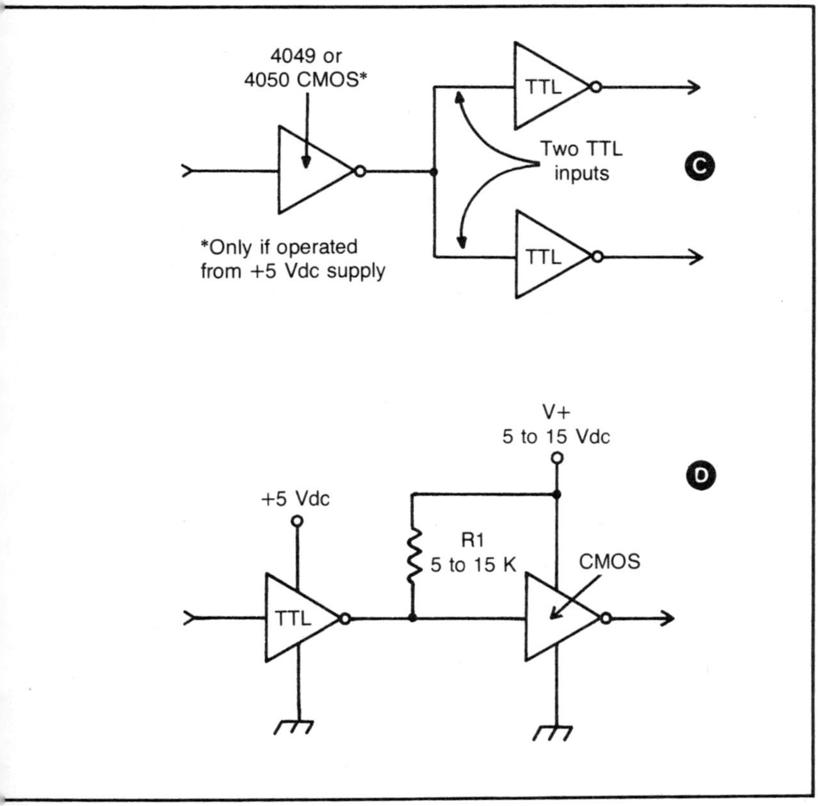

nearly so. The TTL input will see a LOW condition.

We could eliminate transistor Q2, and resistors R3 and R4, if we could live with inversion of the data. Under that condition, a HIGH on the input end of resistor R1 would produce a LOW at the TTL input, and a LOW at the input end of R1 would produce a HIGH at the TTL input. In that circuit, the TTL input will be connected directly to the collector of Q1. This method was once a lot more popular than it is now because "in the old days" there was a lot of equipment around that used mixed logic elements, some of which were totally nonstandard.

Figure 2-6B shows the method of interfacing almost any TTL device with a CMOS input. The constraint on this method is that the CMOS device must be operated from TTL-level power supplies, i.e., +5 volts and ground. A 2.2 kohm pull-up resistor is connected between the interface point and +5 volts in order to form a current source for the TTL output (TTL inputs are 1.8 mA current sources!)

Figure 2-6C shows the special case of the 4050 and 4049 CMOS devices. The 4049 device is a hex inverter, while the 4050 is a hex buffer. Both contain six independent stages; one inverts while the other is noninverting. The special thing about the 4049 and 4050 devices is that they can directly drive up to two standard TTL inputs *provided* that the 4049/4050 is operated from +5 volt dc power supply. In other words, the 4049 and 4050 have a TTL fan-out of two when operated from +5 volts dc. If any other package potential is used, then all bets are off and some other means of interfacing is needed.

Figure 2-6D shows how to interface TTL to CMOS devices that are not operated from +5 volt dc power supplies. The key here is to use open-collector TTL devices such as the 7406, 7407, 7416 and 7417 devices. The 7405 is not eligible because it wants to see a pull-up resistor to +5 volts only. Resistor R1 is used as a pull-up resistor to operate the open-collector output of the TTL stage.

Note that there are other TTL devices other than inverters with open-collector outputs. They also will be useful for this method, but the use of the inverter will allow any TTL device to be interfaced with any CMOS device.

The examples shown in this section are shown using inverter stages. This was a matter of convenience, and the information applies generally to all TTL and CMOS devices, not just inverters. The only exceptions are where open-collector TTL devices are required. In those cases, the selection of devices is constrained to those with open-collector output.

INTERFACING WITH OTHER CIRCUITS

It is possible to interface TTL and CMOS devices with other forms of electronic circuit, for example: relays, LEDs and lamps. In this section we will review some of the more common interface methods. Although certain examples are shown, you should be able to adapt and adopt as needed.

Figure 2-7 shows the methods for interfacing with light-emitting diodes (LEDs). These devices are used extensively as panel lamps on modern electronic equipment. They are longer lasting than incandescent lamps, so increase the reliability of the equipment being designed.

Figure 2-7A shows the method for interfacing an open-collector TTL device with an LED. The LED and its current-limiting resistor (R1) is connected between the output of the TTL device and the +5 volt power supply (or, higher voltages in some cases). If the dc power supply for the LED is +5 volts, then any TTL open-collector device may be used. Otherwise, one must use 7406, 7407, 7416, or 7417. The resistor value shown here is for most ordinary LEDs operated from +5 volts. The assumption is that the LED is happy with a current of 15 mA. If the LED requires a higher or lower current, then the resistor value will have to be adjusted in accordance with Ohm's law. Also, the resistor value will have to be increased for higher supply voltages. Keep in mind that the current must be limited to 15 mA (i.e., 0.015 amperes). For a supply voltage of +12 volts, for example, the resistor should be approximately 12/0.015, or 800 ohms.

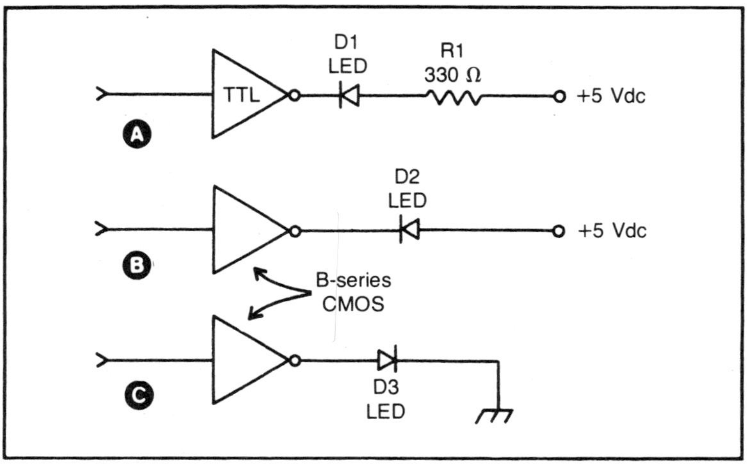

Fig. 2-7. LED interfacing.

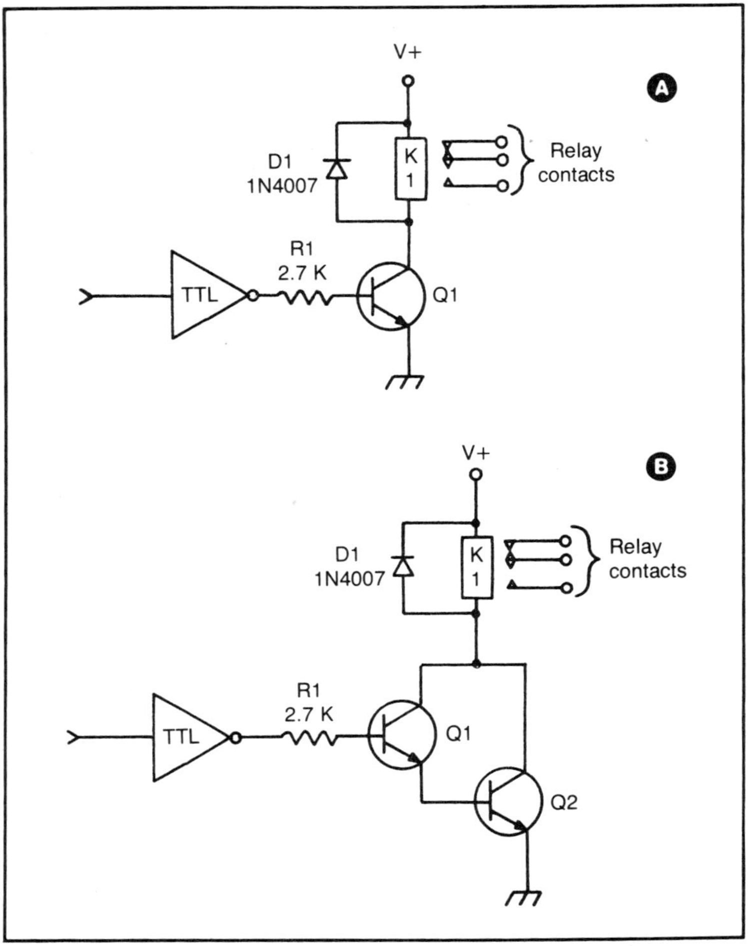

Fig. 2-8. (A) TTL relay interfacing. (B) TTL relay interfacing with Darlington-pair amplifier.

Two cases for CMOS devices are shown in Figs. 2-7B and 2-7C. The B-series CMOS devices have a higher drive current than A-series devices, so can directly drive low-current LEDs without the need for interfacing. The circuit in Fig. 2-7B shows the connection for turning on the LED when the CMOS input is HIGH. Similarly, the version in Fig. 2-7C turns on the LED when the input is LOW.

Figure 2-8 shows two circuits for driving electromechanical relays or other loads from TTL outputs. The TTL output terminal is used to turn on a driver transistor (Q1 in both Fig. 2-8A and 2-8B).

The version in Fig. 2-8A is for low-current relays with coil potentials in the under 30-volts range. The transistor can be any NPN device that will hack the current and potentials of the relay; 2N2222 and 2N3053 are commonly selected. You will have to look at the *beta* of the transistor and the load current in order to determine if this circuit will work properly. The TTL device will have a fan-out of approximately 10, so will produce up to 18 mA of current. We can adjust the series resistor (R1) for higher current levels, but for the value shown 1.5 mA is about right. The beta of the transistor is defined as the ratio of the collector current to the base current (I_c/I_b), so we know that the maximum collector current that can be supported is *beta* × I_b. If the transistor *beta* is 80, for example, the collector current will have to be 120 mA or less. If that is true, then this circuit will operate properly. If the collector current requirements imposed by the load are greater, then it is necessary that either another circuit be used (e.g., Fig. 2-8B) or the transistor be replaced with one that has a higher *beta* rating.

The diode in parallel with the relay coil is used to suppress inductive kick spikes caused by de-energizing the relay. The energy in the magnetic field around the relay coil is dumped back into the circuit when the relay is turned off, and forms a high-voltage spike that can damage the electronic components in the circuit, especially Q1. The diode is a 1-ampere 1000 volt (PIV) device, so will easily clip that spike.

The high-current circuit is shown in Fig. 2-8B. In this circuit the single transistor is replaced with a *Darlington Pair.* The two transistors are connected in the Darlington amplifier configuration, meaning that we treat them as one transistor. The *beta* of the pair is the product of the individual betas. In other words, *beta (Q1)* × *beta (Q2)*. If the two transistors have identical betas, then the total beta will be the square of that figure! It is easy, therefore, to obtain *beta* figures in the 5,000 to 25,000 range. Obviously, a very high current can be driven with the 1.5 mA available from our lowly TTL inverter.

It is common practice when using Darlington pairs to make Q1 a driver transistor and Q2 a power transistor; 2N3053 and 2N3055 are frequently paired. There are also available commercial power Darlington devices that contain both transistors inside of a common TO-3 power transistor package. These devices will make nice power relay drivers.

Chapter 3

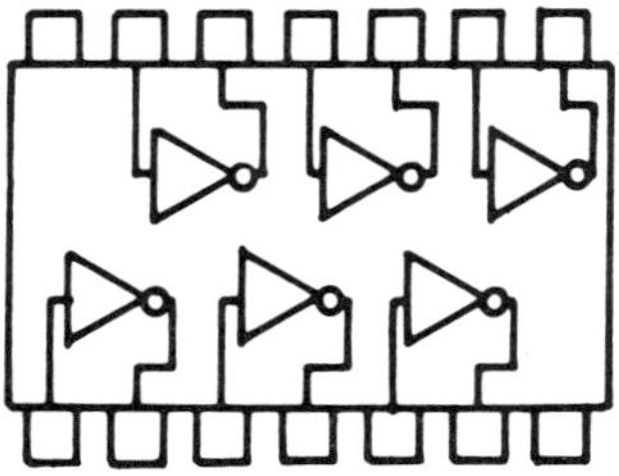

TTL Devices

The TTL (*transistor-transistor-logic*), or, T^2L family of digital IC logic elements is probably the most widely used of all logic families. Even common CMOS devices are sometimes forced into operation in TTL circuits because of the predominance of TTL devices. These logic elements were introduced in IC form in the early to late sixties and are now standard. The TTL IC device lowered the cost of digital circuitry from relatively high costs involved when gates and inverters were made from discrete diodes and transistors, such that now, in quantity, it is possible to buy two-input NAND gates and inverters at prices that are just plain ridiculous compared with earlier costs. A 7400 NAND gate priced for 100-quantities can be purchased for about 20¢, or about a nickel a gate.

TTL devices outstripped other devices in the matter of speed. When RTL was the only kind of IC logic element available, operating speeds tended to be in the 3 to 5 megahertz range. Modern TTL devices now typically operate easily at 18 to 20 megahertz, with certain specially selected specimens operating to 25 or 30 megahertz. It is not difficult to find people who have grossly exceeded the rated speed successfully! Some specially designed devices will operate to 50, 80, or even 120 megahertz.

The specifics of the TTL line were covered in some detail in Chapter 2, under the rubric of "IC logic families." The purpose of that chapter was to acquaint you with the various families, both historic and current, and to give you some appreciation for the differences between them. Of all of those families considered, only

TTL and CMOS (and to some extent ECL) are commonly used in a large range of equipment.

The article on TTL in Chapter 2 used an inverter circuit in order to show the comparison between families (all examples were inverters). Figure 3-1 shows how the inverter becomes a two-input NAND gate, and thus how extra inputs are accommodated in TTL devices. The two inputs are formed as separate emitter regions for the input transistor, Q1 (see also Fig. 2-3A). As many as eight inputs are accommodated in some TTL devices (see spec sheet for 7430 device!)

The TTL device operates from a +5 volt dc regulated power supply. The voltage range tolerated by TTL devices is quite narrow, so it is generally ill-advised to attempt to use nonregulated power supplies.

It is also true that TTL devices are not tolerant of careless accidents like dropping a screwdriver onto a printed circuit board! The old RTL family was immune to such problems, as long as no potential over +4 volts dc was on the board, because no combination of opens or shorts could damage the chip. Not so with TTL devices—grounding some outputs can and will cause immediate destruction of the chip. The TTL output is a current sink operated from a totem-pole output stage. Since the input driven by a TTL output never goes to ground, no provision is made for such operation. A grounded TTL input can mean destruction of the IC!

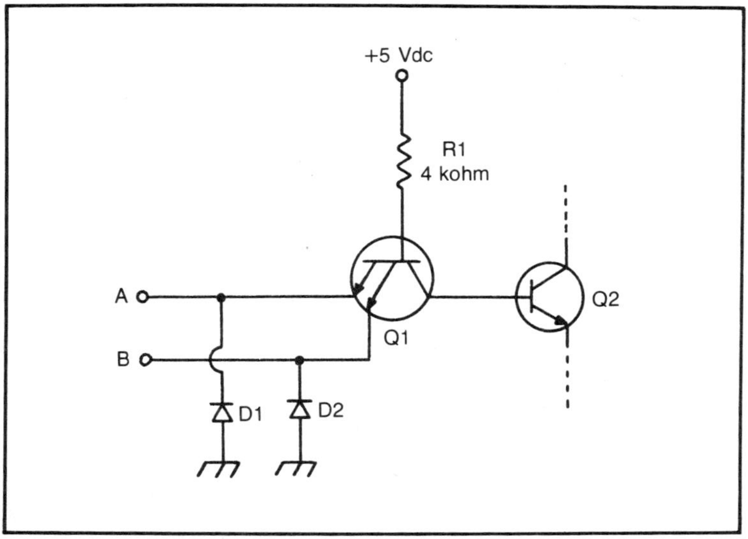

Fig. 3-1. TTL inputs.

A disadvantage of TTL is the immense amounts of current that seem to be needed. It is probably true that 25 milliamperes per chip is needed to operate TTL devices. In a large digital project, such as a computer, there might be anything from dozens to hundreds of chips, so the current requirements often exceed one-ampere. In fact, the current requirements are such that bench power supplies for people who design circuits with TTL devices should be at least 1-ampere (at +5 volts dc, regulated), and more is "nice to have." I personally recommend having a 3 to 5 ampere power supply if at all possible. Chapter 15 of this book gives several dc power-supply circuits that will be more than suitable for most applications. Chapter 15 also gives some of the theory and design criteria for power supplies, so you may well find it easy to design your own should none of the sample projects be satisfactory.

The TTL family of devices come in two forms of packaging: *dual-inline-package* (DIP) and *flat-pack*. The DIP is perhaps the most common form of IC package, and consists of a plastic or ceramic body with a row of pins protruding at 90-degrees from the body. Typical TTL devices are found in quite a variety of DIPs, with pin sets of 14, 16, 20, 24, 32, 36 and 40 pins (14- and 16-pin DIP is the most common, and both have a body that is 0.3 inches wide pinrow to pinrow). All pins are spaced on 0.1 inch centers within the same row.

The flat pack is used in military applications, and consists of a flat ceramic body with pins coming out in the same plan with the body, rather than at 90-degrees as in the DIP. The purpose of the flat pack, despite the fact that they are harder to use, is to have a lower profile, hence allow higher packaging density.

Some TTL devices come in white ceramic, either DIP or FP, and will have a gold-plated metal over the center region of the IC package. These are special military and aerospace ICs that are radiation protected. Most high-quality TTL devices will not carry 74xx or 74xxx numbers, but will carry the 54xx and 54xxx equivalents, denoting MILSPEC ICs. These are a lot more expensive than the commercial 74xx/74xxx equivalents (a 5475 is a 7475 in uniform) because of the greater amount of testing and burn-in carried out for 54xx/54xxx devices. Often, the dies are the same but the 54xx/54xxx devices are burned-in longer. In some hobbyist circles, dealers occasionally sell military and commercial surplus 54xx/54xxx as 74xx/74xxx devices, and this usually is better for the buyer.

TTL SUBFAMILIES

The regular TTL device is not the only type of TTL available, for there are certain special subfamilies that are sometimes seen. The logic elements are pin-for-pin interchangeable, but not always it is possible to actually make the interchange! In other words, the 7475 and 74LS75 are identical as to pinouts, logic function and so-forth, but operate at different frequency ranges and power levels. There are certain circuits where some property of one particular sub-family is critical for design. Stories abound in the microcomputer industry of engineering change orders coming out specifying one or the other TTL subfamily. I can recall my first microcomputer had a problem with a pulse being too short—it would disappear before the rest of the circuit was ready to look for it. The sole change, which fixed the problem incidentally, was to change from a 74LS121 to a 74121 device!

The names of the subfamilies are: *Regular TTL*, *High-power TTL*, *Low-power TTL*, *Schottky TTL* and *Low-power Schottky TTL*. The Regular TTL devices are those which were the subject of the TTL section in Chapter 2, and carry the type numbers with the 74xx/74xxx designations. Typically, the Regular TTL devices will have propagation delay times (i.e., difference between the pulse appearing at the output and its being applied to the input) of 10 to 15 nanoseconds, and operating speeds to 20 MHz (30 - 50 MHz when hand-selected for speed). Power dissipation tends to be on the order of 10 mW per gate/inverter. The High-power TTL device operates to 50 - 60 MHz, with the concomitant increase in power dissipation (25 milliwatts rather than 10). Also, because of the higher operating speeds, the propagation delay of high-power TTL is 6 to 8 nanoseconds. Devices in the High-power TTL subfamily carry type numbers of the sort 74Hxx/74Hxxx (a 74H02 is a High-power TTL 7402).

The Low-power TTL devices, which carry type numbers of the sort 74Lxx/74Lxxx, operate at considerably less power dissipation levels than do regular TTL devices. The typical dissipation is 1 mW rather than 10 mW, which is common for Regular TTL. Of course, we always have the speed-vs-power trade-off, so we find that typical operating speeds for the 74Lxx/74Lxxx devices is on the order of 2.5 to 5 MHz, and propagation delay times between 30 and 50 nanoseconds.

There are two forms of Schottky TTL device. These devices have a Schottky diode across certain critical transistor base-emitter

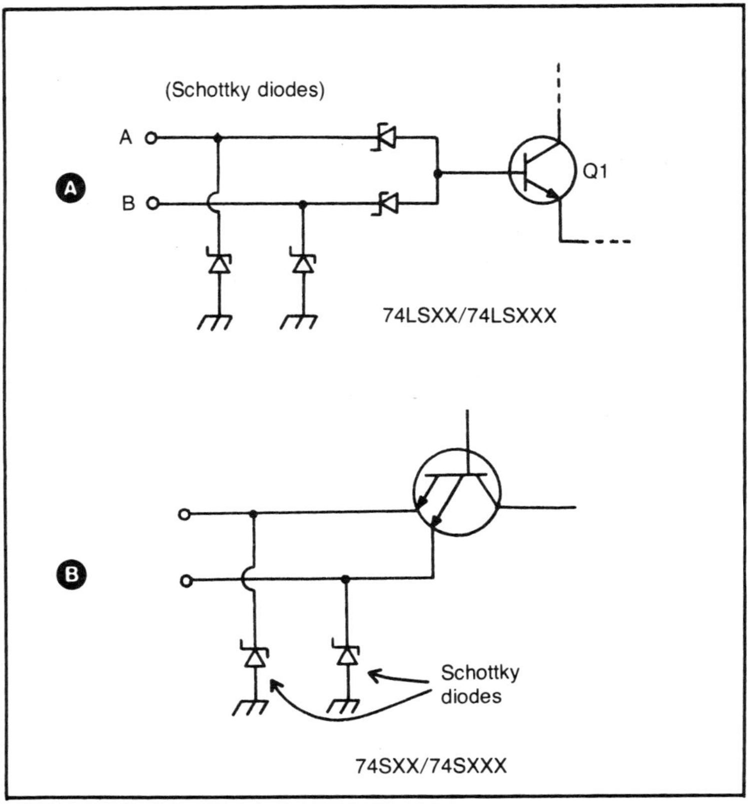

Fig. 3-2. 74LS inputs and 74S inputs.

junctions (see Fig. 3-2) in order to prevent saturation of the base region. If the base is not saturated, then it is easier to sweep all charge out of the base and thereby increase the switching speed. The regular Schottky TTL devices carry type numbers of the sort 74Sxx/74Sxxx, while Low-power Schottky TTL devices have type numbers such as 74LSxx/74LSxxx. The 74Sxx/74Sxxx series devices will operate to 120 MHz, with 3 nanosecond delay times. The power dissipation is typically larger than regular TTL (20 mW instead of 10 mW), but is lower than 74Hxx/74Hxxx devices. The LS-series operate to 40 to 50 MHz, and have typical propagation delay times of 10 nanoseconds. In this latter regard, they are either as fast or faster than regular TTL. The 74LSxx/74LSxxx devices operate to 40 or 50 MHz, yet in seeming contradiction to the normal power-vs-speed rule will dissipate only 2 milliwatts of electrical power.

TTL POWER SUPPLY SPECS AND BYPASSING

The dc power supply used on TTL circuits must have an operating potential of +5 volts dc, and ground. This potential must be regulated to within approximately plus or minus 0.25 volts, for a range of +4.75 to +5.25 volts, officially. Unofficially, it is sometimes found that this range is a little flaky. At the high end, over +5.05 volts, for example, the reliability of the TTL devices will suffer somewhat unless the circuit is well ventilated with forced air! At the lower end, it is found that complex function devices sometimes operate unpredictably or not at all at potentials less than +4.9 volts! Some gates, in contrast, work well down to +4 volts! If a circuit seems to operate "some of the time," and it contains complex function devices such as monostable multivibrators or flip-flops, then it might be wise to check the power supply voltage and see if it is below +4.9 volts dc—I don't care what the spec sheet says!

In the rest of this chapter, we will examine some of the more popular TTL chips (and some that aren't so popular, but are very useful to some readers) on a case-by-case basis. The pinouts given in each case are the standard DIP package, unless otherwise specified in the section on the particular device.

Device Number: 7400

Type of Device: Quad Two-Input NAND Gate

QUADRUPLE 2-INPUT POSITIVE-NAND GATES

positive logic: $Y = \overline{AB}$

00

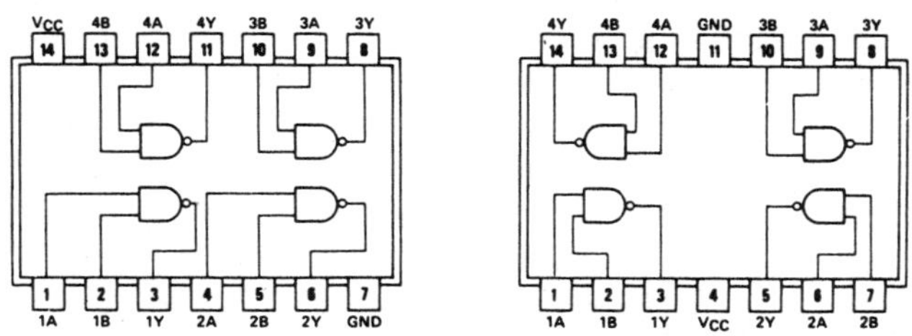

SN5400 (J)	SN7400 (J, N)	SN5400 (W)
SN54H00 (J)	SN74H00 (J, N)	SN54H00 (W)
SN54L00 (J)	SN74L00 (J, N)	SN54L00 (T)
SN54LS00 (J, W)	SN74LS00 (J, N)	
SN54S00 (J, W)	SN74S00 (J, N)	

Operation & Application: The 7400 contains four independent two-input NAND gates. All four gates may be used independently. A LOW on any input will cause the output to be HIGH. If both inputs are HIGH, then the output will be LOW.

Propagation delay: 10-12 nS
Current Requirements per package: 10-15 mA
Power Dissipation:

Device Number: 7401

Type of Device: Quad Two-Input NAND Gates with Open-Collector Outputs

QUADRUPLE 2-INPUT POSITIVE-NAND GATES WITH OPEN-COLLECTOR OUTPUTS

positive logic:
$Y = \overline{AB}$

01

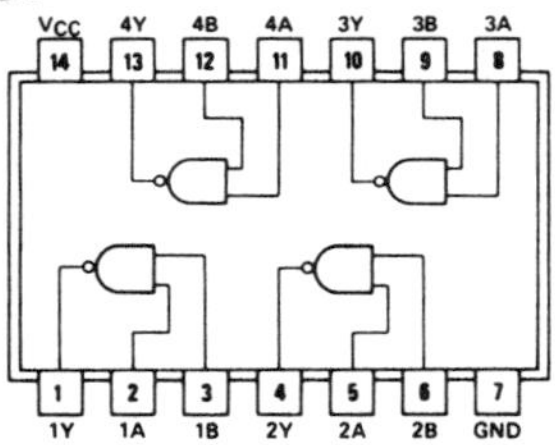

SN5401 (J) **SN7401 (J, N)**
SN54LS01 (J, W) **SN74LS01 (J, N)**

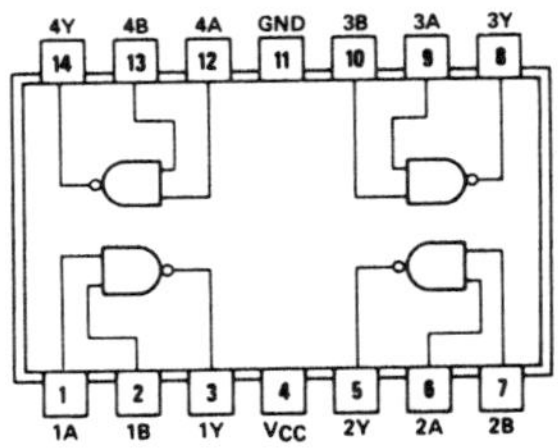

SN5401 (W)
SN54H01 (W)
SN54L01 (T)

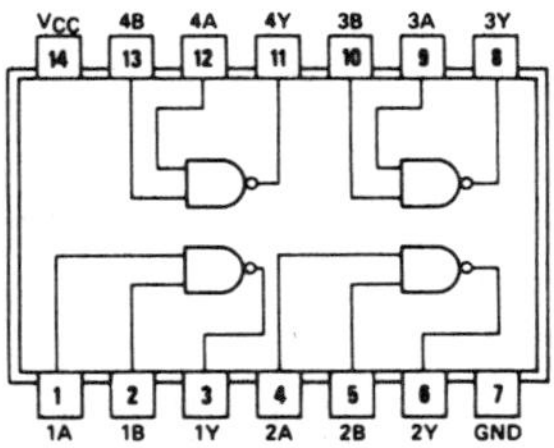

SN54H01 (J) **SN74H01 (J, N)**

Operation & Application: The 7401 contains four independent two-input NAND gates that each have open-collector output stages. A 1.5 to 2.7 kohm pull-up resistor is required between any given output and the +5 volt dc line. Note that the 7401 and 7400 are not pin-for-pin compatible, but are logically similar. A LOW on any input makes the corresponding output HIGH. If both inputs are HIGH, then the output will be LOW.

Propagation delay: 8 to 40 nS (faster in output-LOW condition)
Current Requirements per package: 10 mA
Power Dissipation:

Device Number: 7402

Type of Device: Quad Two-Input NOR Gate

QUADRUPLE 2-INPUT POSITIVE-NOR GATES

positive logic:
$Y = \overline{A+B}$

02

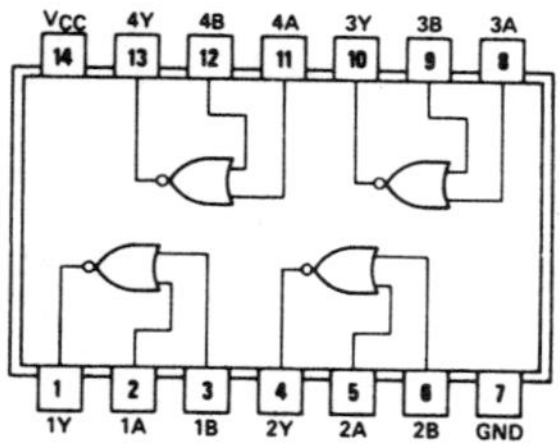

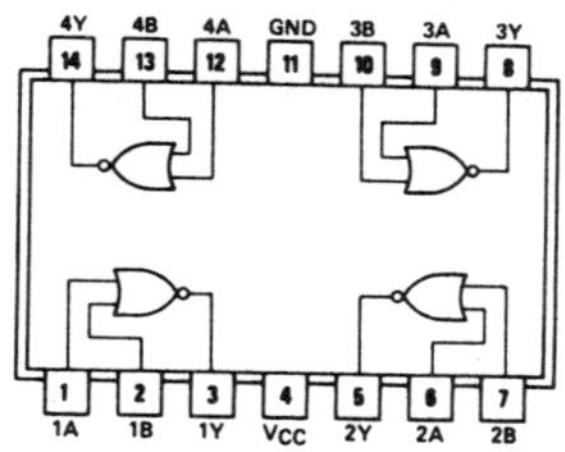

SN5402 (J)
SN54L02 (J)
SN54LS02 (J, W)
SN54S02 (J, W)

SN7402 (J, N)
SN74L02 (J, N)
SN74LS02 (J, N)
SN74S02 (J, N)

SN5402 (W)
SN54L02 (T)

Operation & Application: The 7402 contains four independent NOR gates. A HIGH on any one input will make the output LOW. If both inputs are LOW, then the output will be HIGH. Note that this device is logically the *opposite* of the popular 7400 device.

Propagation delay: 10-12 nS

Current Requirements per package: 10-15 mA

Power Dissipation:

Device Number: 7403

Type of Device: Quad Two-Input NAND Gate with Open-Collector Outputs

QUADRUPLE 2-INPUT POSITIVE-NAND GATES WITH OPEN-COLLECTOR OUTPUTS

03

positive logic:
$Y = \overline{AB}$

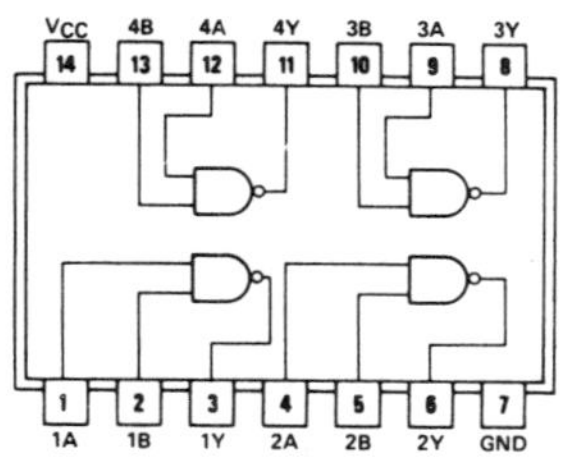

SN5403 (J)
SN54L03 (J)
SN54LS03 (J, W)
SN54S03 (J, W)

SN7403 (J, N)
SN74L03 (J, N)
SN74LS03 (J, N)
SN74S03 (J, N)

Operation & Application: This device is logically the same as the 7400 and has the same pin-outs. The output terminals, however, are open-collector types so a 1.5 K to 2.7 kohm pull-up resistor must be connected between each output and the +5 volt dc line.

Propagation delay: 8-40 nS (faster into LOW output)
Current Requirements per package: 8-10 mA
Power Dissipation:

Device Number: 7404

Type of Device: Hex Inverter

HEX INVERTERS

04

positive logic:
$Y = \overline{A}$

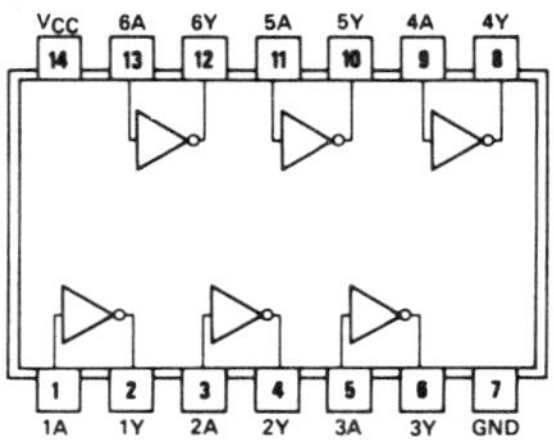

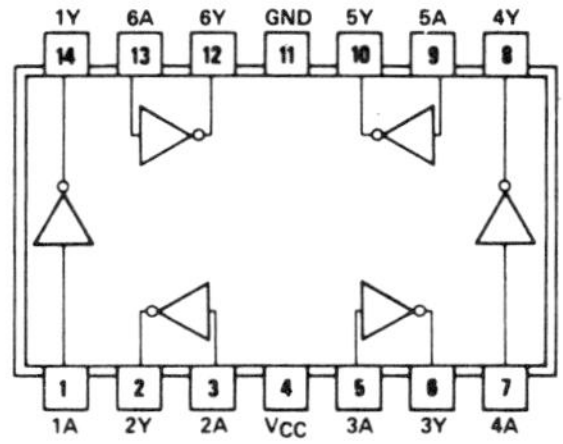

SN5404 (J)	SN7404 (J, N)	SN5404 (W)
SN54H04 (J)	SN74H04 (J, N)	SN54H04 (W)
SN54L04 (J)	SN74L04 (J, N)	SN54L04 (T)
SN54LS04 (J, W)	SN74LS04 (J, N)	
SN54S04 (J, W)	SN74S04 (J, N)	

Operation & Application: A Hex inverter contains six independent inverter circuits. A HIGH on the input will cause a LOW on the output and vice-versa: $A = \overline{A}$.

Propagation delay: 8-12 nS
Current Requirements per package: 10-15 mA
Power Dissipation:

Device Number: 7405

Type of Device: Open-Collector Hex Inverter

HEX INVERTERS WITH OPEN-COLLECTOR OUTPUTS

05

positive logic:
$Y = \overline{A}$

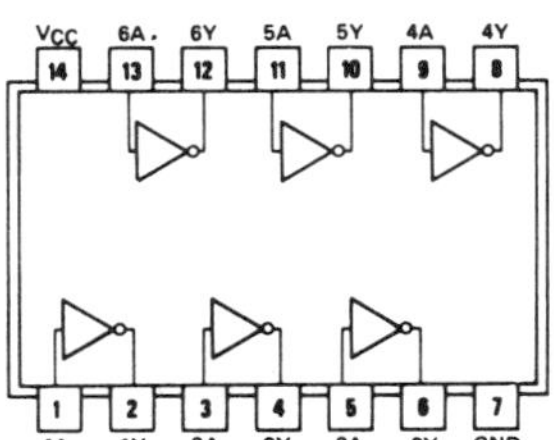

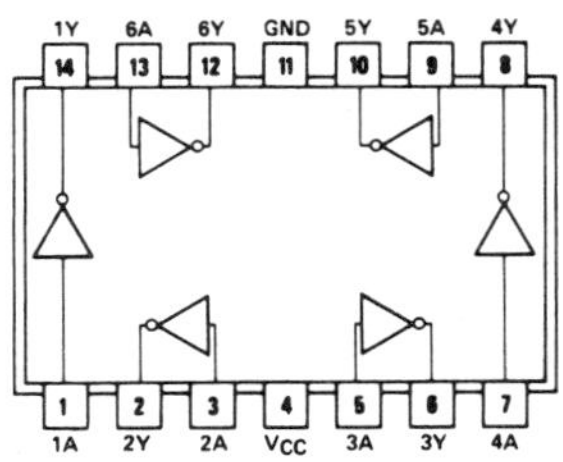

SN5405 (J)	SN7405 (J, N)	SN5405 (W)
SN54H05 (J)	SN74H05 (J, N)	SN54H05 (W)
SN54LS05 (J, W)	SN74LS05 (J, N)	
SN54S05 (J, W)	SN74S05 (J, N)	

Operation & Application: An inverter circuit produces a HIGH output when the input is LOW, and a LOW output when the input is HIGH. In other words, the output is always the *complement* of the input. This IC contains six independent inverter sections that share common +5 Vdc (pin 14) and ground (pin 7) terminals. The 7405 device is intended for use with TTL levels only, and requires a 2000 to 3000 ohm pull-up resistor between each output terminal and +5 volts dc. The open collector type of inverter is often used when we want to drive external loads other than another TTL device. We must keep in mind that the output LOW/input HIGH condition causes current to flow in the pull-up load, while the reverse condition turns off the current in the load. If we were to connect an LED (with 330 to 470 ohm current limiting resistor in series) between the output terminal of any one inverter and +5 Vdc, the LED will turn on when the input is HIGH and off when the input is LOW.

Propagation delay: 7 to 10 nanoseconds when output is LOW, 35 to 45 nanoseconds for output HIGH

Current Requirements per package: 10 to 15 mA

Power Dissipation:

Device Number: 7406

Type of Device: Open-Collector Hex Inverter

HEX INVERTER BUFFERS/DRIVERS WITH OPEN-COLLECTOR HIGH-VOLTAGE OUTPUTS

06

positive logic:
$Y = \overline{A}$

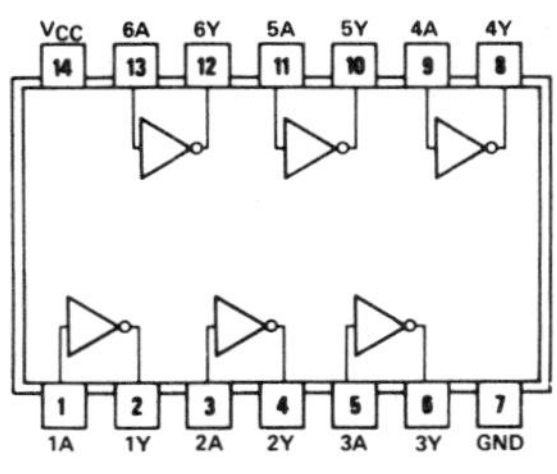

SN5406 (J, W) **SN7406 (J, N)**

Operation & Application: See the description for the 7405. This device is a lot like the 7405 described previously, except that the output transistor for each section can handle up to 30 milliamperes and 30 volts. This feature allows us to drive higher current and higher voltage (e.g., the standard 28 Vdc used in industrial electronics) loads. It is reasonable to use the 7406 device for driving electromechanical relays and certain small lamps, as well as LEDs. Even though the output transistors will handle higher voltages, it is still necessary to keep the package voltage (i.e., that between pins 14 and 7) at the standard TTL level. Only the output transistor is at a higher voltage level.

Propagation delay: 8-12 nS into output HIGH, and 12-18 nS into output LOW

Current Requirements per package: 30 mA, not counting the output transistors

Power Dissipation:

Device Number: 4707

Type of Device: Open-Collector Noninverting Hex Driver

HEX BUFFERS/DRIVERS WITH OPEN-COLLECTOR HIGH-VOLTAGE OUTPUTS

07

positive logic:
Y = A

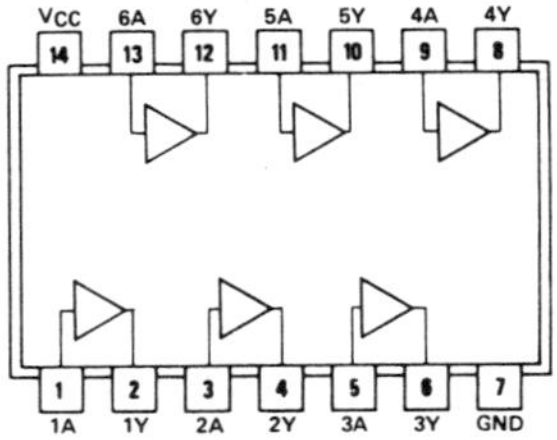

SN5407 (J, W) **SN7407 (J, N)**

Operation & Application: This device is a hex driver and is *noninverting*. The noninverting feature means that a HIGH input results in a HIGH output, and a LOW input results in a LOW output. There is no complementing of the input signal. The noninverting driver can be used in applications where it is necessary to drive more TTL loads than a normal TTL output can handle, and certain non-TTL loads up to 30 volts and 30 mA per section. Like the 7406 device, it can handle higher voltage loads than +5 volts, but the package V+ must remain at +5 volts dc. All six sections of the 7407 are independent of each other except for the common power supply connections.

Propagation delay: 5-8 nS into HIGH output, 18-22 nS into LOW output

Current Requirements per package: 25-30 mA (not counting output transistors)

Power Dissipation:

Device Number: 7408

Type of Device: Quad Two-Input AND Gate

QUADRUPLE 2-INPUT POSITIVE-AND GATES

08

positive logic:
Y = AB

SN5408 (J, W)
SN54LS08 (J, W)
SN54S08 (J, W)

SN7408 (J, N)
SN74LS08 (J, N)
SN74S08 (J, N)

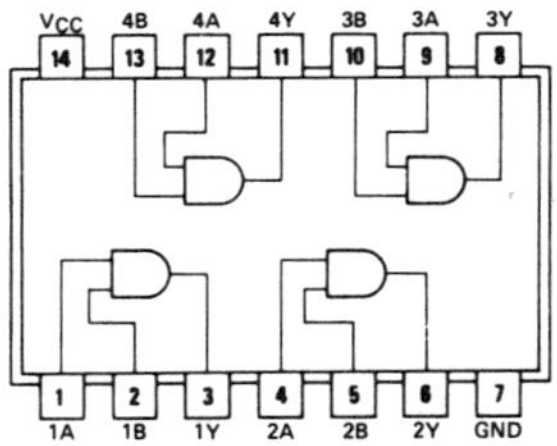

Operation & Application: An AND gate operates such that a LOW on any input causes the output to be LOW. The output is HIGH *only* when *all* inputs are also HIGH. For the 7408 device, all four AND gates operate independently, so a LOW on either input (e.g., pins 1 and 2 for G1) will cause the output to be LOW also. The output goes HIGH only when both inputs are HIGH.

Progagation delay: 12-15 nS

Current Requirements per package: 15-20 mA

Power Dissipation:

Device Number: 7409

Type of Device: Open-collector Quad Two-Input AND gate

QUADRUPLE 2-INPUT POSITIVE-AND GATES WITH OPEN-COLLECTOR OUTPUTS

09

positive logic:

Y = AB

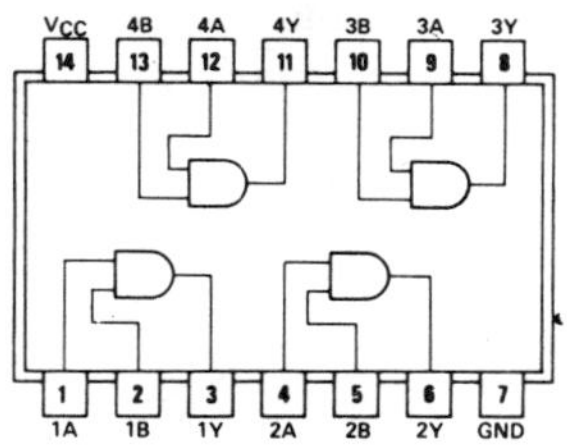

SN5409 (J, W)	**SN7409 (J, N)**
SN54LS09 (J, W)	**SN74LS09 (J, N)**
SN54S09 (J, W)	**SN74S09 (J, N)**

Operation & Application: Open-collector version of the 7408. Device will operate with TTL voltage levels to the output transistor, so must have a 2000-3000 ohm pull-up resistor between each output terminal and +5 Vdc.

Device Number: 7410

Type of Device: Triple Three-Input Positive Logic NAND Gate

TRIPLE 3-INPUT POSITIVE-NAND GATES

10

positive logic:

$Y = \overline{ABC}$

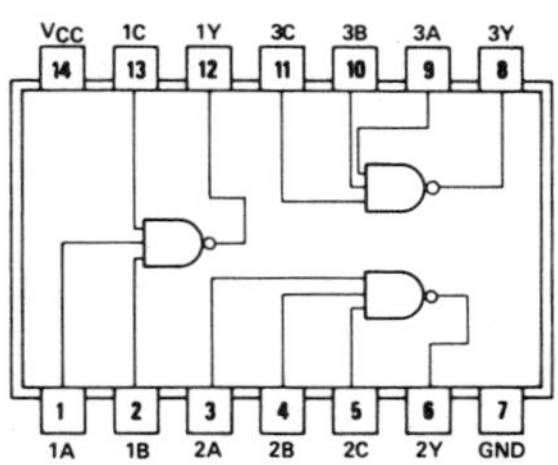

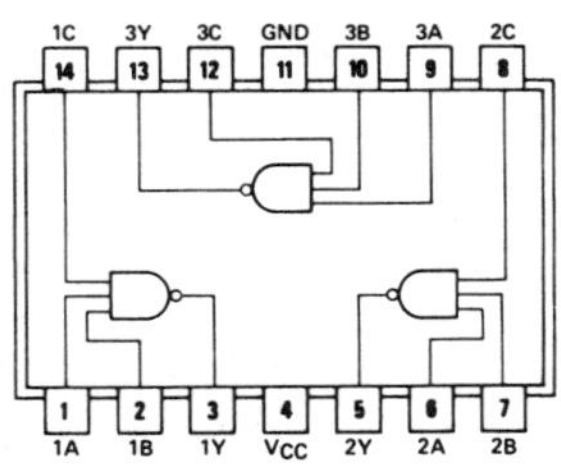

SN5410 (J)	**SN7410 (J, N)**	**SN5410 (W)**
SN54H10 (J)	**SN74H10 (J, N)**	**SN54H10 (W)**
SN54L10 (J)	**SN74L10 (J, N)**	**SN54L10 (T)**
SN54LS10 (J, W)	**SN74LS10 (J, N)**	
SN54S10 (J, W)	**SN74S10 (J, N)**	

Operation & Application: This 14-pin DIP device contains three three-input NAND gates. All three sections are dependent except for the +5 volt dc and ground connections. On any one gate, a low on any one input makes the output HIGH. All three inputs of one gate must be HIGH for the output to be LOW. This device can be used as a replacement for individual sections of the 7400 if one input of the sections being used is tied HIGH permanently. Each gate may also be used as an inverter if all three inputs are tied together . . . something that is true of all TTL NAND gates regardless of the number of inputs.

Propagation delay: 8-10 nS

Current Requirements per package: 6-10 mA

Device Number: 7411

Type of Device: Triple Three-Input AND Gate

TRIPLE 3-INPUT POSITIVE-AND GATES

11

positive logic:
Y = ABC

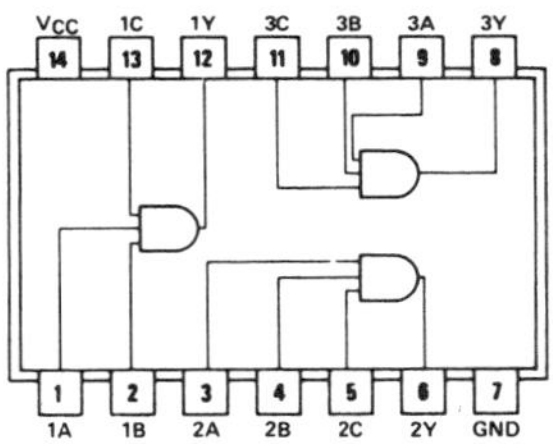

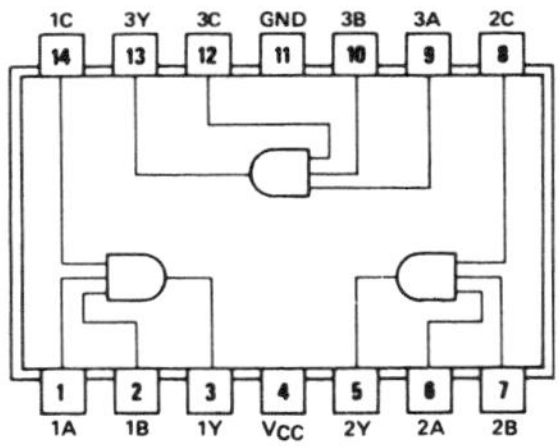

SN54H11 (J) SN74H11 (J, N) SN54H11 (W)
SN54LS11 (J, W) SN74LS11 (J, N)
SN54S11 (J, W) SN74S11 (J. N)

Operation & Application: This gate is similar to the 7410 device, except that the individual sections are AND gates instead of NAND gates. The output will be HIGH only if all three inputs of any given section are also HIGH. The spec sheet for this device lists the logic algebra expression for the 7411 as Y = ABC.

Propagation delay: 8-10 nS

Current Requirements per package: 6-10 mA

Device Number: 7412

Type of Device: Open-Collector Outputs on a Triple Three-Input NAND Gate

TRIPLE 3-INPUT POSITIVE-NAND GATES WITH OPEN-COLLECTOR OUTPUTS

12

positive logic:
Y = $\overline{ABC}$

SN5412 (J, W)
SN54LS12 (J, W)
SN7412 (J, N)
SN74LS12 (J, N)

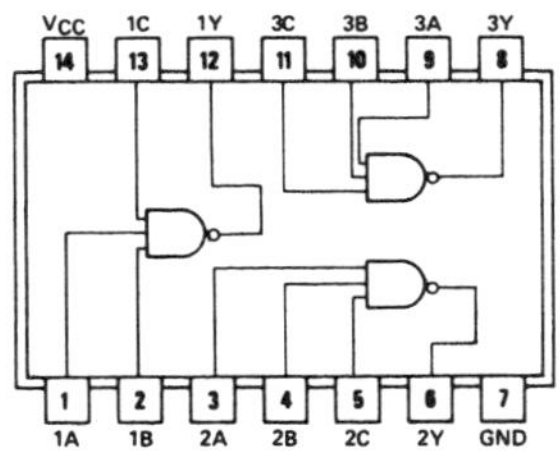

Operation & Application: This device is functionally equivalent (but not socket compatible without circuit modification) with the 7410 device. The difference is that the output terminals of each section are the open-collector type. A 2.7 kohm pull-up resistor is needed between each output terminal and the +5 volt dc power supply line.

Propagation delay: 10 nS

Current Requirements per package: 10 mA

Device Number: 7413

Type of Device: Dual Four-Input Positive-NAND Schmitt Trigger

DUAL 4-INPUT POSITIVE-NAND SCHMITT TRIGGERS

13

positive logic:

$Y = \overline{ABCD}$

SN5413 (J, W)
SN54LS13 (J, W)

SN7413 (J, N)
SN74LS13 (J, N)

NC—No internal connection

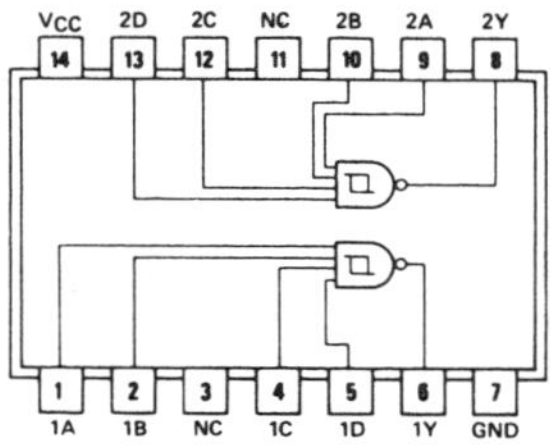

Operation & Application: This device contains two four-input NAND Schmitt triggers. For each device, any one input LOW will make the output HIGH. All four inputs must be HIGH for the output to be LOW. The Schmitt trigger inputs require that the input signal be 1.7 on the positive-going transition for the output state to change. Similarly, the trip point for the negative-going transition is 0.9 volts. The 1.7-0.9 volt hysteresis between positive-going transitions is sometimes very useful. The input impedance of this device is approximately 6 to 10 kohms.

Propagation delay: 22-27 nS

Current Requirements per package: 16 mA

Device Number: 7414

Type of Device: Hex Schmitt-Trigger INverter

HEX SCHMITT-TRIGGER INVERTERS

14

positive logic:

$Y = \overline{A}$

SN5414 (J, W)
SN54LS14 (J, W)

SN7414 (J, N)
SN74LS14 (J, N)

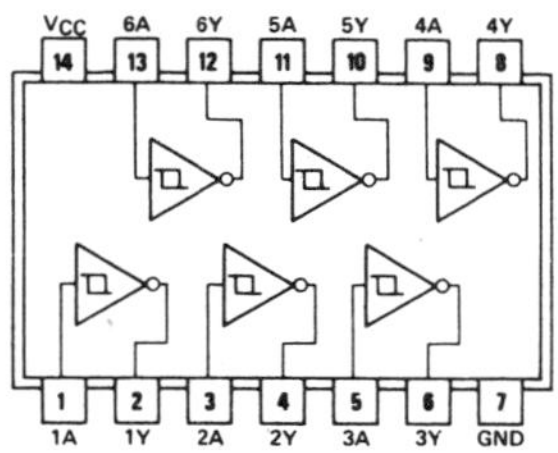

Operation & Application: An inverter is a circuit that will produce a LOW output when the input is high, and a HIGH output when the input is LOW. In other words, the output is always the *complement* of the input. Like the 7413 device, the individual sections of the 7414 are Schmitt triggers. The same trip points apply, so see the discussion for the 7413.

All six sections of the 7414 can be used independently of each other. The only common connections are the +5 volt dc and ground terminals.

Propagation delay: 15-20 nS

Current Requirements per package: 25-35 mA

Device Number: 7415

Type of Device: Triple Three-Input Positive AND Gates with Open-Collector Output Terminals

TRIPLE 3-INPUT
POSITIVE-AND GATES
WITH OPEN-COLLECTOR OUTPUTS

15

positive logic:

Y = ABC

SN54H15 (J, W) SN74H15 (J, N)
SN54LS15 (J, W) SN74LS15 (J, N)
SN54S15 (J, W) SN74S15 (J, N)

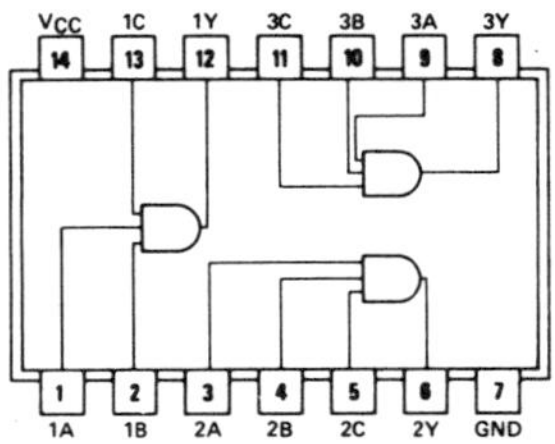

Operation & Application: Device contains three independent three-input AND gates. It is essentially an open-collector version of the 7411 device; hence a pull-up resistor (2 to 3 kohms) is required between each gate output terminal and the +5 volt dc power supply.

Propagation delay: 10 nS

Current Requirements per package: 10 mA

Device Number: 7416

Type of Device: Open-Collector Hex Inverter (15 volts)

HEX INVERTER BUFFERS/DRIVERS
WITH OPEN-COLLECTOR
HIGH-VOLTAGE OUTPUTS

16

positive logic:

$Y = \overline{A}$

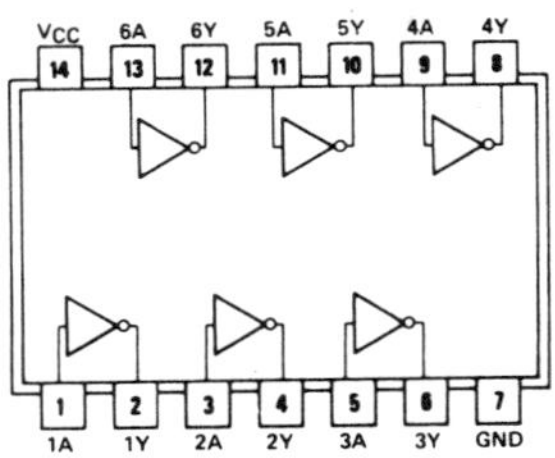

SN5416 (J, W) SN7416 (J, N)

Operation & Application: Six independent inverter/driver stages sharing common power supply and ground terminals. A HIGH applied to any input will produce a LOW on the related output terminal. Since this is an open-collector output device, we must have a 2 to 3 kohm pull-up resistor between each output terminal used and +5 volts dc. Each output can sink up to 40 mA, and will tolerate potentials to +15 volts dc. The package power supply (pins 14 and 7) must, however, be operated at +5 volts dc or damage to the IC will result.

Propagation delay: 10-15 nS

Current Requirements per package: 25-35 mA

Device Number: 7417

Type of Device: Noninverting Hex Driver with Open-Collector Outputs (15 volts)

HEX BUFFERS/DRIVERS WITH OPEN-COLLECTOR HIGH-VOLTAGE OUTPUTS

17

SN5417 (J, W)
SN7417 (J, N)

positive logic:
Y = A

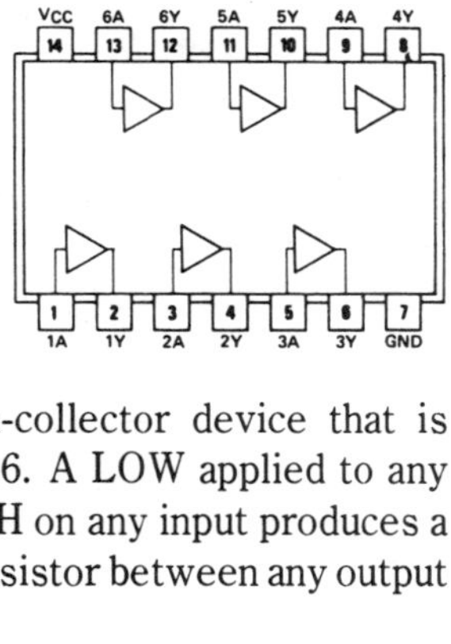

Operation & Application: Open-collector device that is simply a noninverting version of the 7416. A LOW applied to any input produces a LOW output, and a HIGH on any input produces a HIGH output. Use a 2 to 3 kohm pull-up resistor between any output used and +5 volts dc.

Propagation delay: 5-7 nS (output HIGH) or 20-25 nS (output LOW)

Current Requirements per package: 25 mA

Device Number: 7420

Type of Device: Dual Four-Input NAND Gate

DUAL 4-INPUT POSITIVE-NAND GATES

20

positive logic:
$Y = \overline{ABCD}$

SN5420 (J)	**SN7420 (J, N)**
SN54H20 (J)	**SN74H20 (J, N)**
SN54L20 (J)	**SN74L20 (J, N)**
SN54LS20 (J, W)	**SN74LS20 (J, N)**
SN54S20 (J, W)	**SN74S20 (J, N)**

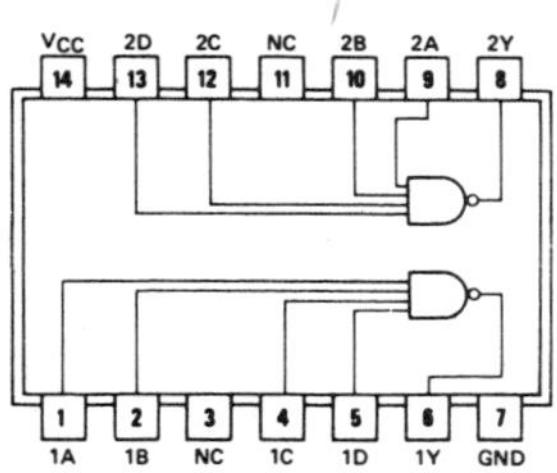

SN5420 (W)
SN54H20 (W)
SN54L20 (T)

NC—No internal connection

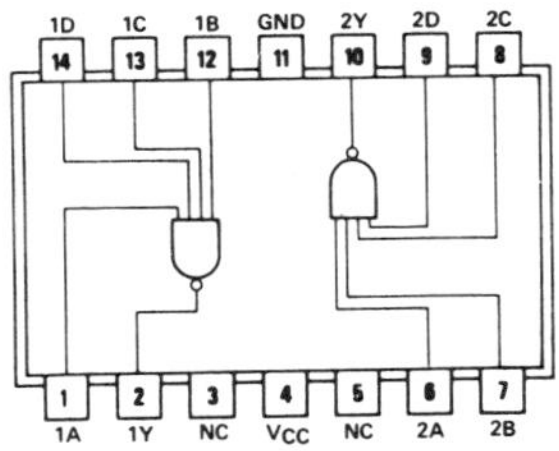

Operation & Application: Device contains two four-input NAND gates. A LOW on any one input of either gate produces a HIGH output; All inputs of any one gate must be HIGH for the output to be LOW. Each gate can be used independently, except for power supply and ground terminals.

Propagation delay: 10-12 nS

Current Requirements per package: 4-8 mA

Device Number: 7421

Type of Device: Dual Four-Input AND Gate

DUAL 4-INPUT POSITIVE-AND GATES

positive logic: Y = ABCD

21

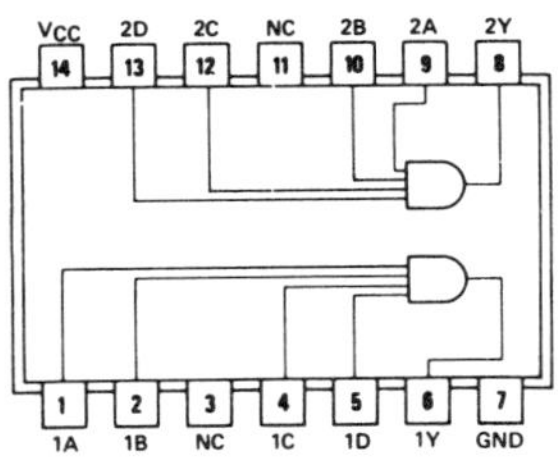

SN54H21 (J) SN74H21 (J, N)
SN54LS21 (J, W) SN74LS21 (J, N)

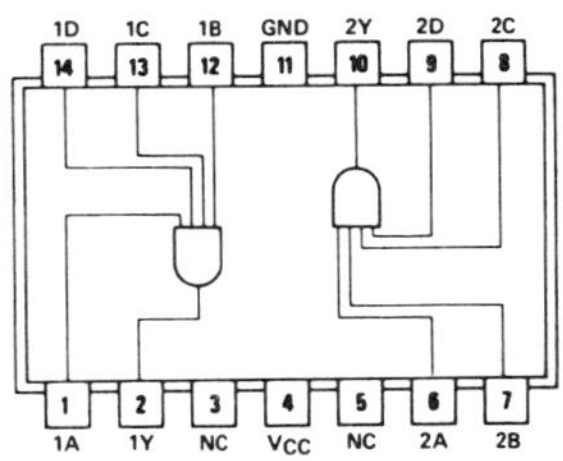

SN54H21 (W)

NC—No internal connection

Operation & Application: Device contains a pair of four-input AND gates. The logic for these gates is Y = ABCD. All four inputs must be HIGH for the output to be HIGH. Both gates are independent except for power supply and ground connections.

Propagation delay: 10 nS

Current Requirements per package: 15 mA

Device Number: 7430

Type of Device: Eight Input NAND Gate

8-INPUT POSITIVE-NAND GATES

positive logic:
Y = $\overline{ABCDEFGH}$

30

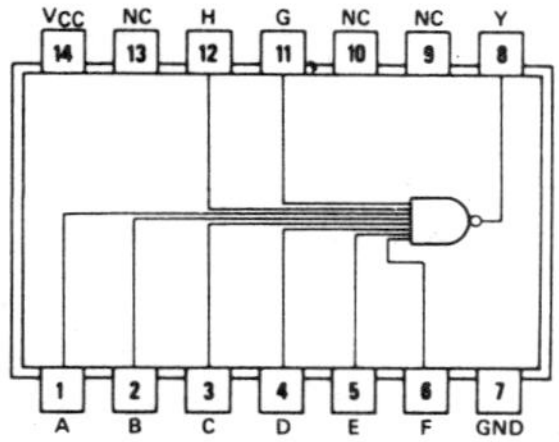

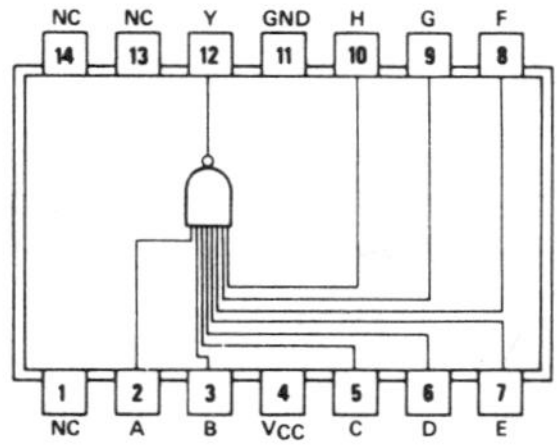

SN5430 (J)	SN7430 (J, N)	SN5430 (W)
SN54H30 (J)	SN74H30 (J, N)	SN54H30 (W)
SN54L30 (J)	SN74L30 (J, N)	SN54L30 (T)
SN54LS30 (J, W)	SN74LS30 (J, N)	
SN54S30 (J, W)	SN74S30 (J, N)	NC—No internal connection

Operation & Application: This TTL device contains a single eight-input NAND gate. If any one input is LOW, then the output will be HIGH. It requires ALL eight inputs HIGH for the output to be LOW. Has applications such as eight-bit decoders for addresses, data values and input/output port numbers in microcomputer circuits.

Propagation delay: 8-12 nS

Current Requirements per package: 2-4 mA

Device Number: 7432

Type of Device: Quad Two-Input OR Gate

QUADRUPLE 2-INPUT POSITIVE-OR GATES

32

positive logic:
Y = A+B

SN5432 (J, W)
SN54LS32 (J, W)
SN54S32 (J, W)

SN7432 (J, N)
SN74LS32 (J, N)
SN74S32 (J, N)

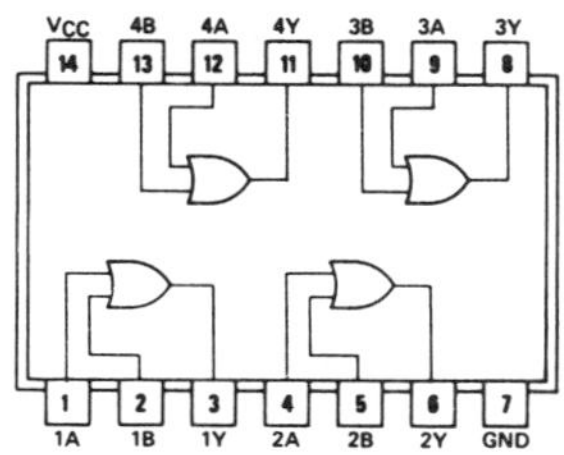

Operation & Application: Device contains four independent (except for power supply and ground) TTL OR two-input gates. The output of any gate will be HIGH when either input is HIGH. The output is LOW only when both inputs are also LOW.

Propagation delay: 10-12 nS

Current Requirements per package: 18-22 mA

Device Number: 7433

Type of Device: Quad Two-Input NOR Gate with Open-Collector Terminals

QUADRUPLE 2-INPUT
POSITIVE-NOR BUFFERS
WITH OPEN-COLLECTOR OUTPUTS

33

positive logic:
Y = $\overline{A+B}$

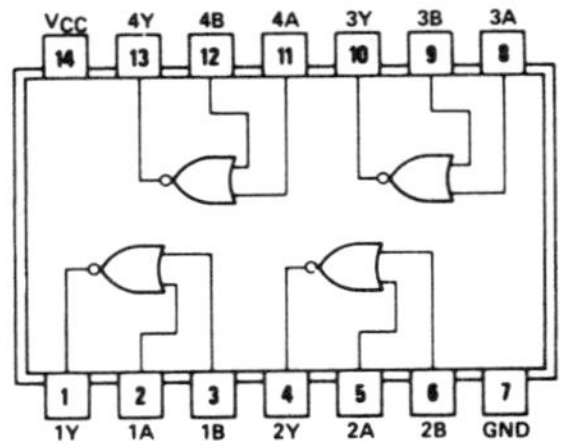

SN5433 (J, W) SN7433 (J, N)
SN54LS33 (J, W) SN74LS33 (J, N)

Operation & Application: This device contains four independent (except for power supply and ground) two-input NOR gates. Each NOR gate has an open collector output terminal, so requires a 2 to 3 kohm pull-up resistor between the output terminal(s) and +5 volt dc. A high on any one input will cause the output to be LOW. It requires both inputs to be LOW for the output terminal to be HIGH.

Propagation delay: 8-12 nS

Current Requirements per package: 20 mA

Device Number: 7437

Type of Device: Quad Two-Input NAND Gate/Buffer

QUADRUPLE 2-INPUT
POSITIVE-NAND BUFFERS

37

positive logic:
Y = $\overline{AB}$

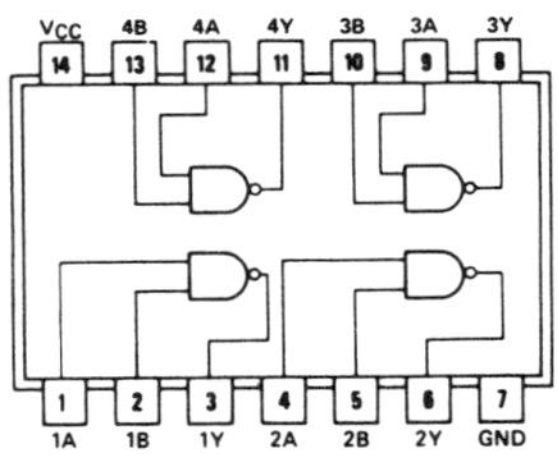

SN5437 (J, W) SN7437 (J, N)
SN54LS37 (J, W) SN74LS37 (J, N)
SN54S37 (J, W) SN74S37 (J, N)

Operation & Application: This device is functionally similar to the 7400 in that it contains four independent two-input NAND gates. If either input on any one gate is LOW, then its associated output is HIGH. It requires both inputs to be HIGH for the output to be LOW. A principle difference between this device and the 7400 is the output drive capacity. The 7400 can drive only a few TTL input terminals from its output, while the 7437 will drive up to 30 TTL input lines; hence the 7437 is called a NAND gate/buffer.

Propagation delay: 10-12 nS

Current Requirements per package: 3-6 mA (all outputs HIGH), 30-40 mA (all outputs LOW)

Device Number: 7438

Type of Device: Quad Two-Input NAND Gate/Buffer

QUADRUPLE 2-INPUT
POSITIVE-NAND BUFFERS
WITH OPEN-COLLECTOR OUTPUTS

38

positive logic:
$Y = \overline{AB}$

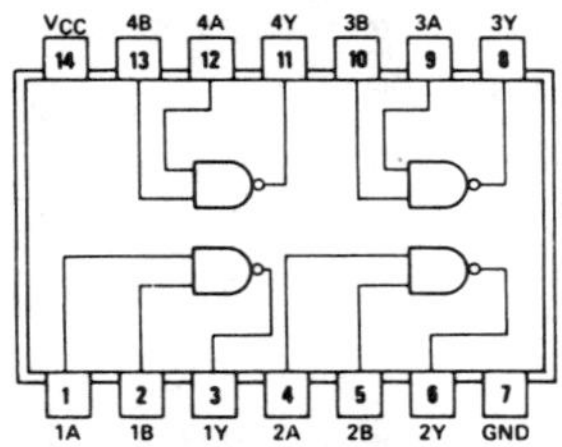

SN5438 (J, W) SN7438 (J, N)
SN54LS38 (J, W) SN74LS38 (J, N)
SN54S38 (J, W) SN74S38 (J, N)

Operation & Application: Same as 7437, except that it requires pull-up resistors between each output terminal and +5 volts dc. See discussion of 7437 for other specs.

Device Number: 7440

Type of Device: Dual Four-Input NAND Gate/Buffer

DUAL 4-INPUT
POSITIVE-NAND BUFFERS

positive logic:
$Y = \overline{ABCD}$

40

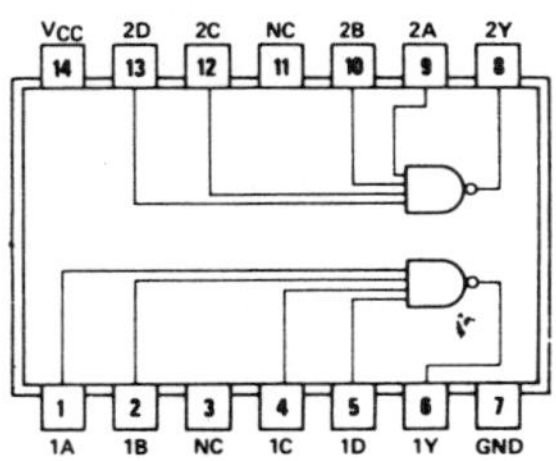

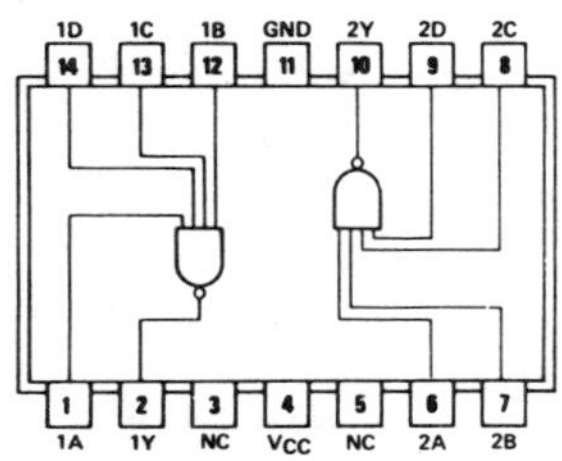

SN5440 (J) SN7440 (J, N)
SN54H40 (J) SN74H40 (J, N)
SN54LS40 (J,W) SN74LS40 (J, N)
SN54S40 (J, W) SN74S40 (J, N)

SN5440 (W)
SN54H40 (W)

NC—No internal connection

Operation & Application: Device contains two independent four-input NAND gates. On either gate, a LOW on any one input will force the output HIGH; a HIGH on both inputs is required to force the output LOW. This device is a buffer as well, meaning that it will drive far more TTL outputs than a normal gate; the 7440 device will drive up to 30 TTL loads.

Propagation delay: 10-12 nS.

Current Requirements per package: 5 mA (both outputs HIGH) to 20 mA (both outputs LOW)

Device Number: 7442

Type of Device: BCD-to-1-of-10 Decoder (TTL)

4 LINE-TO-10-LINE DECODERS

42 BCD-TO-DECIMAL

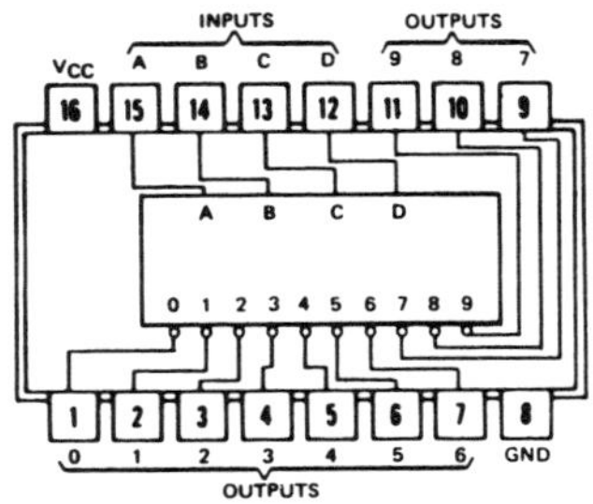

SN5442A (J, W)	SN7442A (J, N)
SN54L42 (J)	SN74L42 (J, N)
SN54LS42 (J, W)	SN74LS42 (J, N)

Operation & Application: This device accepts four BCD (1-2-4-8 code scheme) input lines, and produces a unique output depending upon which of the ten possible BCD words is applied to those inputs. The discrete outputs are active-LOW and can sink 16 milliamperes. Note, however, that the outputs are TTL and cannot be used directly with a high-voltage display as *Nixie*® tubes; an interface transistor is required. All codes over 1001_2 are invalid. The valid codes are:

Decimal	*BCD*	*Output Line LOW*	*Decimal*	*BCD*	*Output Line LOW*
0	0000	1	5	0101	6
1	0001	2	6	0110	7
2	0010	3	7	0111	9
3	0011	4	8	1000	10
4	0100	5	9	1001	11

Propagation delay: 15-20 nS
Current Requirements per package: 30 mA

Device Number: 7443

Type of Device: Excess-3-to-1-of-10 Decoder

4 LINE-TO-10-LINE DECODERS

43 EXCESS-3-TO-DECIMAL

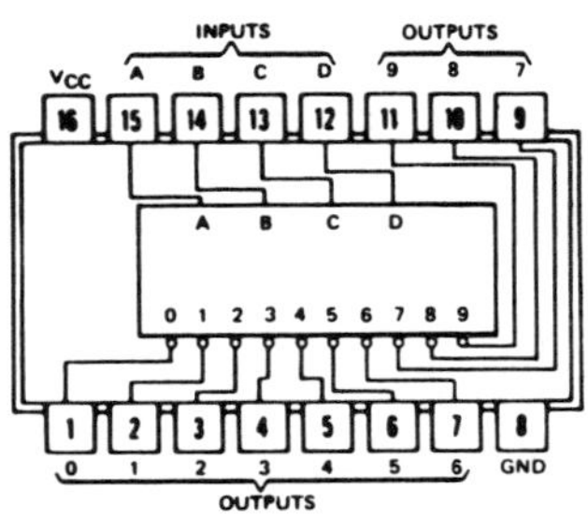

SN5443A (J, W)	SN7443A (J, N)
SN54L43 (J)	SN74L43 (J, N)

Operation & Application: Similar to the 7442, except that, *instead* of BCD the inputs accept "Excess-3 Code". This code is used in some cases, and is simply the BCD code with +3 added. For example, the BCD for zero is 0000, while the Excess-3 code for 0000 is the BCD for +3, or 0011.

Propagation delay: 15-20 nS

Current Requirements per package: 30 mA

Device Number: 7444

Type of Device: Excess-3-Gray-to-1-of-10 Decoder

4 LINE-TO-10-LINE DECODERS

44 EXCESS-3-GRAY-TO-DECIMAL

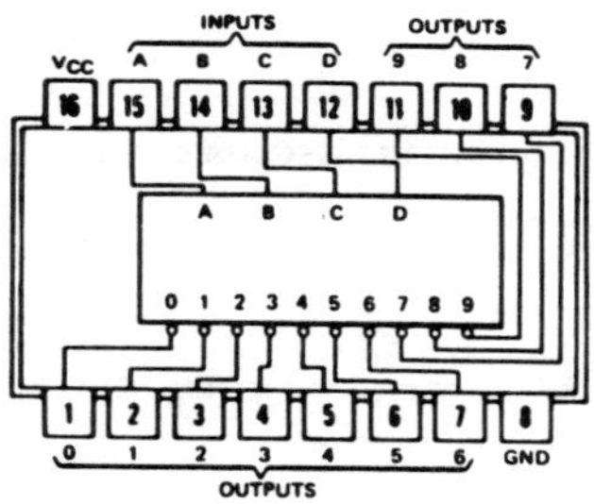

SN5444A (J, W) SN7444A (J, N)
SN54L44 (J) SN74L44 (J, N)

Operation & Application: The Gray code is a four-bit code that only changes by one bit every time it is incremented or decremented. The Excess-3-Gray code, then, is an Excess-3 code that will change only one of the four possible bits when the input value changes. The 7444 device is similar to 7443 and 7442 devices in other respects.

Propagation delay: 15-20 nS

Current Requirements per package: 30 mA

Device Number: 7445

Type of Device: High Power BCD-to-1-of-10 Decoder

BCD-TO-DECIMAL DECODER/DRIVER

45 LAMP, RELAY, OR MOS DRIVER
80-mA CURRENT SINK
OUTPUTS OFF FOR INVALID CODES

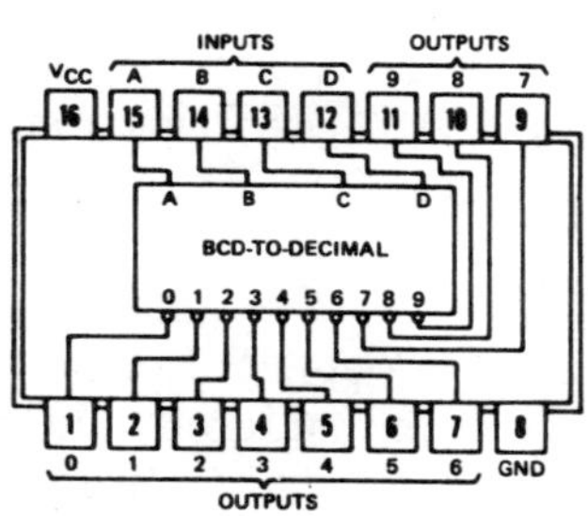

SN5445 (J, W) SN7445 (J, N)

Operation & Application: Device is similar to the 7442 in description, with the exception that the outputs are not limited to TTL levels. The 7445 output lines will handle up to +30 volts dc and sink to 80 mA of current. In other respects, the device is 7442. Some designs use this device in place of the 7442 despite the higher current requirement of the main package and the longer propagation delay.

Propagation delay: 48 nS

Current Requirements per package: 45 mA

Device Number: 7447

Type of Device: BCD-To-Seven-Segment Decoder

BCD-TO-SEVEN-SEGMENT DECODERS/DRIVERS

46 ACTIVE-LOW, OPEN-COLLECTOR, 30-V OUTPUTS

47 ACTIVE-LOW, OPEN-COLLECTOR, 15-V OUTPUTS

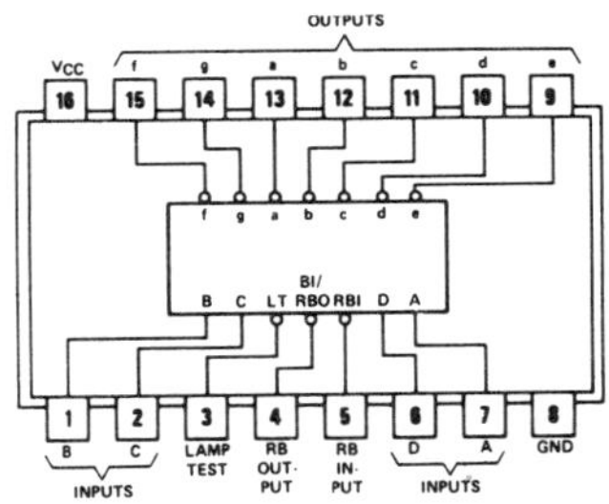

SN5446A (J, W)	SN7446A (J, N)
SN54L46 (J)	SN74L46 (J, N)
SN5447A (J, W)	SN7447A (J, N)
SN54L47 (J)	SN74L47 (J, N)
SN54LS47 (J, W)	SN74LS47 (J, N)

Operation & Application: This device is intended to interface the popular seven segment LED numerical readout with a counter or other source that generates BCD code in the 1-2-4-8 pattern. Each output will sink up to 40 volts, and can tolerate to +30 volts (making it possible to accommodate other-than-LED seven segments readouts, as well). If LED displays are used, then a current limiting resistor (330 to 390 ohms) must be used between the segment terminal on the display and its associated terminal on the 7447; one current-limiting resistor is needed for each segment. The LAMP TEST input (pin no. 3) is kept HIGH normally. But, for purposes of testing the display and 7447, it may be brought LOW. This will cause the display to light up "8," meaning that all segments are lit.

The ripple blanking terminals are used to blank leading (i.e., non-significant zeros. A LOW on the blanking input will extinquish the LED display if a BCD zero (0000) appears on the inputs of the 7447 preceding the device. A LOW on the blanking output appears when the BCD word applied to the input of this device is zero, and is used to blank following displays.

Propagation delay: 40-50 nS

Current Requirements per package: 45 mA (not counting segment current)

Device Number: 7473

Type of Device: Dual Level Triggered J-K Flip-Flop

DUAL J-K FLIP-FLOPS WITH CLEAR

73

'73, 'H73, 'L73
FUNCTION TABLE

INPUTS				OUTPUTS	
CLEAR	CLOCK	J	K	Q	$\bar{Q}$
L	X	X	X	L	H
H	⎍	L	L	Q_0	$\bar{Q}_0$
H	⎍	H	L	H	L
H	⎍	L	H	L	H
H	⎍	H	H	TOGGLE	

'LS73A
FUNCTION TABLE

INPUTS				OUTPUTS	
CLEAR	CLOCK	J	K	Q	$\bar{Q}$
L	X	X	X	L	H
H	↓	L	L	Q_0	$\bar{Q}_0$
H	↓	H	L	H	L
H	↓	L	H	L	H
H	↓	H	H	TOGGLE	
H	H	X	X	Q_0	$\bar{Q}_0$

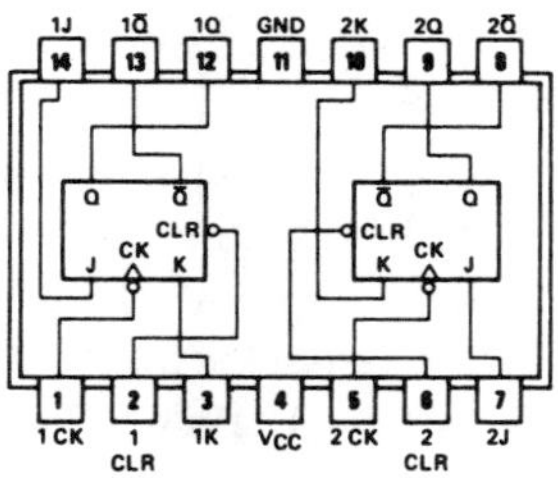

SN5473 (J, W) SN7473 (J, N)
SN54H73 (J, W) SN74H73 (J, N)
SN54L73 (J, T) SN74L73 (J, N)
SN54LS73A (J, W) SN74LS73A (J, N)

Operation & Application: This package contains two independent J-K flip-flops. The rules for the J-K flip-flop are given elsewhere in this book. Note that the FFs have *clear* (CLR) terminals. These active-LOW inputs will cause the FF to go to the state in which Q = LOW and NOT-Q = HIGH. A brief LOW on these terminals, then, will reset their respective flip-flops. CLR is normally held HIGH when not in use. Level triggering means that operation occurs when the clock is at a certain *level*, in this case LOW.

Propagation delay: N/A

Current Requirements per package: 18-22 mA

Device Number: 7474

Type of Device: Dual Edge-Triggered Type-D Flip-Flop

DUAL D-TYPE POSITIVE-EDGE-TRIGGERED FLIP-FLOPS WITH PRESET AND CLEAR

74

FUNCTION TABLE

INPUTS				OUTPUTS	
PRESET	CLEAR	CLOCK	D	Q	$\bar{Q}$
L	H	X	X	H	L
H	L	X	X	L	H
L	L	X	X	H*	H*
H	H	↑	H	H	L
H	H	↑	L	L	H
H	H	L	X	Q_0	$\bar{Q}_0$

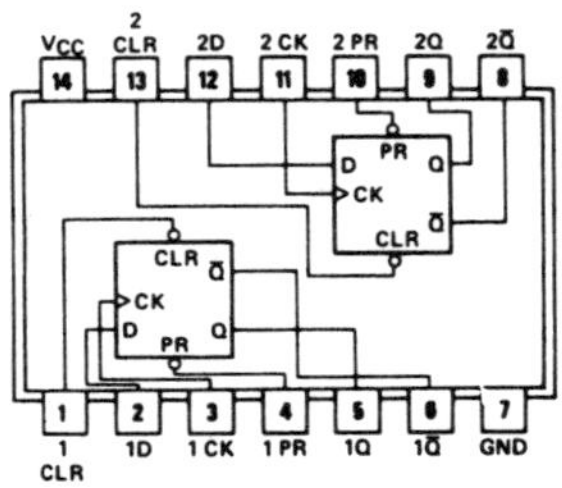

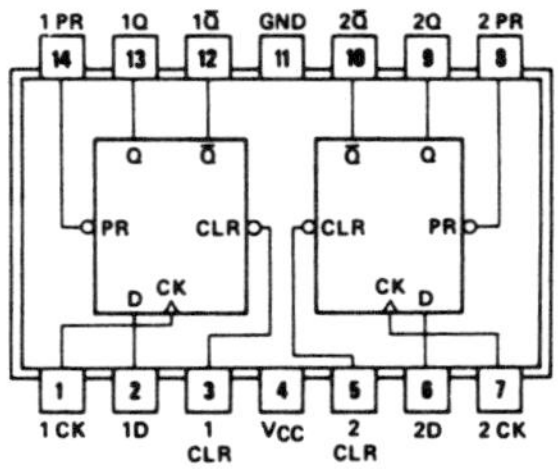

SN5474 (J)	SN7474 (J, N)	SN5474 (W)
SN54H74 (J)	SN74H74 (J, N)	SN54H74 (W)
SN54L74 (J)	SN74L74 (J, N)	SN54L74 (T)
SN54LS74A (J, W)	SN74LS74A (J, N)	
SN54S74 (J, W)	SN74S74 (J, N)	

Operation & Application: This device contains two independent Type-D flip-flops. A type-D FF will cause the data on the D-input to be transferred to the Q output only when the clock input is active. In an edge triggered device, the active period for the clock line is on a transition from HIGH to LOW, or from LOW to HIGH. In the case of the 7474 device, transfers occur on the transition from LOW to HIGH. The data only changes at the output during this period, regardless of changes on the D input. The 7474 is equipped with *clear* and *set* inputs. Since these are active-LOW inputs, they must be tied HIGH when not in use in order to prevent operation on noise pulses. The *set* input on either FF causes the output to go to the state in which Q = HIGH and NOT-Q = LOW. Similarly, the *clear* input will cause the output to go to the state in which Q = LOW and NOT-Q = HIGH. DO NOT EVER ALLOW CLEAR AND SET TO BE LOW SIMULTANEOUSLY.

Propagation delay: N/A

Current Requirements per package: 15-20 mA

Device Number: 7475

Type of Device: Quad Latch

4-BIT BISTABLE LATCHES

75

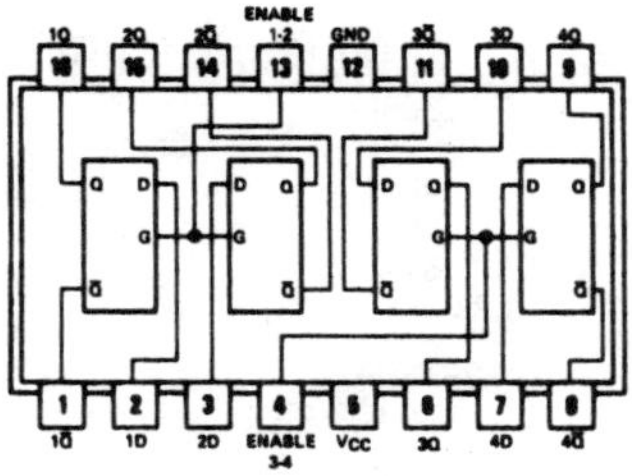

SN5475 (J, W) SN7475 (J, N)
SN54L75 (J) SN74L75 (J, N)
SN54LS75 (J, W) SN74LS75 (J, N)

FUNCTION TABLE
(Each Latch)

INPUTS		OUTPUTS	
D	G	Q	$\bar{Q}$
L	H	L	H
H	H	H	L
X	L	Q_0	$\bar{Q}_0$

H = high level, L = low level, X = irrelevant
Q_0 = the level of Q before the high-to-low transistion of G

Operation & Application: A "latch" is a type-D flip-flop, and is so named because it can remember (i.e., latch) a single bit of data. In the 7475 device, there are four Type-D FFs arranged in two banks of two each. Each pair has its own enable, or strobe, line. When the enable/strobe line is brought HIGH, then data on the D inputs is transferred to the Q outputs. After the enable/strobe input drops LOW again, the last valid data existing on the input will remain on the output indefinitely (as long as power is up and the clock remains LOW). In order to use this "dual two-input" latch in the popular four-bit quad-latch mode simply connect the two enable/strobe inputs together. A single 7475 may be used on BCD and hexadecimal (BCH) lines, while a pair will be useful in eight-bit microcomputer systems. 7475 devices are sometimes used to make latched output ports for microcomputers, but their main use is in digital counter applications.

Propagation delay: 20-25 nS

Current Requirements per package: 35 mA

Device Number: 7476

Type of Device: Dual J-K Flip-Flop (Level-Triggered)

DUAL J-K FLIP-FLOPS WITH PRESET AND CLEAR

76

'76, 'H76
FUNCTION TABLE

INPUTS					OUTPUTS	
PRESET	CLEAR	CLOCK	J	K	Q	$\bar{Q}$
L	H	X	X	X	H	L
H	L	X	X	X	L	H
L	L	X	X	X	H*	H*
H	H	⎍	L	L	Q_0	$\bar{Q}_0$
H	H	⎍	H	L	H	L
H	H	⎍	L	H	L	H
H	H	⎍	H	H	TOGGLE	

'LS76A
FUNCTION TABLE

INPUTS					OUTPUTS	
PRESET	CLEAR	CLOCK	J	K	Q	$\bar{Q}$
L	H	X	X	X	H	L
H	L	X	X	X	L	H
L	L	X	X	X	H*	H*
H	H	↓	L	L	Q_0	$\bar{Q}_0$
H	H	↓	H	L	H	L
H	H	↓	L	H	L	H
H	H	↓	H	H	TOGGLE	
H	H	H	X	X	Q_0	$\bar{Q}_0$

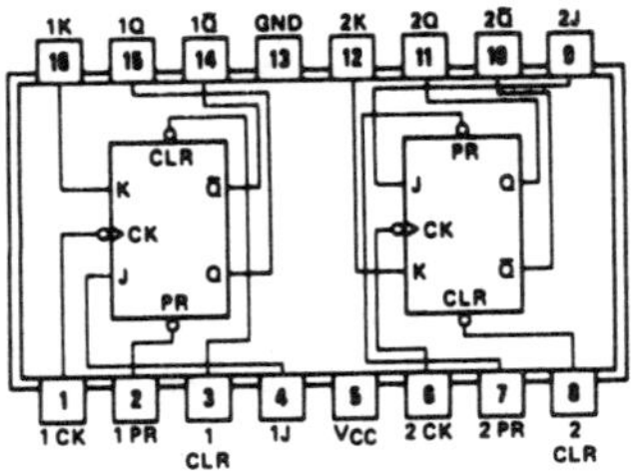

SN5476 (J, W) **SN7476 (J, N)**
SN54H76 (J, W) **SN74H76 (J, N)**
SN54LS76A (J, W) **SN74LS76A (J, N)**

Operation & Application: This device contains two independent J-K flip-flops. The rules for the operation of the J-K flip-flop are given elsewhere in this book. Note that the data applied to the J and K inputs must not change during the clock period (because it is a level triggered device). If the set and clear inputs are not used, then they must be tied to +5 volt through a current limiting resistor of 1 to 2 kohms. The set and clear inputs are active-LOW, so will operate when brought to ground potential. The set causes the FF outputs to go to the state in which Q = HIGH and NOT-Q = LOW. Similarly, the clear input causes the FF outputs to go to the state in which Q = LOW and NOT-Q = HIGH

Propagation delay: N/A

Current Requirements per package: 22 mA

Device Number: 7483

Type of Device: FULL-ADDER with Fast Carry (4-bit)

4-BIT BINARY FULL ADDERS WITH FAST CARRY

83

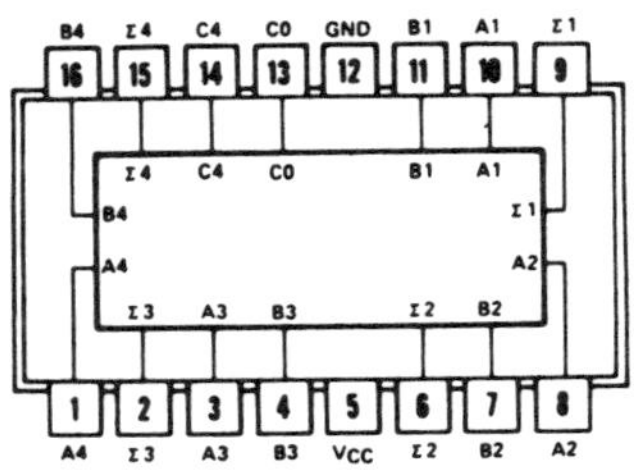

SN5483A (J, W) **SN74 83A (J, N)**
SN54LS83A (J, W) **SN74LS83A (J, N)**

Operation & Application: This device is a four-bit arithmetic unit that will add two four bit numbers that are encoded in the usual 1-2-4-8 system. The two numbers are input on the A and B lines (A1/B1 = 1, A2/B2 = 2, A3/B3 = 4 and A4/B4 = 8), and the summation is output on the "sigma" lines using the same weighting (1248). If the output number overflows (i.e., is larger that FF_{16}, then the carry bit (C4) is made HIGH. The C.O. line (pin no. 13) is used to cascade to the C4 line of the next least significant stage when 8 or longer bit words are used. In the case where only a four bit data word is used, the C.O. is grounded, indicating a LOW condition.

Propagation delay: 15-20 nS

Current Requirements per package: 65 mA

Device Number: 7484

Type of Device: 16-bit Random Access Memory

16-BIT RANDOM-ACCESS MEMORIES

84

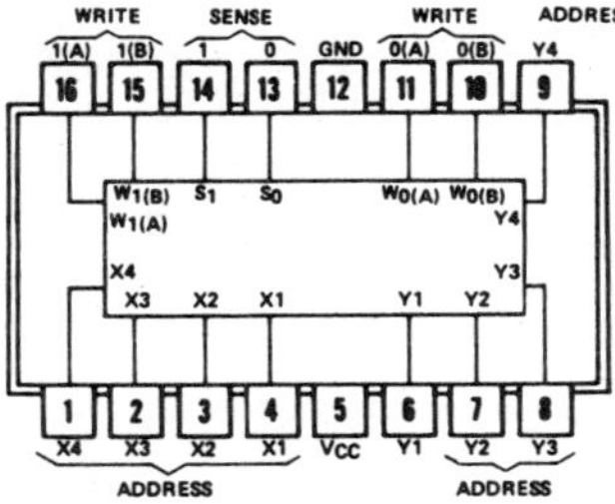

SN5484A (J, W) SN7484A (J, N)

Operation & Application: This device is used to form small scratch-pad memories in digital circuits and small-scale microprocessor projects. It contains a 4×4 matrix of 16 flip-flops and two write amplifiers. It features direct addressing and non-destructive readout.

Propagation delay: 18-30 nS

Current Requirements per package: 60 mA

Device Number: 7485

Type of Device: Four-Bit Magnitude Comparator

4-BIT MAGNITUDE COMPARATORS

85

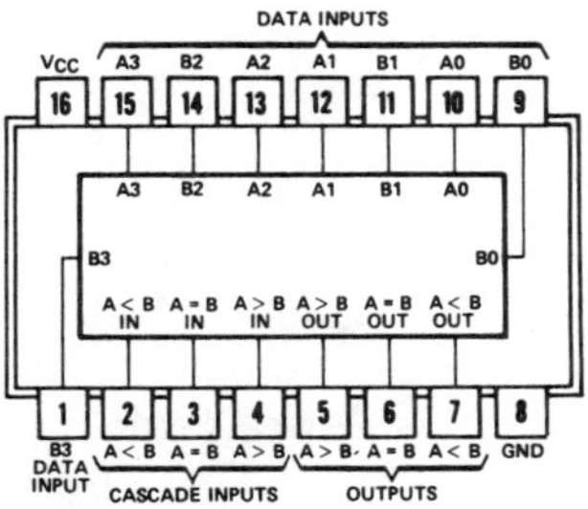

SN5485 (J, W) SN7485 (J, N)
SN54LS85 (J, W) SN74LS85 (J, N)
SN54S85 (J, W) SN74S85 (J, N)

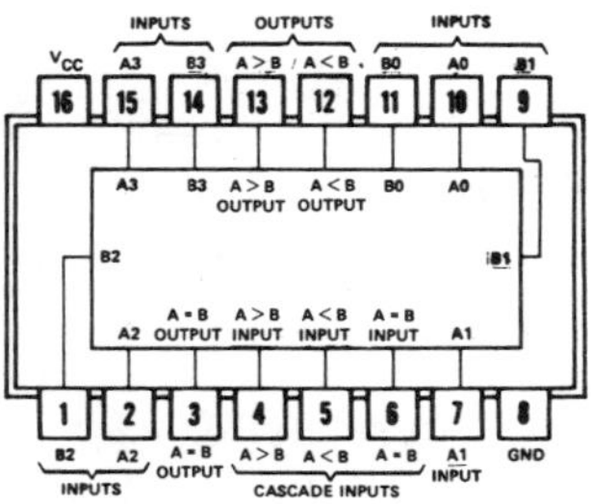

SN54L85 (J) SN74L85 (J, N)

Operation & Application: The 7485 device compares two four-bit binary words (1-2-4-8 coding) and issues one of three unique active-HIGH outputs: A is less than B, A is greater than B, and A equals B. The 7485 device has cascade inputs to sense the state of the previous (less significant) stage. The output of the most significant stage indicates the state of the entire N-bit word applied to the cascaded comparators. The cascade inputs of the least significant stage must be treated such that A=B is connected permanently HIGH and the other two are connected permanently LOW.

Propagation delay: 25 nS

Current Requirements per package: 60 mA

Device Number: 7486

Type of Device: Quad Exclusive-OR Gate

QUADRUPLE 2-INPUT EXCLUSIVE-OR GATES

86 $Y = A \oplus B = \overline{A}B + A\overline{B}$

FUNCTION TABLE

INPUTS		OUTPUT
A	B	Y
L	L	L
L	H	H
H	L	H
H	H	L

H = high level, L = low level

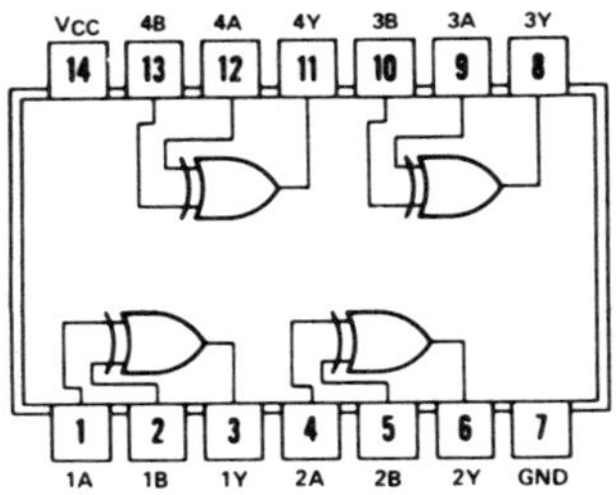

SN5486 (J, W) SN7486 (J, N)
SN54LS86 (J, W) SN74LS86 (J, N)
SN54S86 (J, W) SN74S86 (J, N)

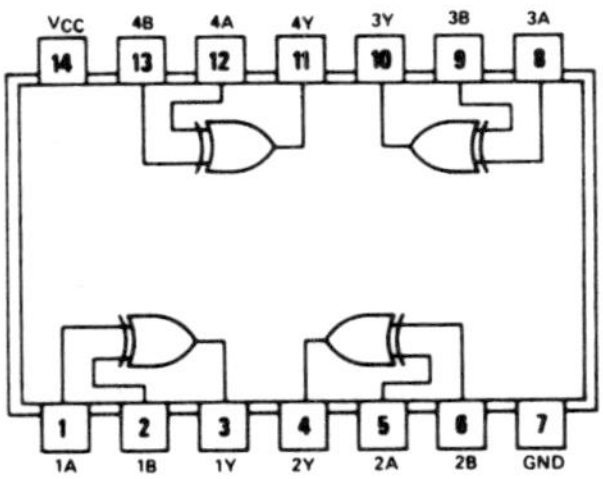

SN54L86 (J) SN74L86 (J, N)

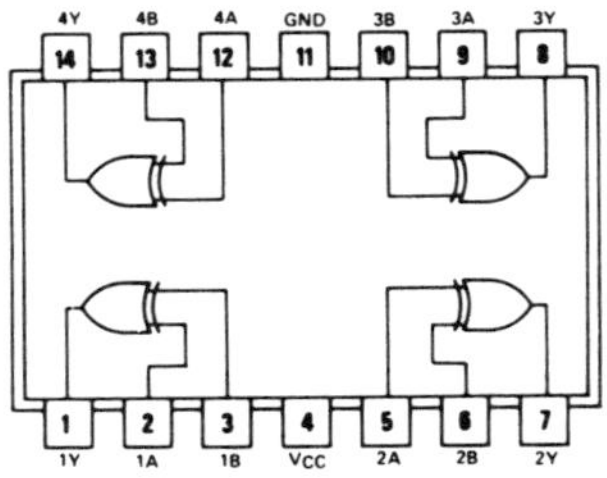

SN54L86 (T)

Operation & Application: This device contains four independent Exclusive-OR (XOR) gates. The output of each gate is HIGH only when one—and only one—input is HIGH. If both inputs are HIGH, or, if both inputs are LOW, then the output is LOW also.

Any time the same level is applied to both inputs of an XOR gate the output will be LOW.

Propagation delay: 20 nS
Current Requirements per package: 30 mA

Device Number: 7489

Type of Device: 16 × 4 (64-bit) Memory Element

64-BIT READ/WRITE MEMORIES

89 16 4-BIT WORDS

See Bipolar Microcomputer Components Data Book, LCC4270

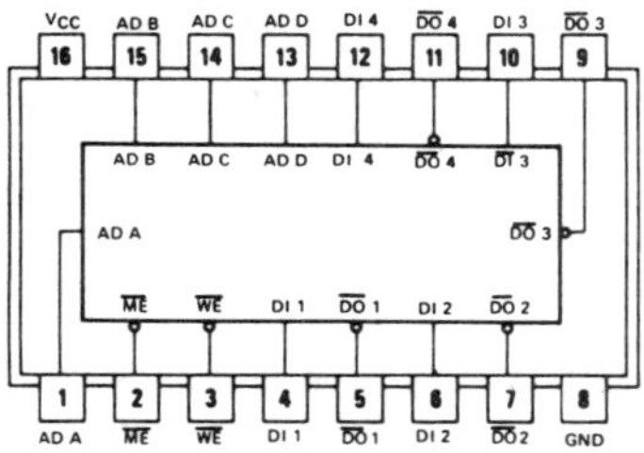

SN7489 (J, N)

Operation & Application: The 7489 device is a 64-bit memory arranged in a 4-bit array of 16 half-bytes. Data is input to the 7489 through four lines labeled DI1 through DI4. The address of the memory location where the data is stored is selected by a four-bit data bus consisting of the AD A through AD D lines (pins 1, 13, 14 and 15). There are two active-LOW *enable* lines: $\overline{WE}$ (write enable) and $\overline{ME}$ (memory enable). Normally, the $\overline{WE}$ is left HIGH and will be brought LOW when data is being written to the memory. The data will be stored at the location specified on the address bus at the instant $\overline{WE}$ is brought LOW. The $\overline{ME}$ line is normally left LOW; during this condition the complement of the word stored in the memory location selected by the address bus will appear on the output lines ($\overline{DO1}$ thru $\overline{DO4}$). The outputs will go tri-state (i.e., high impedance to both ground and V+) when both $\overline{ME}$ and $\overline{WE}$ are made HIGH.

Write time: Approximately 50 nanoseconds
Read time: Approximately 35 nanoseconds
Current Requirements per package: 75 mA

Device Number: 7490

Type of Device: Biquinary Divide-by-10 (decade) Counter with Decodable BCD Outputs

DECADE COUNTERS

90 DIVIDE-BY-TWO AND DIVIDE-BY FIVE

SN5490A (J, W)	SN7490A (J, N)
SN54L90 (J, T)	SN74L90 (J, N)
SN54LS90 (J, W)	SN74LS90 (J, N)

NC – No internal connection

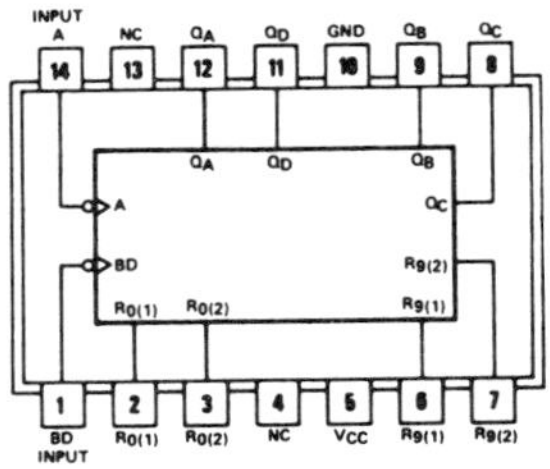

Operation & Application: This device contains two counters, one a divide-by-2 and the other a divide-by-5, hence its name *biquinary*. Although the two counters can be used separately, the usual method is to cascade the counters to make a divide-by-10 decade counter. The ABCD outputs (weighted 1-2-4-8) are decodable into decimal digits. In normal service, the clock is applied to input A (pin 14), and pins 1 and 12 are strapped making the device a cascade decade counter. The count increments on the negative-going transition of the input pulse, a reflection of the J-K flip-flops used internally. The counter can be reset to either 0 or 9. Pins 2-3 are the zero-reset lines; bringing either or both terminals momentarily HIGH will reset the counter to zero. Pins 6-7 are similarly able to reset the counter to 9 (1001); bringing either 6 or 7 momentarily HIGH will do the trick. Contrary to normal practice in the TTL line, +5-volt power is applied to pin no. 5, and power supply ground is pin no. 10.

Maximum operating frequency: 18-20 MHz

Current Requirements per package: 34 mA

Device Number: 7491

Type of Device: 8-Bit Serial-In Serial-Out (SISO) Shift Register with Gated Input Line

8-BIT SHIFT REGISTERS

91 SERIAL-IN, SERIAL-OUT GATED INPUT

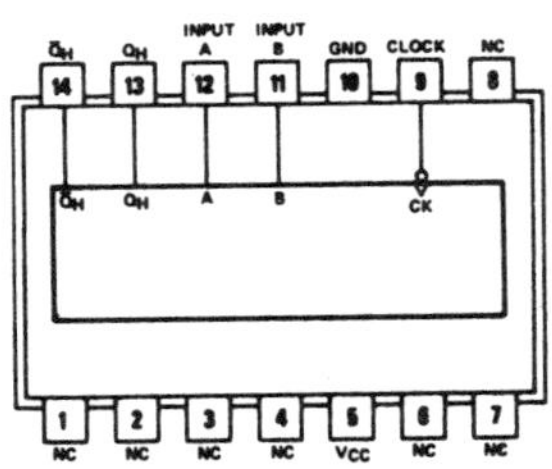

SN5491A (J)	SN7491A (J,N)
SN54L91 (J)	SN74L91 (J, N)
SN54LS91 (J)	SN74LS91 (J, N)

FUNCTION TABLE

INPUTS AT t_n		OUTPUTS AT t_{n+8}	
A	B	Q_H	$\bar{Q}_H$
H	H	H	L
L	X	L	H
X	L	L	H

H = high, L = low
X = irrelevant
t_n = Reference bit time, clock low
t_{n+8} = Bit time after 8 low-to-high clock transitions

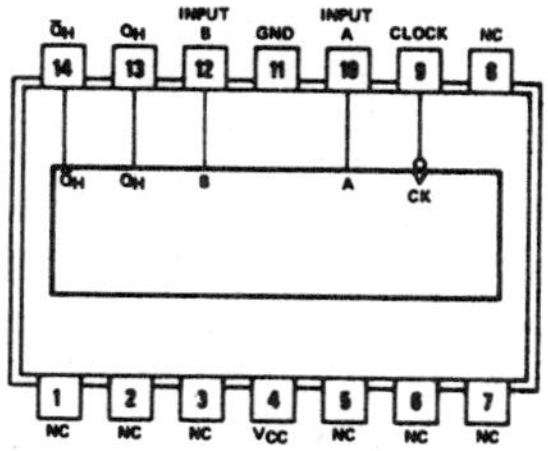

SN5491A (W)
SN54L91 (T)
SN54LS91 (W)

NC – No internal connections

Operation & Application: This device transfers data on the A and B inputs toward the Q output according to the rule given in the truth table shown in the illustration.

Device Number: 7492

Type of Device: Four-Bit Divide-by-2 and Divide-by-6 Counter (Base-12)

DIVIDE-BY-TWELVE COUNTERS

92 DIVIDE-BY-TWO AND DIVIDE-BY-SIX

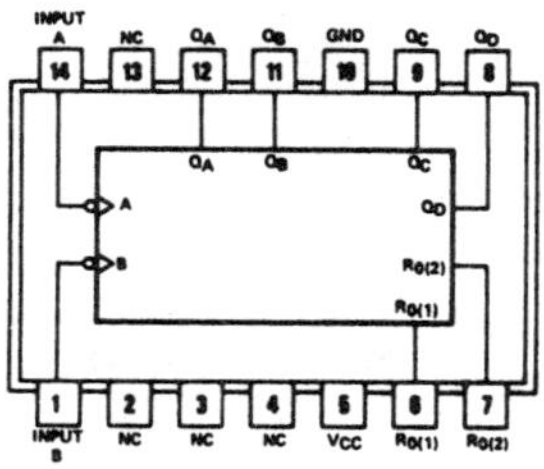

SN5492A (J, W) SN7492A (J, N)
SN54LS92 (J, W) SN74LS92 (J, N)

NC—No internal connection

Operation & Application: This divide-by-12 counter consists of two separate counters, one a divide-by-2 and the other divide-by-6. When the two counters are operated in cascade, the outputs will be ABCD', weighted 1-2-4-6. The $R_{0(1)}$ and $R_{0(2)}$ inputs must be kept at ground potential for normal base-12 counting. To reset the counter to zero, bring either $R_{0(1)}$ or $R_{0(2)}$ momentarily HIGH, or both simultaneously HIGH.

Operating Frequency: 18-20 MHz

Current Requirements per package: 34 mA

Device Number: 7493

Type of Device: Hexadecimal (Base-16) Four-Bit Counter arranged in Bioctal Form

93 DIVIDE-BY-TWO AND DIVIDE-BY-EIGHT

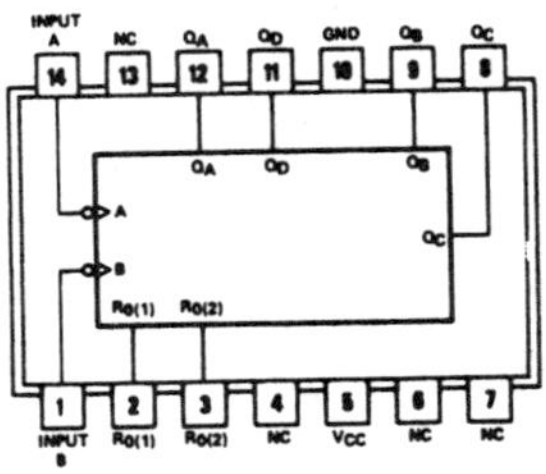

SN5493A (J, W) SN7493A (J, N)
SN54LS93 (J, W) SN74LS93 (J, N)

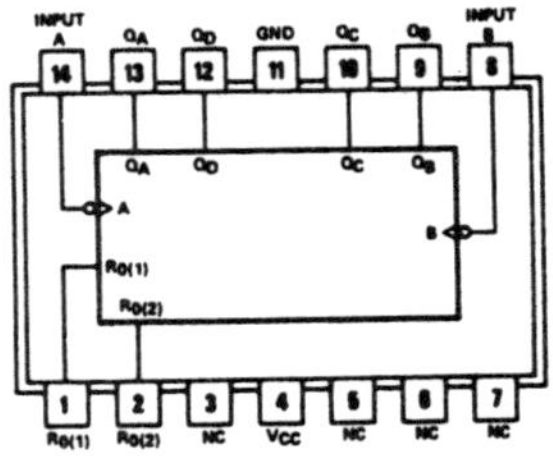

SN54L93 (J, T) SN74L93 (J, N)

NC—No internal connection

Operation & Application: This counter contains two separate counters, one a divide-by-2 and the other divide-by-8. When operated in cascade, these counters form a base-16 counter with four-bit outputs. The counter outputs are weighted in the standard 1-2-4-8 system (like the 7490, but this counter counts all the way to 16 (1111). Pins 2 and 3 are $R_{0(1)}$ and $R_{0(2)}$ terminals, and must be grounded for normal counting. The counter will reset to zero when either 2 or 3 or both are brought momentarily HIGH.

Maximum Operating Frequency: 18-20 MHz

Current Requirements per package: 34 mA

Device Number: 7495

Type of Device: 4-Bit Shift Register (PIPO), SR/SL

4-BIT SHIFT REGISTERS

95 PARALLEL IN/PARALLEL OUT SHIFT RIGHT, SHIFT LEFT SERIAL INPUT

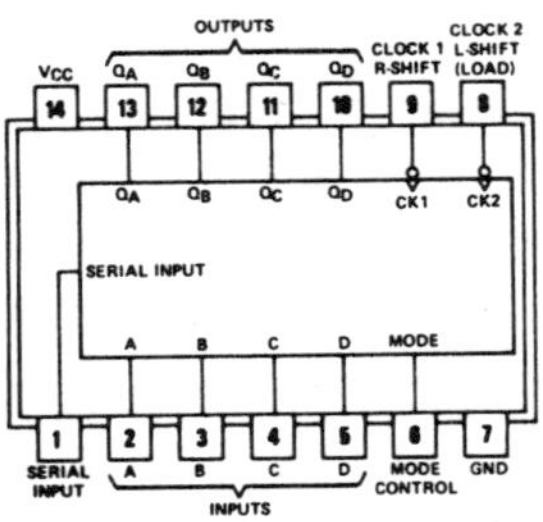

SN5495A (J, W) SN7495A (J, N)
SN54LS95B (J, W) SN74LS95B (J, N)

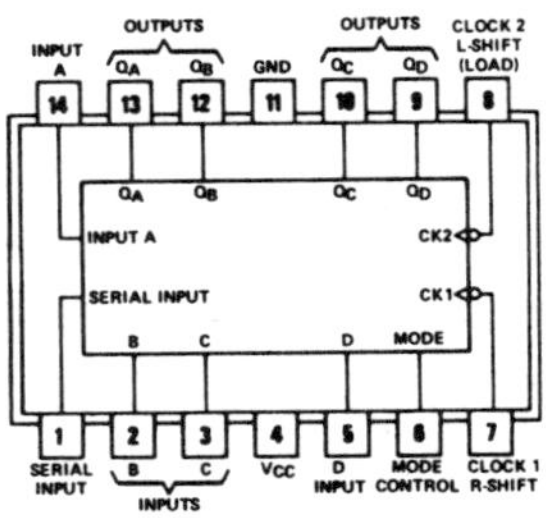

SN54L95 (J, T) SN74L95 (J, N)

Operation & Application: The 7495 device is a parallel-in parallel-out 4-bit shift register, and also includes a serial input (pin no. 1). Each stage in the 7495 has its own output stage because this is a parallel output shift register. The device also has *shift right* and *shift left* terminals (pins 9 and 8, respectively) so may be used in certain multiplication applications.

Maximum Operating Frequency: 30 MHz
Current Requirements per package: 43 mA

Device Number: 7496

Type of Device: 5-Bit Asynchronous Preset Shift Register (PIPO)

5-BIT SHIFT REGISTERS

96 ASYNCHRONOUS PRESET

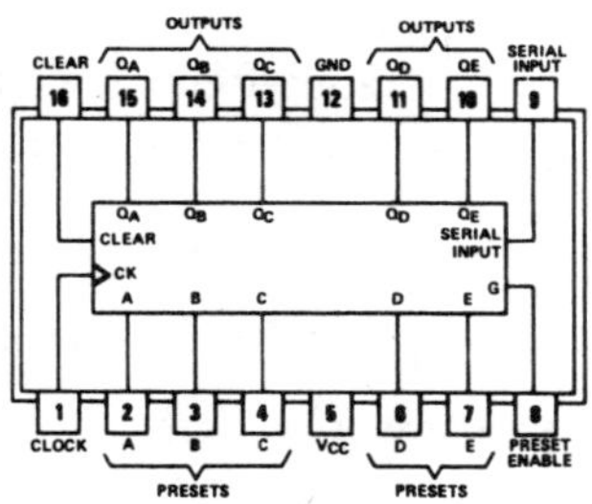

SN5496 (J, W) **SN7496 (J, N)**
SN54L96 (J) **SN74L96 (J, N)**
SN54LS96 (J, W) **SN74LS96 (J, N)**

Operation & Application: The 7496 device is a 5-stage parallel-in parallel-out shift register. The states of the stages can be preset by loading data on the A-B-C-D-E inputs. Each stage of the 7496 has its own discrete output (because it is PIPO), so this data will be reflected on the A-B-C-D-E outputs. A serial input is also provided, so the device can be operated as a SIPO register. There are three control inputs: *preset enable, clear* and *clock*. The preset enable input is left normally HIGH, and brought LOW when the register is to be set to zero. To load, therefore, one must first clear the register. The clock line will shift the data one place to the right every time it is brought from LOW to HIGH (positive-edge operated).

Maximum Operating Frequency: 10-12 MHz
Current Requirements per package: 50 mA

Device Number: 7497

Type of Device: Synchronous 6-Bit (i.e., Base-64) Binary Rate Multiplier

SYNCHRONOUS 6-BIT BINARY RATE MULTIPLIERS

97

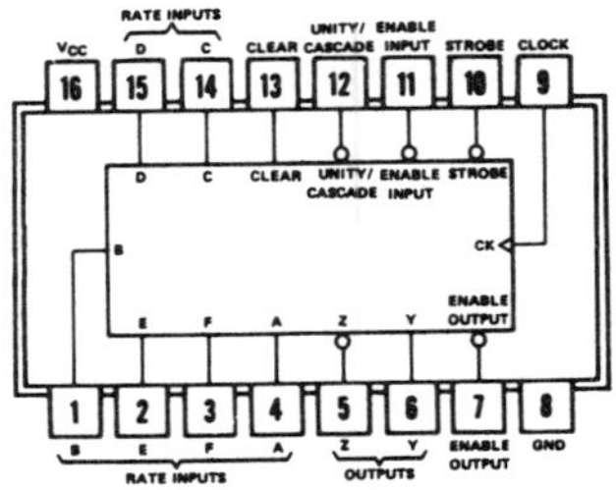

SN5497 (J, W) SN7497 (J, N)

Operation & Application: Six "rate" inputs select the number of output pulses that will be produced for every 64 input pulses. Rates from 0 to 63 are selectable.

Operating frequency: 18-20 MHz

Current Requirements per package: 70 mA

Device Number: 74100

Type of Device: Dual Four-Bit Data Latches

8-BIT BISTABLE LATCHES

100

FUNCTION TABLE
(Each Latch)

INPUTS		OUTPUTS	
D	G	Q	$\bar{Q}$
L	H	L	H
H	H	H	L
X	L	Q_0	$\bar{Q}_0$

H = high level, X = irrelevant
Q_0 = the level of Q before the high-to-low transition of G

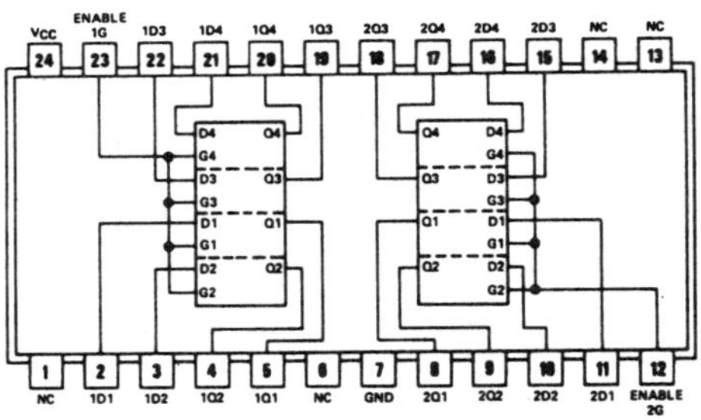

SN54100 (J, W) SN74100 (J, N)

NC — No internal connection

Operation & Application: The 74100 device contains a pair of four-bit data latches (i.e., type-D flip-flops). Each data latch can be operated separately, or, the two can be paralleled for 8-bit operation. The 74100 is thus used extensively in 8-bit microcomputer systems. When the strobe line (pin 12 or 23) is brought HIGH, the latch is enabled. Under that condition, data appearing on the inputs (1D1 - 1D4 or 2D1 - 2D4) will be transferred to the Q outputs. When the strobe line is brought LOW again, the Q outputs will remain at the same data that existed the instant the strobe dropped LOW. From then on the outputs will retain the data. If pins 12 and 23 are connected together, then the device will operate as an eight-bit data latch.

Device Number: 74103

Type of Device: Dual J-K Flip-Flop (Negative Edge-Triggered) with Clear

DUAL J-K NEGATIVE-EDGE-TRIGGERED FLIP-FLOPS WITH CLEAR

103

FUNCTION TABLE

INPUTS				OUTPUTS	
CLEAR	CLOCK	J	K	Q	$\bar{Q}$
L	X	X	X	L	H
H	↓	L	L	Q_0	$\bar{Q}_0$
H	↓	H	L	H	L
H	↓	L	H	L	H
H	↓	H	H	TOGGLE	
H	H	X	X	Q_0	$\bar{Q}_0$

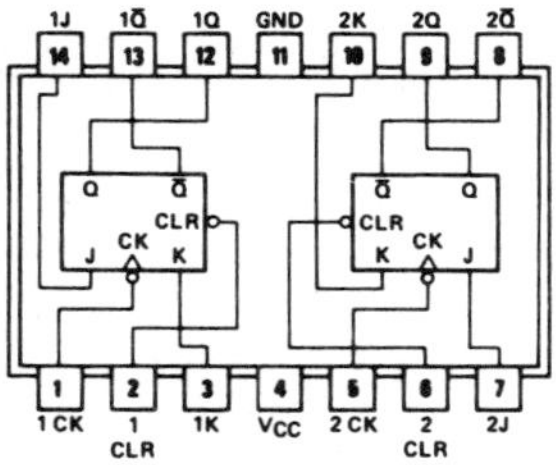

SN54H103 (J, W) SN74H103 (J, N)

Operation & Application: The 74103 contains a pair of independent J-K flip-flops that are negative edge triggered (i.e., output transitions in clocked mode occur when the clock line drops from HIGH to LOW). Only one Direct Mode operation is permissible: *clear.* If the clear input is brought momentarily LOW, the outputs go to the condition where Q = LOW and NOT-Q = HIGH. In the clocked mode, the outputs follow the truth table given above.

Maximum Toggle Frequencies: 18-20 MHz

Current Requirements per package: 22 mA

Device Number: 74107

Type of Device: Dual J-K Flip-Flop (Level Triggered) with Clear

DUAL J-K FLIP-FLOPS WITH CLEAR

107

'107

FUNCTION TABLE

INPUTS				OUTPUTS	
CLEAR	CLOCK	J	K	Q	$\bar{Q}$
L	X	X	X	L	H
H	⎍	L	L	Q_0	$\bar{Q}_0$
H	⎍	H	L	H	L
H	⎍	L	H	L	H
H	⎍	H	H	TOGGLE	

'LS107A

FUNCTION TABLE

INPUTS				OUTPUTS	
CLEAR	CLOCK	J	K	Q	$\bar{Q}$
L	X	X	X	L	H
H	↓	L	L	Q_0	$\bar{Q}_0$
H	↓	H	L	H	L
H	↓	L	H	L	H
H	↓	H	H	TOGGLE	
H	H	X	X	Q_0	$\bar{Q}_0$

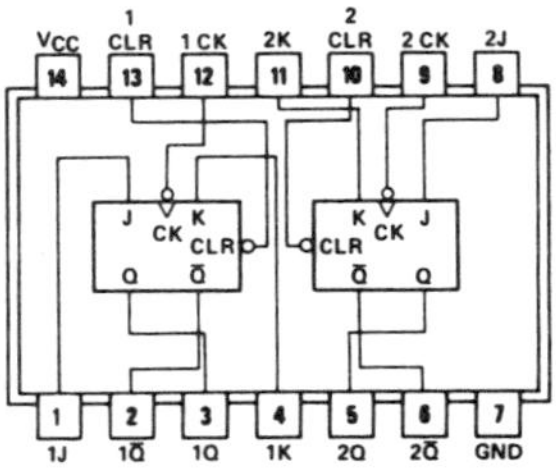

SN54107 (J) SN74107 (J, N)
SN54LS107A (J) SN74LS107A (J, N)

Operation & Application: The 74107 contains a pair of J-K flip-flops that are level-triggered as opposed to the usual negative edge triggering associated with J-K flip-flops. The only Direct Mode terminal on this device is the clear. When the clear is brought momentarily LOW, the outputs go to the condition where Q = LOW and NOT-Q= HIGH. For other operation, the 74107 follows the truth table.

Device Number: 74121

Type of Device: Non-retriggerable Monostable Multivibrator (one-shot)

MONOSTABLE MULTIVIBRATORS

121

FUNCTION TABLE

INPUTS			OUTPUTS	
A1	A2	B	Q	$\overline{Q}$
L	X	H	L	H
X	L	H	L	H
X	X	L	L	H
H	H	X	L	H
H	↓	H	⊓	⊔
↓	H	H	⊓	⊔
↓	↓	H	⊓	⊔
L	X	↑	⊓	⊔
X	L	↑	⊓	⊔

NOTES: 1. An external capacitor may be connected between C_{ext} (positive) and R_{ext}/C_{ext}.
2. To use the internal timing resistor, connect R_{int} to V_{CC}. For improved pulse width accuracy and repeatability, connect an external resistor between R_{ext}/C_{ext} and V_{CC} with R_{int} open-circuited.

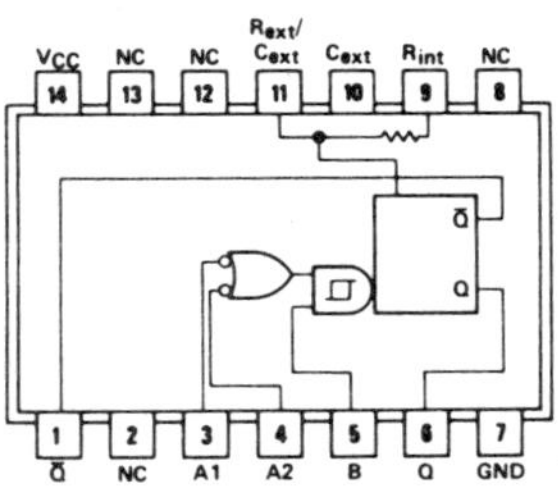

SN54121 (J, W) SN74121 (J, N)
SN54L121 (J, T) SN74L121 (J, N)
'121 . . . R_{int} = 2 kΩ NOM
'L121 . . . R_{int} = 4 kΩ NOM
NC—No internal connection

Operation & Application: The 74121 device is a monostable multivibrator. In response to a pulse on a trigger input will cause the output to snap to the condition Q = HIGH for a preset period of time. Timing of the output pulse is set by the combination of a resistor and capacitor connected to the device (C between pins 10 & 11, R between pin 11 and +5 Volts). The approximate output period is given by 0.7RC. There are several combinations of the A1, A2 and B trigger inputs that provide different modes of triggering, or inhibit status. Both negative-going and positive-going trigger pulses can be accommodated. For example, if A1 = A2 + LOW, then B will trigger the output pulse when a positive-going trigger signal is received. Either A1 or A2 can be made into a negative-going trigger pin if B and the other "A" trigger input are held HIGH.

Current Requirements per package: 25 milliamperes

Device Number: 74122

Type of Device: Retriggerable Monostable Multivibrator (one-shot)

RETRIGGERABLE MONOSTABLE MULTIVIBRATORS WITH CLEAR

122 FUNCTION TABLE

INPUTS					OUTPUTS	
CLEAR	A1	A2	B1	B2	Q	$\bar{Q}$
L	X	X	X	X	L	H
X	H	H	X	X	·L	H
X	X	X	L	X	L	H
X	X	X	X	L	L	H
H	L	X	↑	H	⎍	⊔
H	L	X	H	↑	⎍	⊔
H	X	L	↑	H	⎍	⊔
H	X	L	H	↑	⎍	⊔
H	H	↓	H	H	⎍	⊔
H	↓	↓	H	H	⎍	⊔
H	↓	H	H	H	⎍	⊔
↑	L	X	H	H	⎍	⊔
↑	X	L	H	H	⎍	⊔

NOTES: 1. An external timing capacitor may be connected between C_{ext} and R_{ext}/C_{ext} (positive).

2. For accurate repeatable pulse widths, connect an external resistor between R_{ext}/C_{ext} and V_{CC} with R_{int} open-circuited.

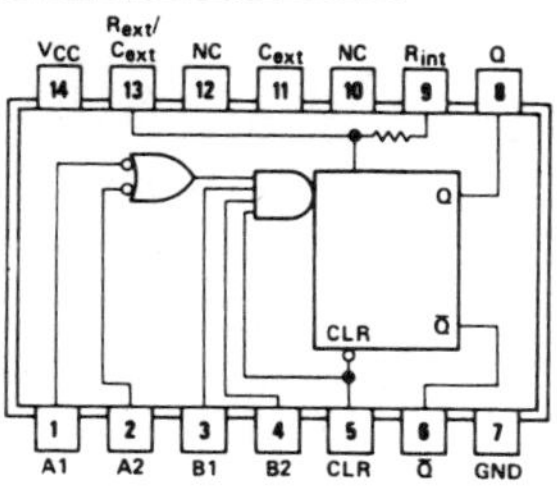

SN54122 (J, W) **SN74122 (J, N)**
SN54L122 (J, T) **SN74L122 (J, N)**
SN54LS122 (J, W) **SN74LS122 (J, N)**
'122 . . . R_{int} = 10 kΩ NOM
'L122 . . . R_{int} = 20 kΩ NOM
'LS122 . . . R_{int} = 10 kΩ NOM

NC—No internal connection

Operation & Application: The 74122 device is similar to the 74121 described previously, except that it has a slightly different trigger-input arrangement, and it is retriggerable. A normal one-shot cannot be retriggered to produce an extended output period until after the present pulse is completed (some also require a slight refractory period after time-out!). The retriggerable one-shot, however, can be retriggered while the output pulse is still present. The total Q = HIGH time will be extended for one 0.7RC time period each time a trigger pulse is applied. In addition to the triggering differences, the 74122 also has a clear input that will normally remain HIGH, but when it is brought LOW will force the output to the state Q = LOW.

Current Requirements per package: 25 milliamperes

Device Number: 74123

Type of Device: Dual Retriggerable Monostable Multivibrator (one-shot)

DUAL RETRIGGERABLE MONOSTABLE MULTIVIBRATORS WITH CLEAR

123

FUNCTION TABLE

INPUTS			OUTPUTS	
CLEAR	A	B	Q	$\bar{Q}$
L	X	X	L	H
X	H	X	L	H
X	X	L	L	H
H	L	↑	⊓	⊔
H	↓	H	⊓	⊔
↑	L	H	⊓	⊔

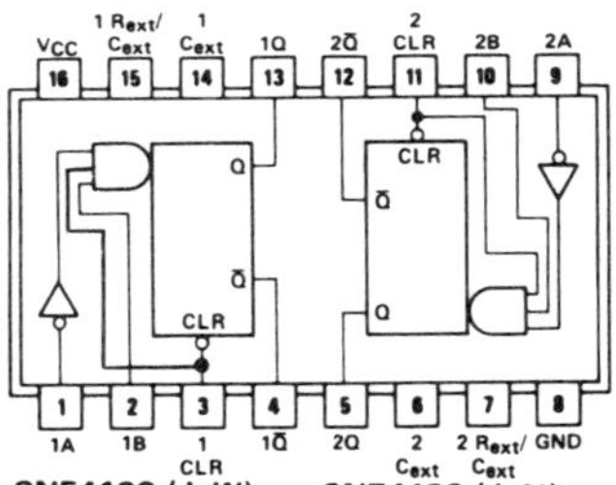

SN54123 (J, W) SN74123 (J, N)
SN54L123 (J) SN74L123 (J, N)
SN54LS123 (J, W) SN74LS123 (J, N)

Operation & Application: The 74123 device contains two independent one-shot stages, each of which is similar to the 74122 device. Each one-shot has its own clear terminal that forces Q = LOW; clear is an active-LOW input. Both one-shots also have two trigger inputs which can be configured to either negative-going or positive-going pulses. Positive-going triggering is accomplished by holding the A-input LOW, and applying the positive-going pulse to the B-input. Similarly, if the trigger pulse is to be negative-going, then hold B HIGH and apply the trigger pulse to the A-input. Like the other forms of one-shot device, the resistor and capacitor used to time the 74123 should have a value of 5 kohms to 25 kohms for the resistor and more than 10 picofarads for the capacitor.

Current Requirements per package: 48 milliamperes

Device Number: 74125

Type of Device: Quad Bus Buffer Gates with Tri-State Outputs

QUADRUPLE BUS BUFFER GATES WITH THREE-STATE OUTPUTS

125

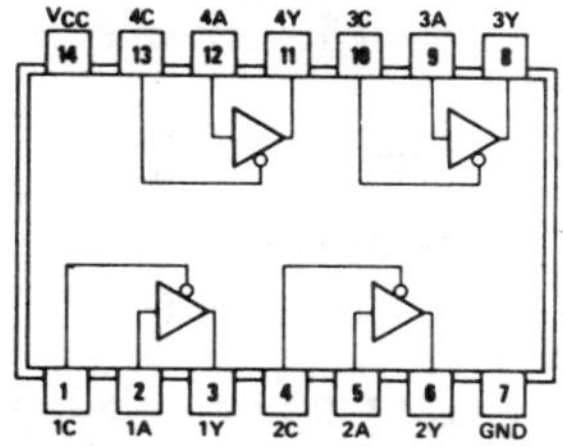

positive logic:

Y = A

Output is off (disabled) when C is high.

SN54125 (J, W) SN74125 (J, N)
SN54LS125A (J, W) SN74LS125A (J, N)

Operation & Application: The 74125 device contains four noninverting buffer stages, all of which can be operated independently from the others in the package. This device is often used in microcomputers and other digital applications as bus drivers; two 74125's can be used to drive an eight-bit microcomputer bus. The outputs of the 74125 are tri-state. If the active-LOW enable pin of any buffer stage (pins, 1, 4, 10 and 13) is HIGH, then the outputs of the buffers are disconnected from their respective output terminals. The output terminal is therefore at a high impedance to either ground or +5 volts dc. If the enable terminal is made LOW, however, the stage will operate as a normal TTL noninverting buffer: a) if the input is HIGH, then the output is also HIGH, b)if the input is LOW, then the output is also LOW. The 74125 device is similar to the 74126 device.

Current Requirements per package: 20 milliamperes

Device Number: 74126

Type of Device: Quad Bus Buffer with Tri-State Outputs

QUADRUPLE BUS BUFFER GATES WITH THREE-STATE OUTPUTS

126

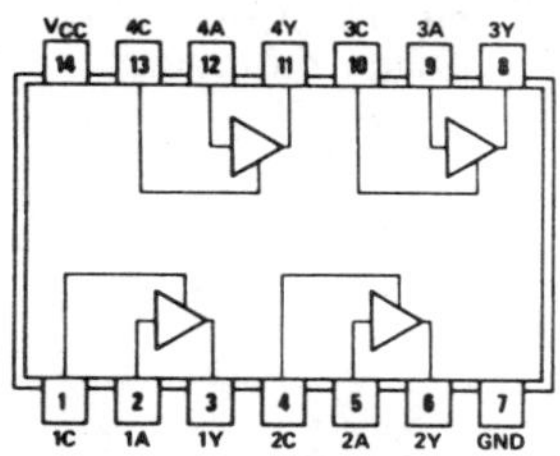

positive logic:

Y = A

Output is off (disabled) when C is low.

SN54126 (J, W) SN74126 (J, N)
SN54LS126A (J, W) SN74LS126A (J, N)

Operation & Application: The 74126 device contains four noninverting buffers, and is essentially the same as the 74125 device described previously. The 74125, however, uses active-LOW enable or "control" inputs, while the 74126 uses active-HIGH inputs.

Current Requirements per package: 20 milliamperes

Device Number: 74140

Type of Device: Dual Four-Input Positive-Logic NAND Gate with 50-ohm Line Driver Outputs

DUAL 4-INPUT POSITIVE-NAND 50-OHM LINE DRIVERS

140

positive logic:
Y = $\overline{ABCD}$

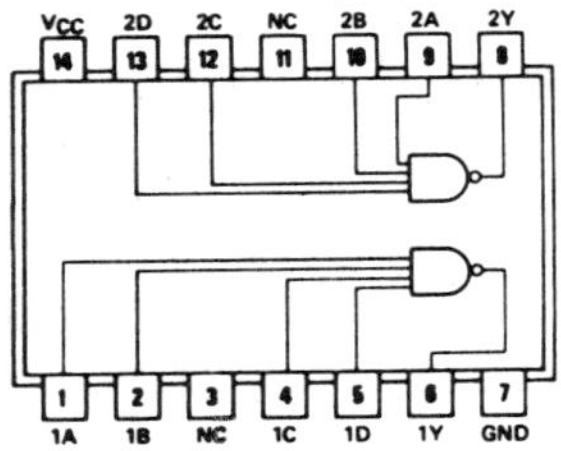

SN54S140 (J, W) SN74S140 (J, N)

NC—No internal connection

Operation & Application: The 74140 device contains two independent four-input NAND gates. These NAND gates are special in that the outputs are low-impedance, and will directly drive 50-ohm lines and input impedances.

Propagation delay: 18 nS

Current Requirements per package: 50 milliamperes

Device Number: 74141

Type of Device: BCD-to-Decimal Decoder/Driver (1-of-10)

BCD-TO-DECIMAL DECODER/DRIVER

141 DRIVES COLD-CATHODE INDICATOR TUBES

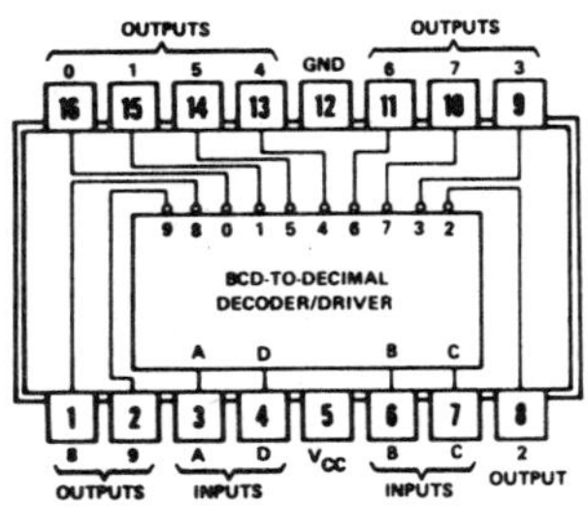

SN74141 (J, N)

Operation & Application: The 74141 device is intended to decode Binary Coded Decimal (BCD) four-bit code that is encoded in 1-2-4-8 weighting. The device produces one LOW output line for each input code. The outputs will sink 7-mA at potentials to 60-volts. Hence, the 74141 device can be used to drive *Nixie®* tubes, neon indicators and other high voltage digital displays that require one-of-ten decoding.

Note: The neon display devices use high voltage. Under appropriate conditions, the 74141 device will handle up to +180 volt lamp potentials. It is extremely important that you be careful when working with these potentials. For one thing, the potentials can be dangerous to humans. the very least that can happen is a painful shock, and it may be very hazardous. Secondary, a short between any output pin and any other pin on the device will destroy the IC instantly.

Current Requirements per package: 18 mA

Device Number: 74142

Type of Device: Decimal Counter/Decoder/Driver

COUNTER/LATCH/DECODER/DRIVER

142 DIVIDE-BY-10 COUNTER
4-BIT LATCH
4-BIT TO 7-SEGMENT DECODER
NIXIE ‡ TUBE DRIVER

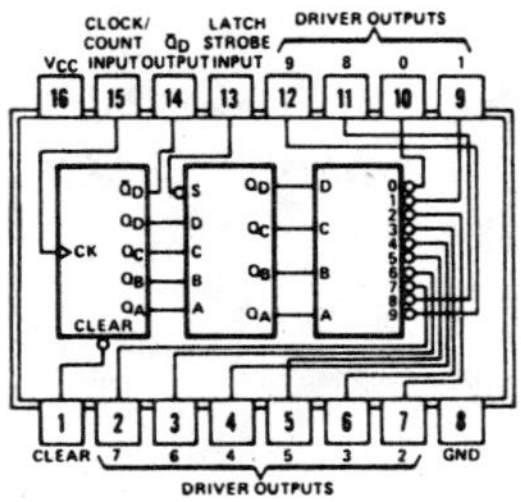

SN74142 (J, N)

Operation & Application: The 74142 device contains all stages necessary to form a decimal (i.e., decade or base-10) counter and display. The device contains a decade counter, four-bit latch, BCD-to-one-of-ten decoder. The clock input on the 74142 responds to positive-going pulses, and will advance the counter one state for each pulse received. The clear input is used to reset the counter; it is normally held HIGH, and is momentarily grounded to clear the counter. The $\overline{Q}_D$ output is used to cascade units—connect the $\overline{Q}_D$ output to the clock input of the next stage. The latch-strobe input is used to hold the display data. When the latch input is LOW, then the output display follows the clock/counter state, but if the latch input is HIGH then the display will remain at the state that was present the instant the latch input went HIGH. The outputs of the 74142 are similar in nature and function to the outputs of the 74141 device, and will accommodate *Nixie*® tubes.

Operating Frequency: 20 MHz

Current Requirements per package: 70 mA

Device Number: 74143

Type of Device: Four-bit Decade Counter/Latch/Seven-Segment Decoder

143

15 mA CONSTANT CURRENT
1- TO 5-V OUTPUT RANGE

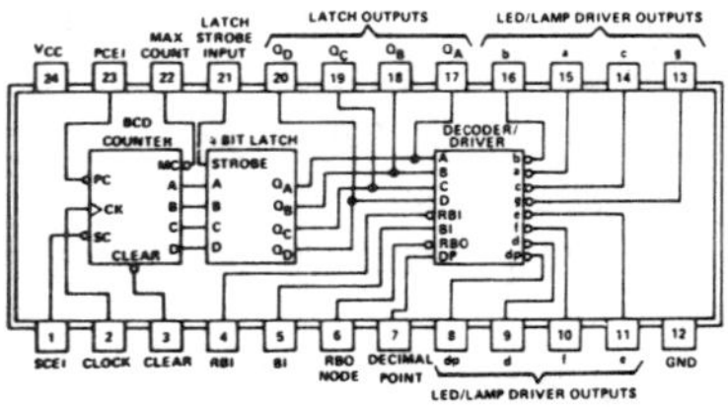

SN54143 (J, W) **SN74143 (J, N)**

Operation & Application: The 74143 device contains in a single 24-pin DIP IC package all of the circuitry needed to form a decade counter stage with an external seven-segment light emitting diode decimal display device. The 74143 outputs are designed to maintain a constant 15 milliamperes over the supply voltage range of 1 to 5 volts. The functions of the 74143 pins are given in the Table below.

FUNCTION	PIN NO.	DESCRIPTION
CLEAR INPUT	3	When low, resets and holds counter at 0. Must be high for normal counting.
CLOCK INPUT	2	Each positive-going transition will increment the counter provided that the circuit is in the normal counting mode (serial and parallel count enable inputs low, clear input high).
PARALLEL COUNT ENABLE INPUT (PCEI)	23	Must be low for normal counting mode. When high, counter will be inhibited. Logic level must not be changed when the clock is low.
SERIAL COUNT ENABLE INPUT (SCEI)	1	Must be low for normal counting mode, also must be low to enable maximum count output to go low. When high, counter will be inhibited and maximum count output will be driven high. Logic level must not be changed when the clock is low.
MAXIMUM COUNT OUTPUT	22	Will go low when the counter is at 9 and serial count enable input is low. Will return high when the counter changes to 0 and will remain high during counts 1 through 8. Will remain high (inhibited) as long as serial count enable input is high.
LATCH STROBE INPUT	21	When low, data in latches follow the data in the counter. When high, the data in the latches are held constant, and the counter may be operated independently.
LATCH OUTPUTS (Q_A, Q_B, Q_C, Q_D)	17, 18, 19, 20	The BCD data that drives the decoder can be stored in the 4-bit latch and is available at these outputs for driving other logic and/or processors. The binary weights of the outputs are: $Q_A = 1$, $Q_B = 2$, $Q_C = 4$, $Q_D = 8$.
DECIMAL POINT INPUT	7	Must be high to display decimal point. The decimal point is not displayed when this input is low or when the display is blanked.
BLANKING INPUT (BI)	5	When high, will blank (turn off) the entire display and force RBO low. Must be low for normal display. May be pulsed to implement intensity control of the display.
RIPPLE-BLANKING INPUT (RBI)	4	When the data in the latches is BCD 0, a low input will blank the entire display and force the RBO low. This input has no effect if the data in the latches is other than 0.
RIPPLE-BLANKING OUTPUT (RBO)	6	Supplies ripple blanking information for the ripple blanking input of the next decade. Provides a low if BI is high, or if RBI is low and the data in the latches in BCD 0; otherwise, this output is high. This pin has a resistive pull-up circuit suitable for performing a wire-AND function with any open-collector output. Whenever this pin is low the entire display will be blanked; therefore, this pin may be used as an active-low blanking input.
LED/LAMP DRIVER OUTPUTS (a, b, c, d, e, f, g, dp)	5, 16, 14, 9 11, 10, 13, 8	Outputs for driving seven-segment LED's or lamps and their decimal points. See segment identification and resultant displays on following page.

Device Number: 74144

Type of Device: Four-Bit Decade Counter/Latch/Seven-Segment Decoder

COUNTERS/LATCHES/DECODERS/DRIVERS

144

UP TO 15-V INDICATORS
UP TO 25 mA
OPEN-COLLECTOR OUTPUT

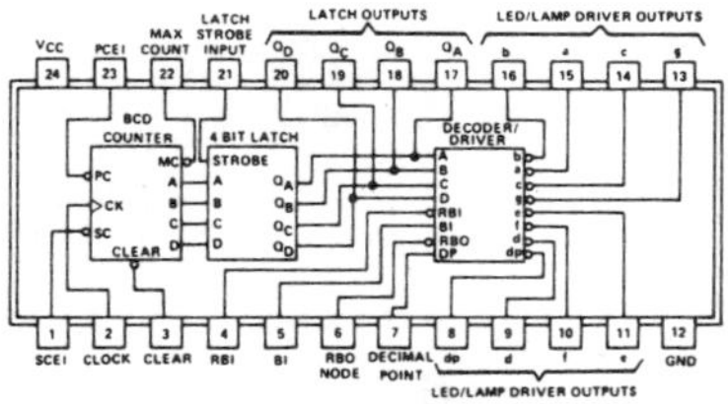

SN54144 (J, W) SN74144 (J, N)

Operation & Application: The 74144 device is the same as the 74143 device, except that the outputs are capable of sinking up to 25 milliamperes. For description of the pinout functions see the 74143 above.

Device Number: 74148

Type of Device: 8-Line-to-3-Line-Octal-Priority-Encoder

8-LINE-TO-3-LINE OCTAL PRIORITY ENCODERS

148

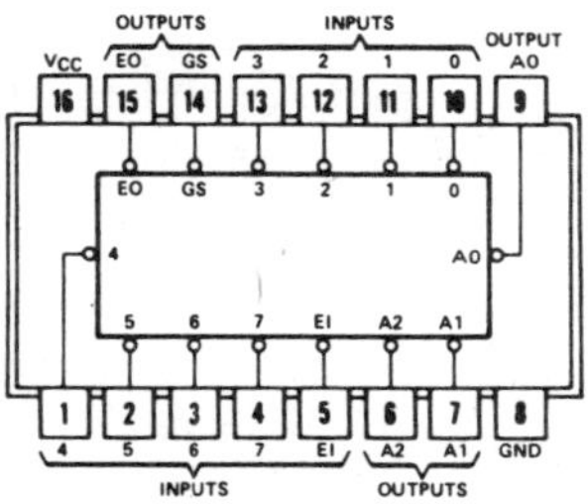

SN54148 (J, W) SN74148 (J, N)
SN54LS148 (J, W) SN74LS148 (J, N)

Operation & Application: The 74148 device permits us to prioritize eight inputs according to order of importance. Applications include keyboard encoding, range selections, N-bit encoding, code conversion and code generation. The *E1* enable line must be LOW for the chip to operate. Cascading is possible by linking *E1* and *E0* terminals on successive devices. Check the specification sheet for the 74148 device for the exact truth table which defines operation.

Propagation delay: 15 nanoseconds
Current Requirements per package: 40 mA

Device Number: 74150

Type of Device: 1-of-16 Data Selector/Multiplexer

1-OF-16 DATA SELECTORS/MULTIPLEXERS

150

SN54150 (J, W) SN74150 J, N)

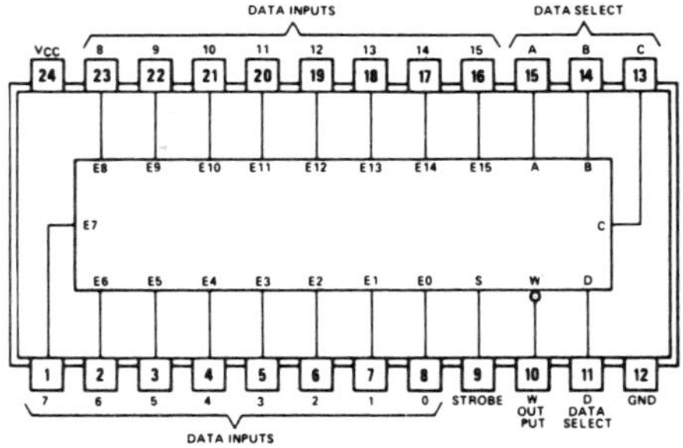

Operation & Application: The 74150 device accepts a four-bit binary word at the address inputs (i.e., *Data Select*). The data input line associated with the selected address is connected to the output terminal through an inverter. Hence, the output data will be the complement of the data applied to the selected input. The sixteen input lines are selected by codes 0000 to 1111. The *strobe* line functions as an active-LOW chip-enable line. Making *strobe* Low will allow proper operation, while making *strobe* HIGH will make the output (pin no. 10) HIGH regardless of the data applied to the selected input.

Propagation delay: 11 nanoseconds

Current Requirements per package: 25 milliamperes

Device Number: 74151

Type of Device: 1-of-8 Octal Data Selector with Complementary Outputs

1-OF-8 DATA SELECTORS/MULTIPLEXERS

151

SN54151A (J, W) SN74151A (J, N)
SN54LS151 (J, W) SN74LS151 (J, N)
SN54S151 (J, W) SN74S151 (J, N)

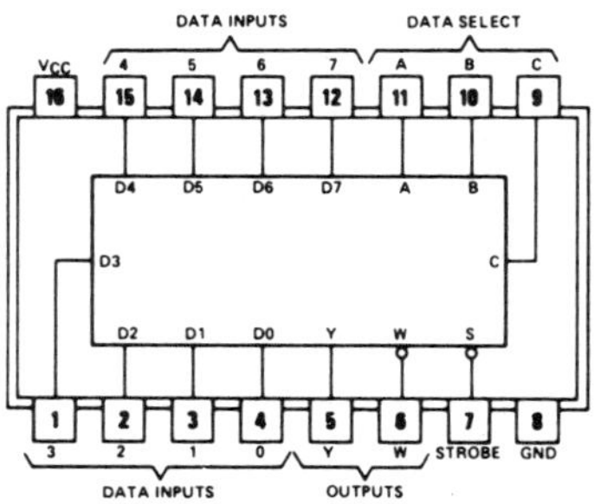

Operation & Application: The 74151 device examines a three-bit input address (*Data Select*) and will connect the input line selected by the three-bit code (000-111) to two outputs. The Y output will exhibit the same data as the selected input, while the W output will exhibit the complement of the data on the selected input.

The *strobe* line acts as an active-LOW chip enable. This line is made LOW for normal operation, and HIGH when the chip is not selected. If a HIGH is applied to the strobe input, then the W output goes HIGH and the Y output goes LOW, regardless of the data present on the selected input line.

Propagation delay: 8 nanoseconds

Current Requirements per package: 25 milliamperes

Device Number: 71452

Type of Device: 1-of-8 Octal Data Selector with Single Output

1-OF-8 DATA SELECTORS/MULTIPLEXERS

152

SN54152A (W)
SN54LS152 (W)

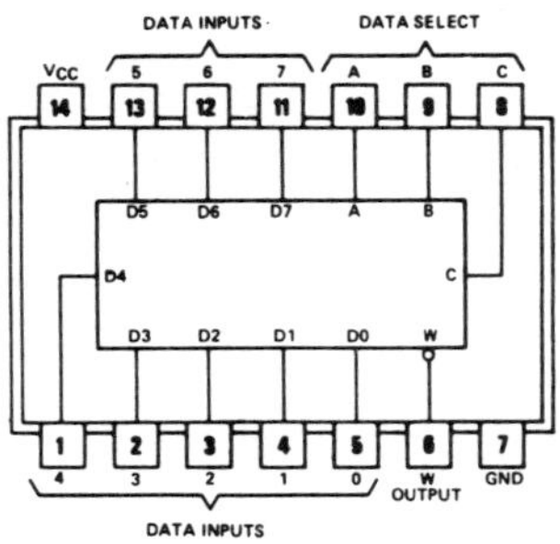

Operation & Application: The 74152 device examines a three-bit input address (*Data Select*) and will connect the input line selected by the three-bit code (000 - 111) to the W output through an inverter. Hence, the output line will exhibit the complements of the data applied to the selected input. The 74152 is very similar to the 74151 device, except that there is only one output, W, rather than complementary outputs as on the 74151. The *strobe* line acts as an active-LOW chip enable. This line is made LOW for normal operation, and HIGH when the chip is not selected. If a HIGH is applied to the strobe input, then the W output goes HIGH regardless of the data applied to the selected input.

Propagation delay: 8 nanoseconds

Current Requirements per package: 25 milliamperes

Device Number: 71453

Type of Device: Dual 1-of-4 Data Selectors/Multiplexers

153

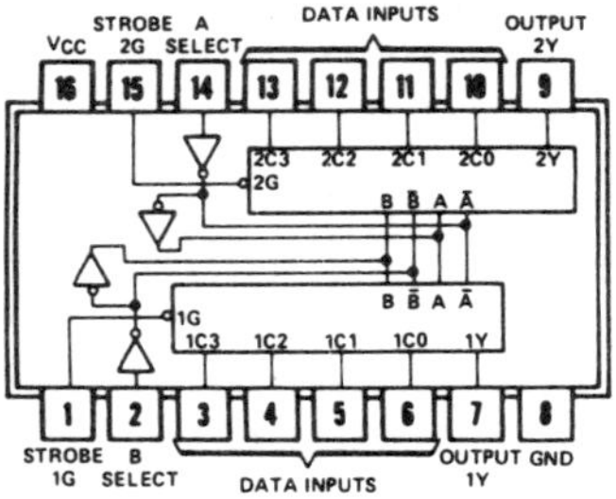

SN54153 (J, W) SN74153 (J, N)
SN54L153 (J) SN74L153 (J, N)
SN54LS153 (J, W) SN74LS153 (J, N)
SN54S153 (J, W) SN74S153 (J, N)

Operation & Application: The 74153 device examines a 3-bit input address (A-select and B-select) and connects the selected inputs of each half to their respective outputs. Note that the address lines are common to both halves, so the two selectors are not independent of each other as to address selection. The inputs and outputs are independent, however. The data on the two output lines is not inverted, as in the case of other selectors, but is the same as the data applied to the selected inputs. The strobe inputs act as active-LOW chip enable lines, and each half of the 74153 may be independently strobed on or off. When the strobe line is LOW, then the output will follow the data applied to the selected input. But, when the strobe line is HIGH, the output associated with that line will drop LOW and remain LOW regardless of data applied to the input.

Propagation delay: 14 to 44 nanoseconds depending upon operation

Current Requirements per package: 25 milliamperes

Device Number: 74154

Type of Device: Four-line-to-sixteen-Line-Decoder/Demultiplexer

4-LINE TO 16-LINE DECODERS/DEMULTIPLEXERS

154

SN54154 (J, W) SN74154 (J, N)
SN54L154 (J) SN74L154 (J, N)

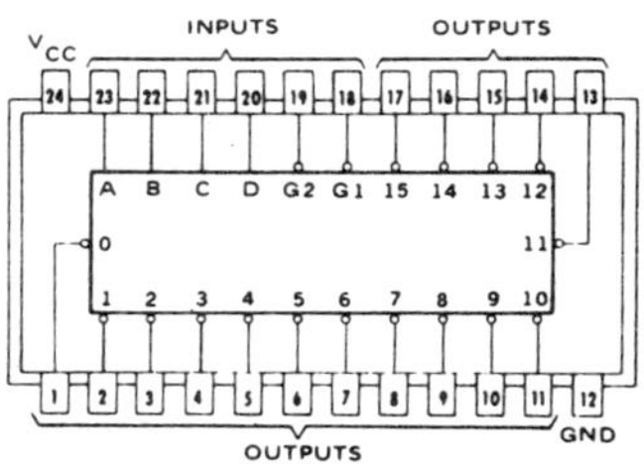

Operation & Application: The 74154 device selects one output according to the four-bit binary word applied to the ABCD inputs (normal 1-2-4-8 weighting). The data input is applied to the G1 terminal, and will be reflected on the selected output line. Hence, if a HIGH is applied to G1, and the ABCD inputs see 0010, then output no. 2 (pin no. 3) will also be HIGH. The data on the selected output will follow the data applied to G1. Input G2 functions as a chip-enable line, and is active-LOW. Applications for the 4154 device include input/output and memory address decoding and bank-select functions in microcomputers, demultiplexing for systems with up to sixteen channels of digital data, and any form of 1-and-only-1 line selection.

Propagation delay: 46 nanoseconds

Current Requirements per package: 20 milliamperes

Device Number: 74155

Type of Device: Dual-1-of-4-Data-Decoder/Demultiplexer/Distributer

DECODERS/DEMULTIPLEXERS

DUAL 2- TO 4-LINE DECODER
DUAL 1- TO 4-LINE DEMULTIPLEXER
3- TO 8-LINE DECODER
1- TO 8-LINE DEMULTIPLEXER

155 TOTEM-POLE OUTPUTS

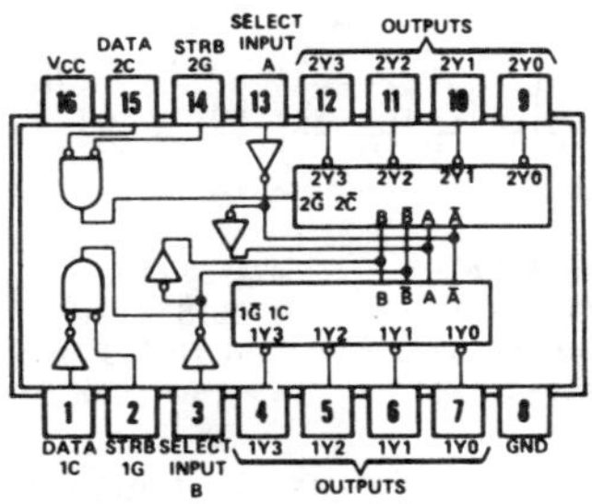

SN54155 (J, W) SN74155 (J, N)
SN54LS155 (J, W) SN74LS155 (J, N)

Operation & Application: Like some other chips in this series, the 74155 device contains two data distributors that are addressed by a common two-bit address bus (Select-A and Select-B). On each half, the data applied to the respective Data inputs are routed to the output selected by the two-bit address lines. Thus, the 74155 device will work wither as a data distributor/demultiplexer (approximately the same thing!), or, will function as a 1-HIGH-of-4, or, 1-LOW-of-4 decoder. The *strobe* lines act as an active-LOW chip enable, and each half has its own strobe so can be turned on and off independently. Be aware that there is a significant difference between the two halves of this circuit! The *Data 1C* input inverts the data, while the *Data 2C* input does not. Hence, one side complements the input, while the other does not; the two halves may not be used interchangeably unless some means is provided to account for this difference.

Propagation delay: 32 nanoseconds

Current Requirements per package: 25 milliamperes

Device Number: 74157

Type of Device: Quadruple Two-Line-to-Four-Line Decoder/Demultiplexer

QUAD 2- TO 1-LINE DATA SELECTORS/MULTIPLEXERS

157 NONINVERTED DATA OUTPUTS

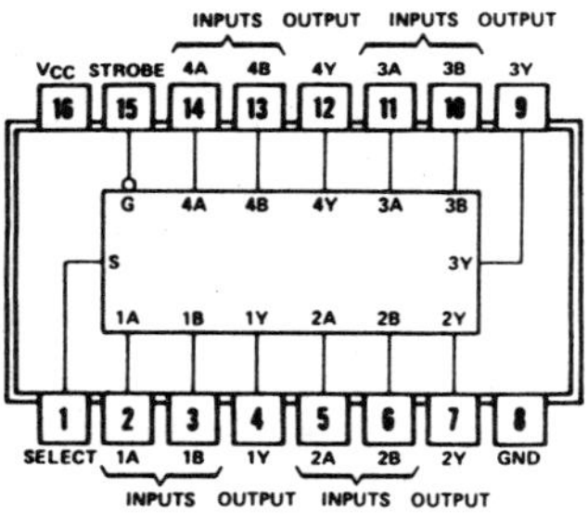

SN54157 (J, W)	**SN74157 (J, N)**
SN54L157 (J)	**SN74L157 (J, N)**
SN54LS157 (J, W)	**SN74LS157 (J, N)**
SN54S157 (J, W)	**SN54S157 (J, N)**

Operation & Application: This device is related to the 74155. It provides four semi-independent single-pole double-throw (SPDT) data switches. If the strobe line is LOW, then the device is selected; if strobe is HIGH then all outputs are LOW regardless of the data applied to the inputs. Assuming the strobe is LOW, then all four switches are controlled by a common *select* input (pin no. 1). If *select* is HIGH, then the outputs will follow the A-inputs, but if *select* is LOW then the outputs will follow the B-inputs. The data on the outputs is not inverted compared with the inputs. For an inverting version see the 74158 device.

Current Requirements per package: 25 milliamperes

Device Number: 74158

Type of Device: Quadruple Two-Line-to-Four-Line Decoder/Demultiplexer with Inverted Outputs

QUAD 2- TO 1-LINE DATA SELECTORS/MULTIPLEXERS

158 INVERTED DATA OUTPUTS

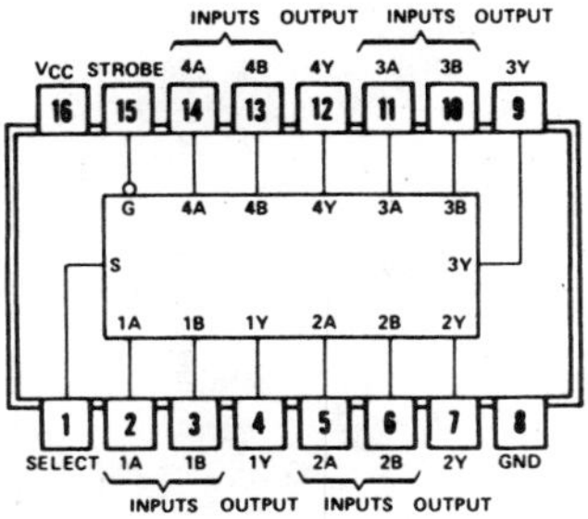

SN54LS158 (J, W) SN74LS158 (J, N)
SN54S158 (J, W) SN74S158 (J, N)

Operation & Application: This device is exactly like the 74157 device, except that the output signals are complements of the selected input (i.e., ate inverted).

Device Number: 74160

Type of Device: Synchronous 4-bit Decade Counter with Direct Clear

SYNCHRONOUS 4-BIT COUNTERS

160 DECADE, DIRECT CLEAR

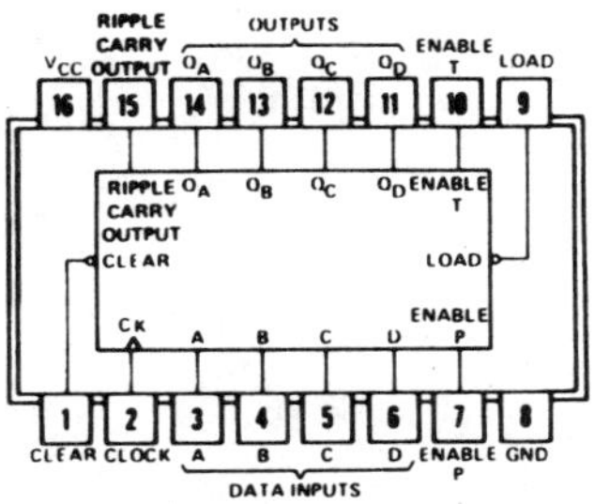

SN54160 (J, W) SN74160 (J, N)
SN54LS160A (J, W) SN74LS160A (J, N)

Operation & Application: The 74160 device is a presettable up counter that is synchronous, rather than of the ripple counter type. The direct *clear* (pin no. 1) is made HIGH for normal counting operation, and is brought momentarily LOW to clear the counter. The 74160 can be preset to some count state by programming the preset point into the ABCD Data inputs (pins 3-6), and then bringing the *load* (pin no. 9) terminal momentarily LOW.

There are two enable lines, marked P and T. For normal counting, the P and T inputs are made HIGH. When the devices are cascaded, then we will connect the *ripple carry output* of the least significant digit stages to the Enable-T input of the next stage. The clock lines of all cascaded stages must be connected together. The

74160 toggles on positive edges of the input signals.

The four output lines are weighted in the usual 1-2-4-8 Binary Coded Decimal (BCD) scheme.

Maximum Frequency: 25 MHz

Current Requirements per package: 35 milliamperes

Device Number: 74161

Type of Device: Synchronous 4-Bit Binary (i.e. Base-16) Synchronous Counter with Direct Clear

SYNCHRONOUS 4-BIT COUNTERS

161 BINARY, DIRECT CLEAR

SN54161 (J, W) SN74161 (J, N)
SN54LS161A (J, W) SN74LS161A (J, N)

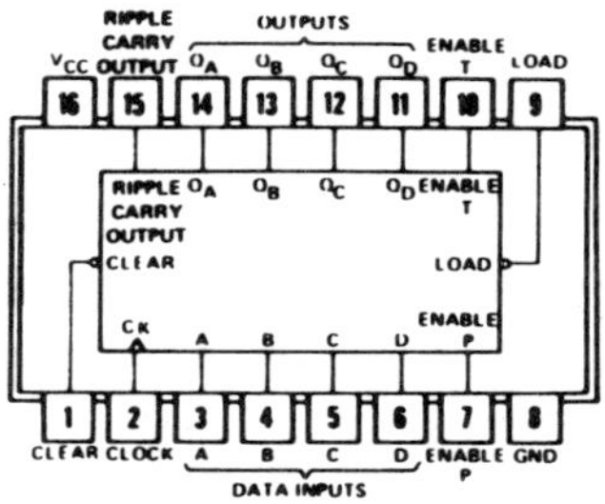

Operation & Application: This device is a base-16 version of the base-10 counter described as the 74160. Refer to the 74160 discussion for details, with the understanding that the two differ according to the table below (assuming preset = 0000):

Pulse No.	74160	74161
0	0000	0000
1	0001	0001
2	0010	0010
3	0011	0011
4	0100	0100
5	0101	0101
6	0110	0110
7	0111	0111
8	1000	1000
9	1001	1001
10	0000	1010
11	0001	1011
12	0010	1100
13	0011	1101
14	0100	1110
15	0101	1111

Device Number: 74162

Type of Device: Synchronous 4-Bit Decade (Base-10) Counter with Synchronous Clear

SYNCHRONOUS 4-BIT COUNTERS

162 DECADE, SYNCHRONOUS CLEAR

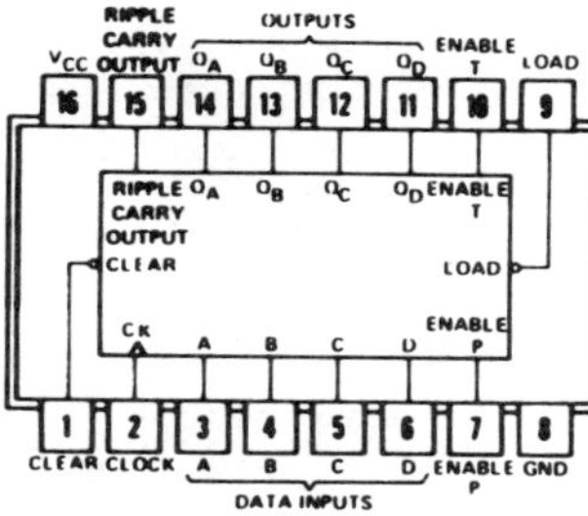

SN54162 (J, W)	**SN74162 (J, N)**
SN54LS162A (J, W)	**SN74LS162A (J, N)**
SN54S162 (J, W)	**SN74S162 (J, N)**

Operation & Application: This device is exactly like the 74160, except that the clear function is synchronized with the clock. See spec sheets for 74160.

Device Number: 74163

Type of Device: Synchronous 4-Bit Binary (Base-16) Counter with Synchronous Clear

SYNCHRONOUS 4-BIT COUNTERS

163 BINARY, SYNCHRONOUS CLEAR

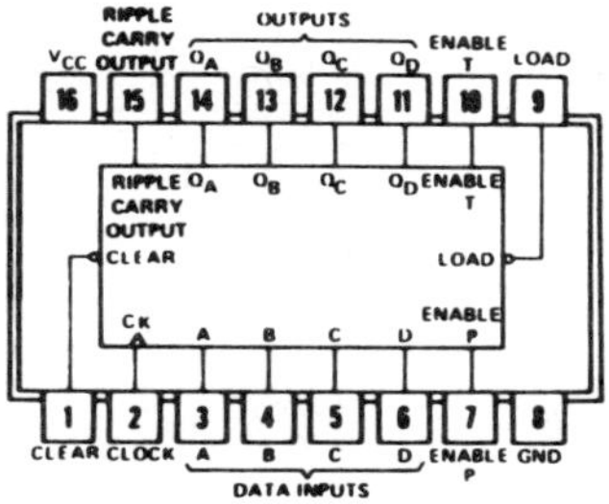

SN54163 (J, W)	**SN74163 (J, N)**
SN54LS163A (J, W)	**SN74LS163A (J, N)**
SN54S163 (J, W)	**SN74S163 (J, N)**

Operation & Application: This device is exactly like the 74161 device, except that the clear function is synchronous with the clock. Consult spec sheets for the 74160 and 74161 devices cited previously in this chapter.

Device Number: 74164

Type of Device: Eight-Bit Parallel Output Serial Input Shift Register with Asynchronous Clear

8-BIT PARALLEL OUTPUT SERIAL SHIFT REGISTERS

164 ASYNCHRONOUS CLEAR

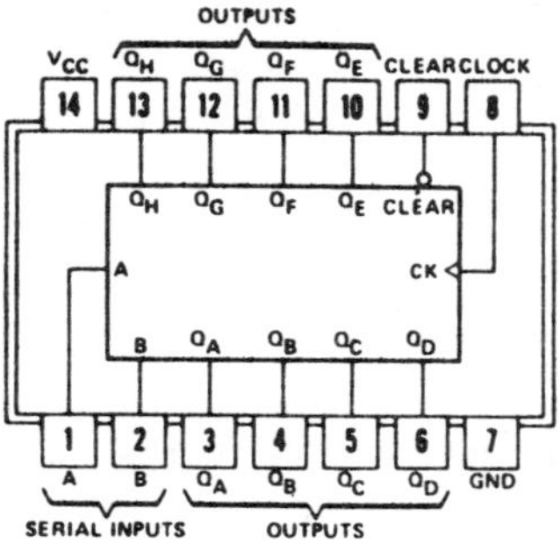

SN54164 (J, W)	**SN74164 (J, N)**
SN54L164 (J, T)	**SN74L164 (J, N)**
SN54LS164 (J, W)	**SN74LS164 (J, N)**

Operation & Application: This eight-bit shift register may be used as either SIPO or SISO shift registers (see text of book). There are two serial inputs, A and B. For normal operation one of these will be held HIGH while data is applied to the other. The *clear* terminal is active-LOW, so will normally be held HIGH; momentarily bringing clear LOW will reset all outputs to LOW. Data will shift one stage to the right on the positive edge of each clock pulse.

Maximum Frequency: 35 MHz

Current Requirements per package: 40 mA

Device Number: 74165

Type of Device: Eight-Bit Parallel Load (i.e., Input) with Complementary Serial Outputs

PARALLEL-LOAD 8-BIT SHIFT REGISTERS WITH COMPLEMENTARY OUTPUTS

165

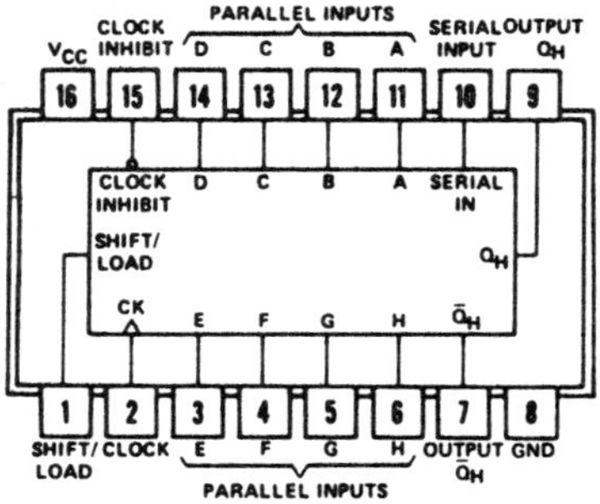

SN54165 (J, W) SN74165 (J, N)
SN54LS165 (J, W) SN74LS165 (J, N)

Operation & Application: This shift register allows simultaneous loading of all eight stages through inputs A through H. In normal operation, the *clock inhibit* line is held LOW and the *shift/ load* is held HIGH. The data loaded will shift right one stage every time the clock line produces a positive-going transition (i.e., it is positive edge triggered). Two outputs are provided: Q and NOT-Q, which are complements of each other. Loading is provided by applying the desired data to inputs A - H, and then bringing the shift/load line momentarily LOW. The clock inhibit is not a true chip-enable because it only prevents further shifting of the data; it does not turn off the outputs or make them tri-state.

Maximum Frequency: 24 MHz

Current Requirements per package: 40 mA

Device Number: 74167

Type of Device: Synchronous Decade Rate Multiplier

SYNCHRONOUS DECADE RATE MULTIPLIERS

167

SN54167 (J, W)   SN74167 (J, N)

NC — No internal connection

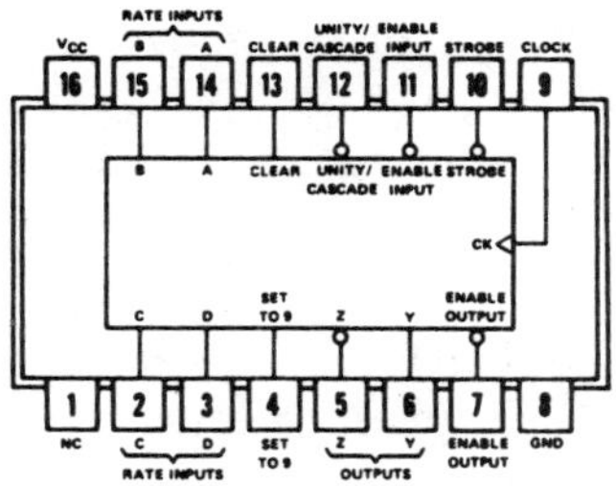

Operation & Application: The rate multiplier will produce an output pulse train from 0 to 9 pulses for each input pulse. The number of pulses produced per input depends on the BCD word applied to the ABCD inputs. The *enable* output is an active-LOW output that produces an output pulse train that is 1/10 the frequency of the input pulse train. Normal operation requires that the strobe, clear and enable lines be held LOW; the input signal square wave is applied to the clock input. Pins 5 and 6 and outputs that are complementary.

Maximum Frequency: 25 MHz

Current Requirements per package: 75 mA

Device Number: 74174

Type of device: Hex Type-D Flip-Flop with Single-Rail Outputs and Common Clear

HEX D-TYPE FLIP-FLOPS

174 SINGLE RAIL OUTPUTS
COMMON DIRECT CLEAR

SN54174 (J, W)   SN74174 (J, N)
SN54LS174 (J, W   SN74LS174 (J, N)
SN54S174 (J, W)   SN74S174 (J, N)

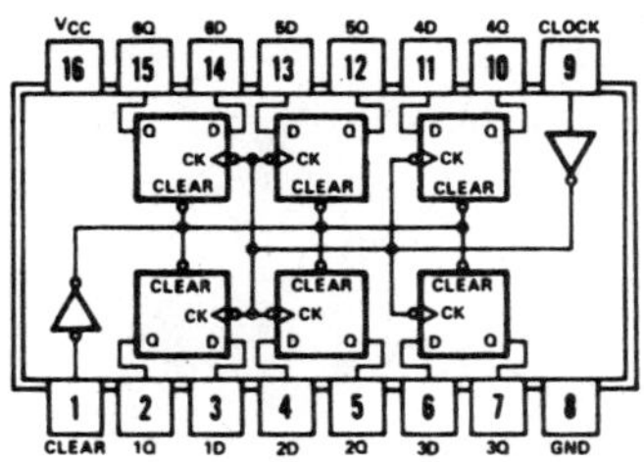

Operation & Application: This hex Type-D flip-flop contains six individual Type-D flip-flops, but they are not independent. The flip-flops contain only one output each, a noninverting "Q" output (which makes it "single-rail" output). There is also a common clear line for all six flip-flops. The clear line is normally kept HIGH, but is dropped LOW momentarily when it is desired that all six flip-flops be reset. When the clear is made active, the flip-flops all go to the condition of Q = LOW. The common clock line is positive-edge triggered. The data present on each input is transferred to its respective Q output on the positive-going edge of the clock pulse.

Maximum Frequency: 36 MHz

Current Requirements per package: 40 mA

Device Number: 74175

Type of Device: Quad Type-D Flip-Flops with Complementary Outputs and Common Clear

QUAD D-TYPE FLIP-FLOPS

175 COMPLEMENTARY OUTPUTS
COMMON DIRECT CLEAR

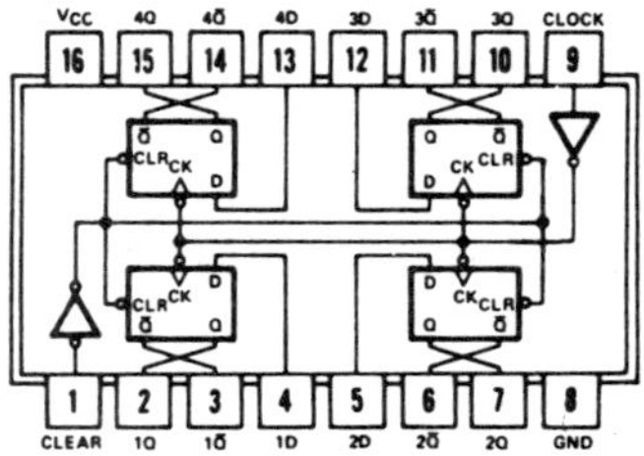

SN54175 (J, W) SN74175 (J, N)
SN54LS175 (J, W) SN74LS175 (J, N)
SN54S175 (J, W) SN74S175 (J, N)

Operation & Application: This device is similar to the 74174, except that there are only four Type-D flip-flops in the package, and each flip-flop has two outputs which are complementary: Q and NOT-Q. The clear input is active-LOW, so is normally held HIGH. When it is desired to clear the flip-flops, the common clear line will be momentarily brought LOW. Like the 74174 device, the clock line is common to all stages. Data present on the inputs will be transferred to the outputs on the positive edge of the clock pulses.

Maximum Frequency: 35 MHz

Current Requirements per package: 32 mA

Chapter 4

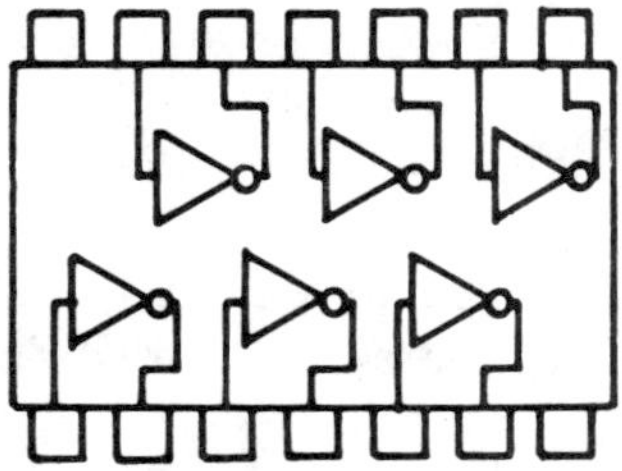

CMOS Devices

The *complementary metal-oxide silicon* (CMOS) family of IC logic devices is not as old as the TTL and some of the other families, but they have certain distinct advantages over TTL devices. One of the most important advantages of CMOS devices is the extremely low power requirements. Where the typical TTL device draws one or more dozen milliamperes, the CMOS device package current requirements are in *microamperes*. As a result of the low current requirements, we can now build devices such as digital watches and calculators that run for a couple of years on hearing aid batteries!

Another advantage of CMOS devices is the generally improved immunity to noise. Most TTL devices, which have a narrow noise immunity band, will respond to high frequency (i.e., short-duration, fast risetime) pulses from power lines and other sources. These pulses can wreak havoc with digital circuits. Even when they are not destructive to the circuitry, they can disrupt the circuits and cause them to act in an unpredictable manner. CMOS devices, on the other hand, are a lot more immune to those pulses for two reasons. One reason is that CMOS devices are low frequency devices, operating to 4 or 5 MHz rather than 20 MHz as in TTL devices. Another is that CMOS devices can have a lot wider band of noise immunity than CMOS. The immunity band for TTL devices is only about 1.6 volts. The CMOS device, however, toggles at the mid-point between V+ and V−, so there can exist a wide band of noise immunity if these voltages are wide apart. In many designs, in fact, the CMOS device has replaced the HNIL logic formerly used.

The heart of the CMOS device is the *metal-oxide semiconductor field-effect transistor* (MOSFET), also called the *insulated-gate field-effect transistor* (IGFET). Figure 4-1 shows the basic form of the MOSFET transistor. This device is essentially a triode, so it has three electrodes. One electrode is the *gate*, while the other two are the *source* and *drain*. These latter electrodes are at either end of a semiconductor structure called the channel. The channel in Fig. 4-1 is made of N-type semiconductor material, hence the MOSFET is an *N-channel* type. If the channel were made from P-type material, then the MOSFET would be a P-type MOSFET. In both cases, the substrate of the MOSFET is of the opposite type material from the channel.

The symbols for the N-channel and P-channel MOSFETs are shown in Figs. 4-2A and 4-2B, respectively. Note that the difference is the direction that the arrow (substrate) points. Very often, the arrow is not printed in CMOS circuit diagrams, but it is in there nonetheless.

The action of the MOSFET, the phenomenon on which it operates, is the change of ohmic resistance of the channel with voltage applied to the gate terminal. The N-channel and P-channel MOSFETs are complementary in this respect, and obey the following rules:

	P-Channel	N-Channel
Gate LOW (i.e., zero or neg)	Low-R	High-R
Gate HIGH (i.e., V+)	High-R	Low-R

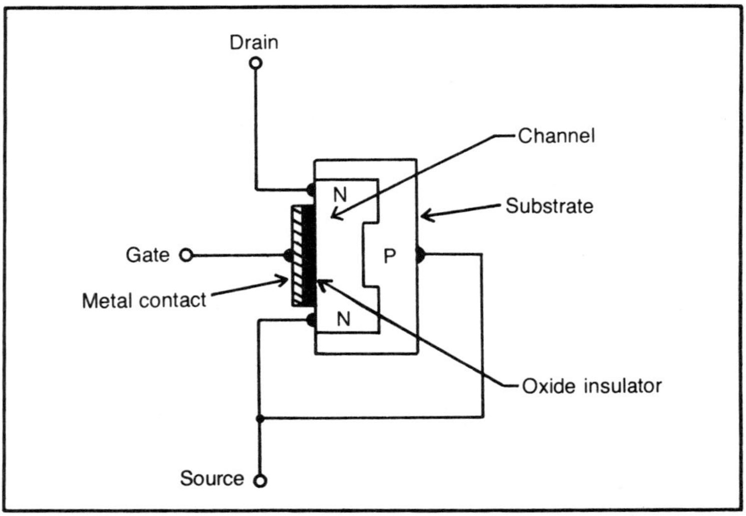

Fig. 4-1. MOSFET transistor.

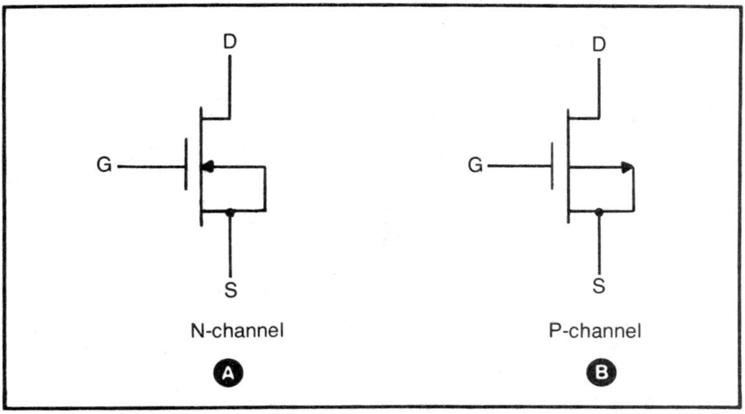

Fig. 4-2. MOSFET circuit symbols (A) N-channel, (B) P-channel.

By varying the gate-source voltage we can thereby vary the channel resistance. This resistance change is reflected as a change in drain current, so we can state that the MOSFET calls for an output current change caused by an input voltage change. When the MOSFET is used in a linear amplifier circuit, therefore, the resultant amplifier will be either a *transconductance amplifier*, or, (if a series resistor is used in the drain circuit), a *voltage amplifier*.

We considered the CMOS family of devices briefly in Chapter 2, so will only briefly discuss the basics here. The typical CMOS output stage consists of an N-channel MOSFET in series with a P-channel MOSFET. The result is a series circuit as shown in Fig. 4-3A. When the gate is LOW, we find that the channel of the P-channel device is low in resistance, so the output terminal is essentially connected to V+ through a resistance of around 200 ohms. At the same time, again with the gate LOW, the N-channel MOSFET has a high resistance channel (megohms). In this case, therefore, the terminal will see an extremely high resistance to ground and a low resistance to V+: the output is therefore HIGH.

Figure 4-3B shows the opposite situation. In this case the input is HIGH, so each gate is also HIGH. Here we find the P-channel resistance very high, and the N-channel resistance very low (again, about 200 ohms). Therefore, the output terminal is essentially grounded so is LOW.

At all times, the path from one power-supply terminal to the other (i.e., *V+* and either *V−* or *ground*, as the case might be) is a pair of resistors in series, i.e., the channel resistances of the two transistors. If the input is static, i.e., at either HIGH or LOW and not in a state of transition between these two states, then there will

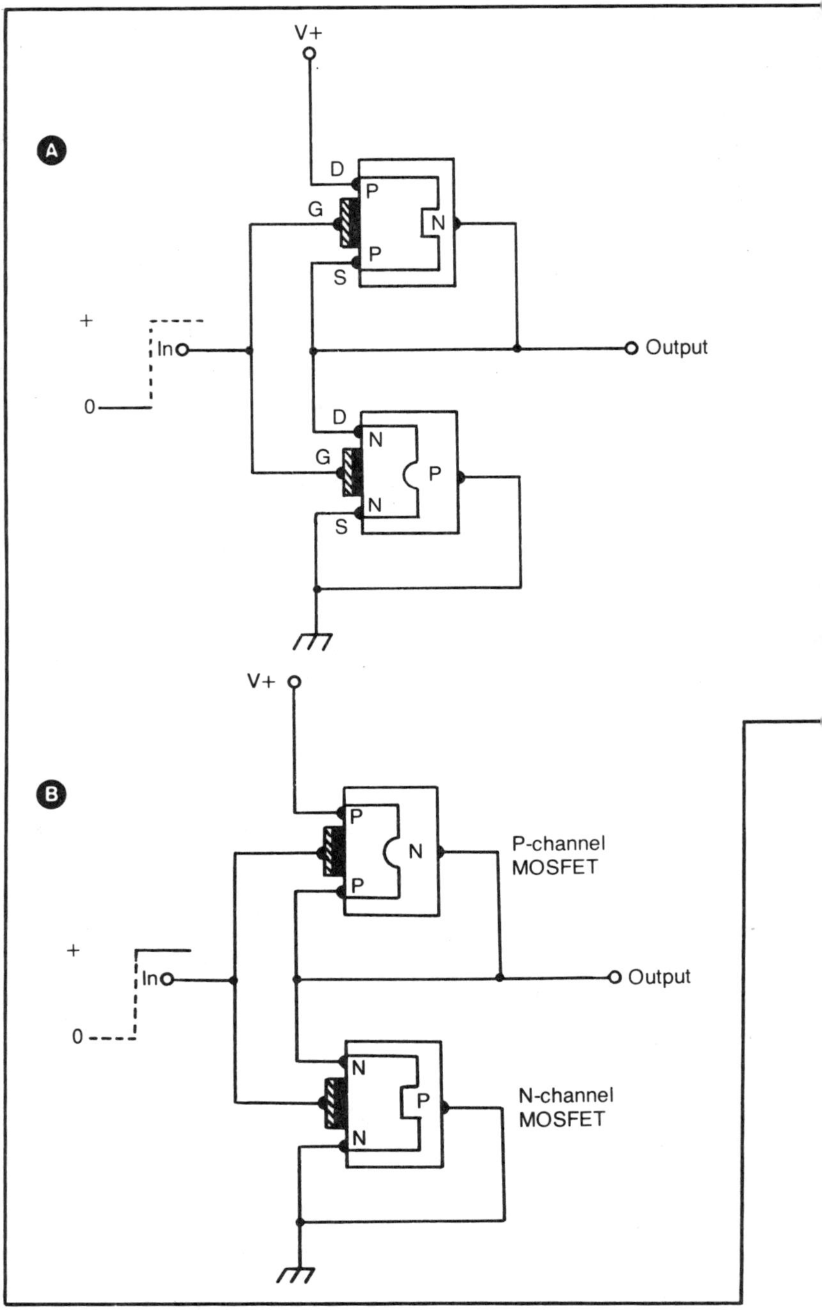

Fig. 4-3. (A) CMOS inverter (input LOW), (B) CMOS inverter (input HIGH), (C) CMOS device transfer function (typical).

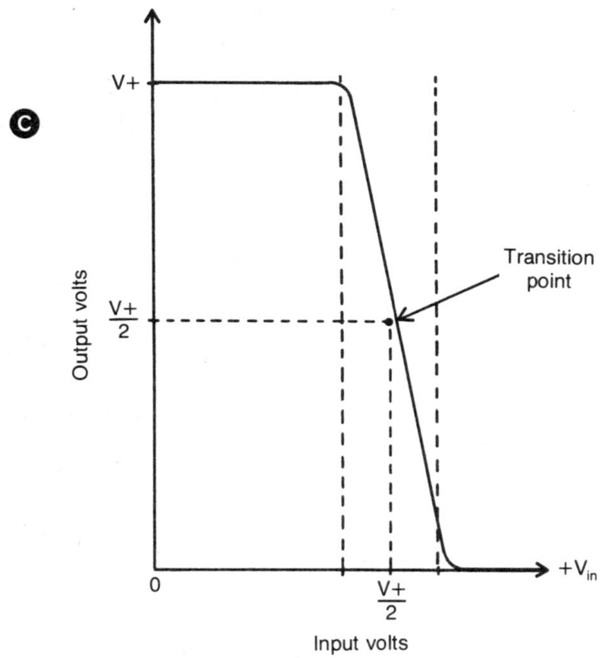

always be a very low resistance in series with a very high resistance. Therefore, we will always have a high resistance across the power supply! This phenomenon is the reason why the CMOS device only draws microamperes.

The only time that the CMOS device draws appreciable current is during transition between HIGH and LOW (i.e., LOW-to-HIGH or HIGH-to-LOW) states. At one point, for a very brief instant, there will be low channel resistances in both transistors. Even so, the low state is very brief so the average current does not increase much.

Figure 4-3C shows the transfer characteristic for a CMOS inverter (which is the simplest form of digital devices). Here, we have plotted *output-versus-input* voltage. When the input voltage (plotted along the horizontal axis) is LOW (i.e., zero in this case, it could be minus in some cases), then the output voltage will be V+, or very nearly V+. When the input voltage goes HIGH (i.e., to V+), however, the output voltage drops LOW. The point where the HIGH-to-LOW transition takes place is ½(V+). If the LOW voltage is not zero, then it will be a minus voltage. In that case, the transition point will be mid-way between V− and V+; for equal V− and V+ power-supply circuits, then the transition point is zero.

CMOS PROBLEMS

The CMOS advantages seem very attractive, and, with the exception of speed, seem to be superior to TTL. But there are limits to CMOS operation that might color what you might decide when selecting a logic family. The drive power of CMOS devices is generally limited, except when driving other CMOS inputs. If one has to drive items like relays, LEDs, lamps and so forth, then one of the interface methods given earlier will have to be used.

The major problem with CMOS devices, however, is that they are sensitive to static electricity. The very thin gate structure has a low breakdown voltage. The ohmic contact for the gate is separated from the channel through a thin layer of metal oxide that has a dielectric strength of 80 to 100 volts. Unfortunately, the static electricity potentials that build up on your body due to normal motion can be in the multi-kilovolt range. If the static potential is high enough to cause a spark, or give you a jolt, when you touch a grounded object, then it is far too high for CMOS devices. The problem is exacerbated in dry sections of the world, or, during dry seasons of the year. Whenever you feel increased static electricity shocks, then it is also a period of greater danger for CMOS devices.

Figure 4-4 shows what happens to a CMOS device if zapped by static electricity. When the static hits the CMOS gate (Fig. 4-4A), a hole is punched in the insulated gate layer (Fig. 4-4B). In some cases, the hole is sufficient to immediately cause a short circuit between the metal ohmic contact of the gate and the semiconductor material of the channel. This is the mechanism of immediate failure. In some cases, inexperienced workers have been known to blow perfectly good CMOS devices while installing them! In a moment, we will consider the mechanism by which this can be prevented.

The static will not necessarily cause an immediate catastrophic failure of the CMOS device. If the hole in the insulated gate layer is very tiny, then there might not be any immediate damage. But, as time goes on, the hole will fill with metal ions from the gate and semiconductor ions from the channel. Eventually the two will meet and a short circuit will exist. This is a source of premature failures of the CMOS device! Improper handling during construction might well be at the base of a failure a long time later.

The B-series (i.e., 4014B) CMOS devices contain zener gate-protection diodes (see Fig. 4-5) that will shunt the dangerous

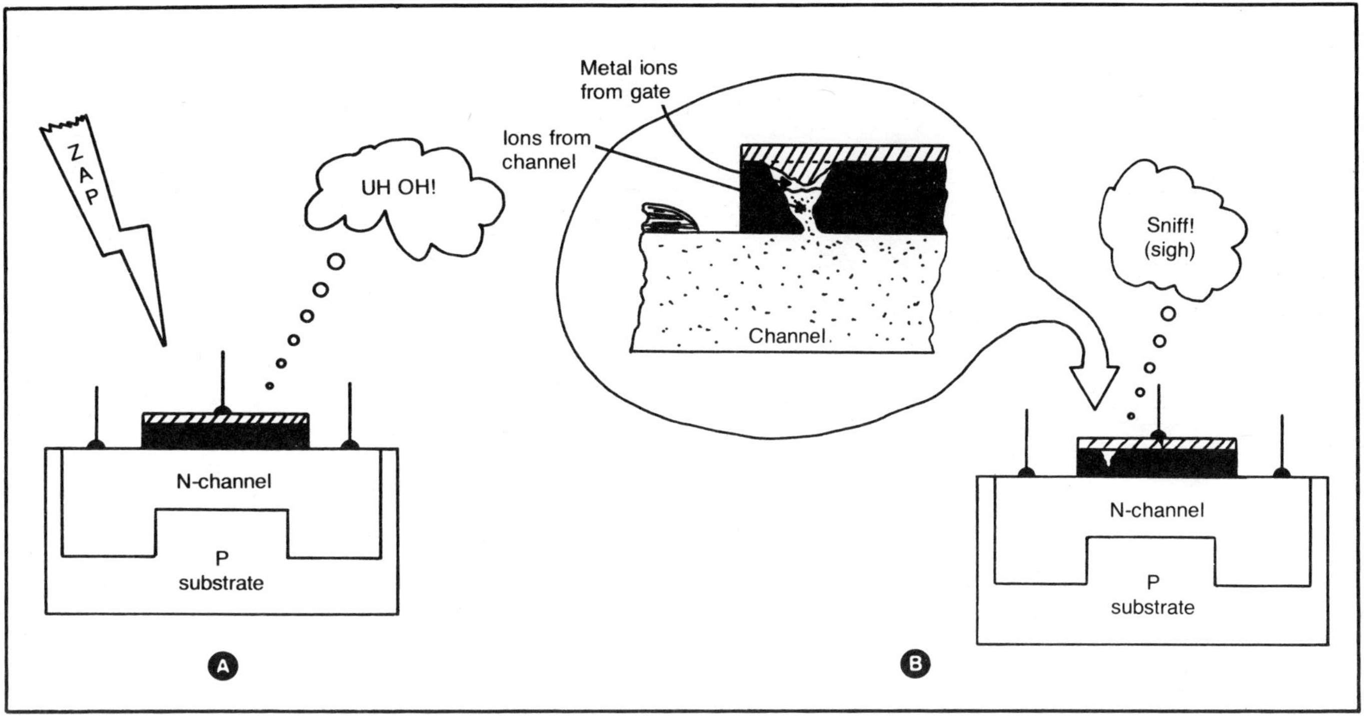

Fig. 4-4. (A) Mechanism of disaster, (B) a ruined CMOS IC.

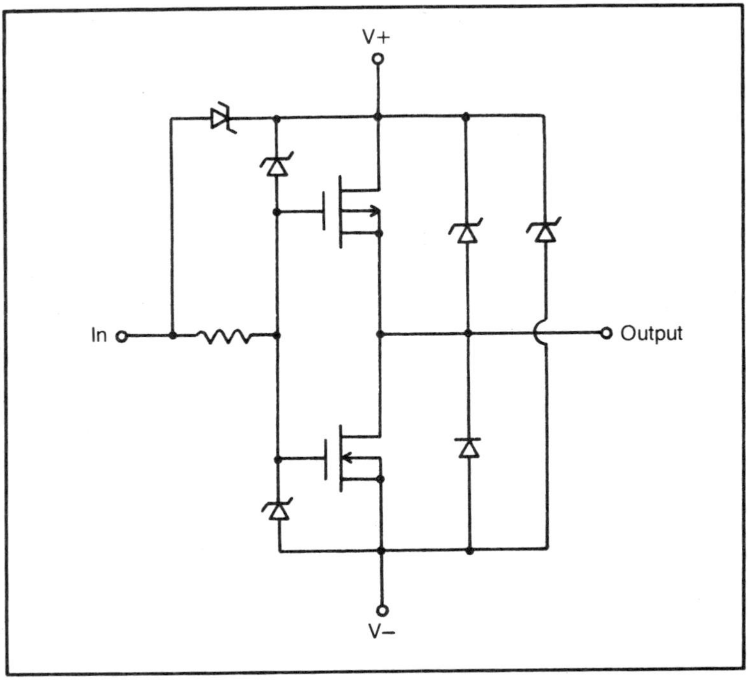

Fig. 4-5. Diode-protected inverter (CMOS).

currents produced by the static charge around the delicate gate insulator. In general, the B-series devices are less sensitive to static, but still must be handled with a little respect. Note that Schottky TTL and certain other IC devices share this problem with the CMOS digital IC devices.

Perhaps the best protection for the CMOS device is proper handling procedures. The main object of these procedures is to keep all pins of the IC at the same potential, all of the time until the device is actually installed in a circuit. Toward that end, the CMOS device will be shipped from the manufacturer in one of several popular protective containers. Some are mounted on a black conductive foam rubber carrier. The carbonized foam will keep all pins at the same potential. Similarly, there are plastic containers (sleeve form) that are conductive and will therefore keep the pins at the same potential. Finally, some individual CMOS devices are shipped in aluminum foil wrappers. This method is especially common among hobby parts suppliers who will buy the devices in wholesale quantities on either foam or in sleeves, and then individually wrap them for customers. Whether or not this is alright depends upon the care taken by the retailer in repackaging the devices.

Printed circuit boards or small exposed projects that contain CMOS or other sensitive devices are often shipped in conductive plastic bags, or, wrapped in aluminum foil. Regardless of whether dealing with individual CMOS devices, a carton full of CMOS devices, or, a printed wiring board containing CMOS devices, do not attempt to defeat the protection. Curiosity can kill the device—stifle yourself and keep the darn wrapper intact!

Installing the CMOS device can be a moment of danger. The idea is to keep everything at the same potential, including yourself! First, when you pick up the CMOS device carrier, touch ground to leak off any built-up static electricity. Also do the same for the circuit in which the device will be installed. Only then should you transfer the device . . . it will be safe then.

Figure 4-6 shows a work station idea that is used by some. Ideally, the ground plane on the work bench will be conductive foam pad, but an aluminim sheet will also work. If the aluminum is used, then ground the surface through a 1 megohm, 2-watt resistor. This protection is not for the CMOS but for you! If a catastrophic short occurs, e.g., to 110 volts ac, and you are working on a grounded surface, then bye-bye: you will be *electrocuted*! The 1 megohm resistor limits the current to a quasi-safe value. Note the wattage rating of the resistor. I don't seriously expect two watts to flow, but there is still a good reason for using a 2-watt resistor. It seems that resistors have *voltage ratings!* The two-watt resistor has the higher rating of the carbon composition resistors, so is the safest to use in this circuit.

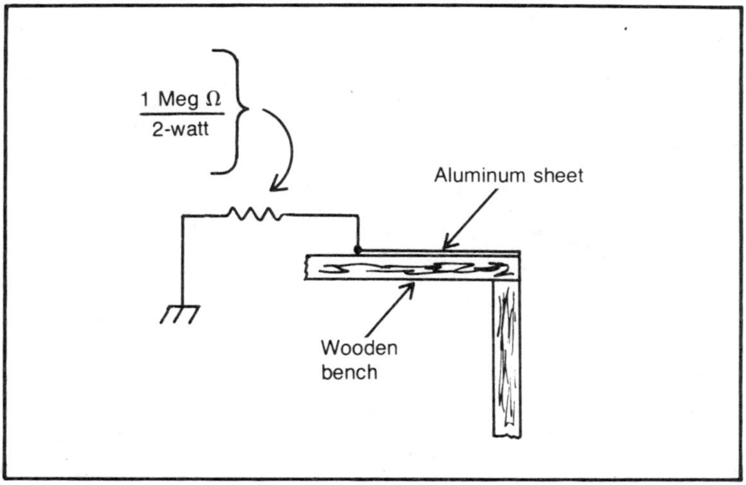

Fig. 4-6. CMOS-safe work table.

It is interesting to note that we can often leave the CMOS device mounted to its foam when solder-tacking it into the circuit for test purposes or troubleshooting. The foam resistance is so high that it won't usually offend the low-voltage circuits in which CMOS devices operate!

Finally, make sure that you continually discharge your body static, and please avoid wearing static-prone garments: wool and some synthetics are particular offenders. Garments are so offensive where static electricity is concerned that explosions have occurred because of the wrong kind of underwear! Wherever explosive gases are used it is not wise to wear static prone garments for this reason. In operating rooms (surgery) once-upon-a-time the anesthetic gases (ether, cyclopropane, etc.) were explosive. As a result, all of the staff had to wear only cotton garments in order to avoid static build-up. Even the floor was conductive (e.g., about 5 megohms to ground) . . . it seems that a hospital operating room was ideal for use with CMOS devices! The lesson here, I suppose, is to build your laboratory in an operating room (*gleep*!). Also, make sure your tools are grounded. The soldering iron should be a type that has a grounded tip, or a ground provided externally where applicable.

CMOS DEVICES

In the pages to follow we will find a catalog of the more common CMOS digital IC devices. The format is similar to that of the TTL device lists in the previous chapter.

4000 SERIES DEVICES

Device Number: 4000

Type of Device: Dual Three-Input NOR plus Spare Inverter

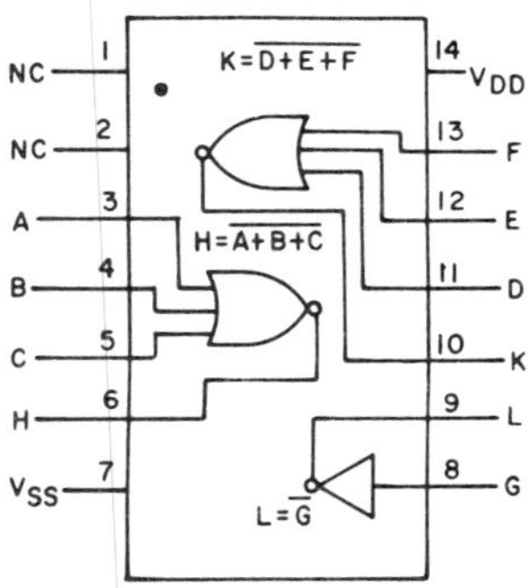

CD4000A
CD4000B
CD4000UB

Dual 3-Input NOR Gate Plus Inverter

Operation & Application: *Inverter Section.* A HIGH input produces a LOW output, and a LOW input produces a HIGH output. *NOR Gate Sections.* A HIGH on any of three inputs produces a LOW output. All three inputs must be LOW for the output to be HIGH. All three sections of this device can be operated either independently, or, in combination with each other. Only the power-supply terminals are common to all three sections.

Device Number: 4001

Type of Device: Quad Two-Input NOR Gate

Quad 2-Input NOR Gate

CD4001A
CD4001B
CD4001UB

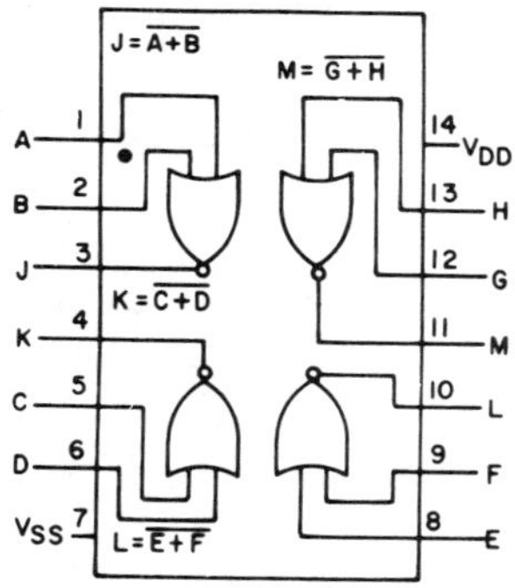

Operation & Application: This device is similar in function to the TTL 7402 device. It contains four independent two-input NOR gates. A HIGH on any one input will cause the related output to be LOW; both inputs on any one section must be LOW for the output to be HIGH. Only the power-supply connections are common to all four sections of the 4001 device.

Device Number: 4002

Type of Device: Dual Four-Input NOR Gate

Quad 4-Input NOR Gate

CD4002A
CD4002B
CD4002UB

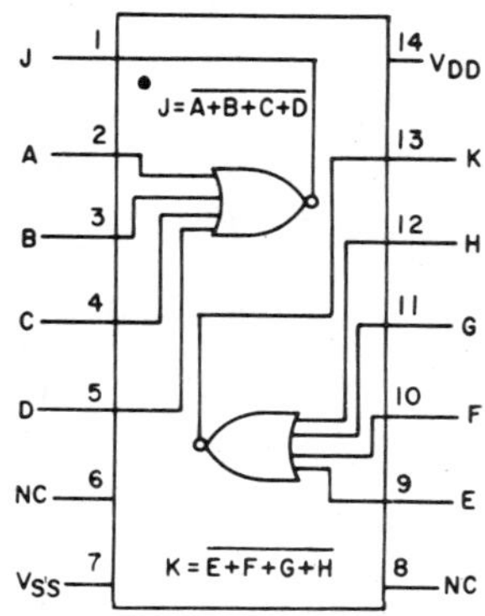

Operation & Application: The 4002 device contains two four-input NOR gates. On either section, a HIGH on any of the four inputs will cause the output to be LOW. All four inputs must be LOW for the output to be HIGH. Both sections can be operated independently, or in conjunction with other devices. Only the power-supply terminals are common to both sections.

Device Number: 4006

Type of Device: CMOS 18-Stage Static Shift Register with Variable Length Capability

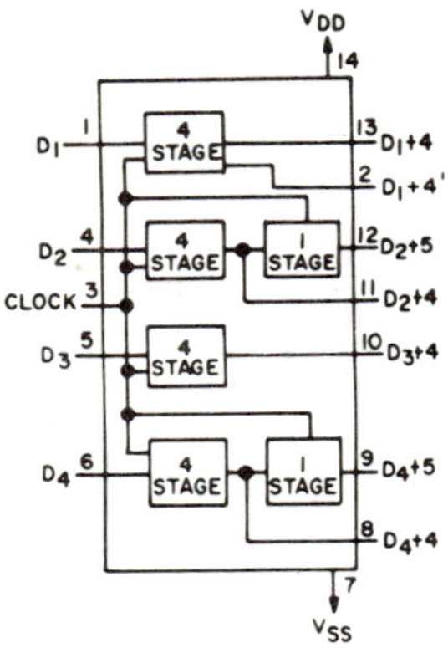

18-Stage Static Shift Register

CD4006A
CD4006B

Operation & Application: The 4006 shift register contains four separate shift registers: two four-stage sections and two five-stage sections. Each section has an independent data path. The five-stage sections have an output tap at the fourth stage. All stages operate from a common clock line; all data are shifted one place to the right on the negative-going transition of the clock pulse. Depending upon external connections, we can make the 4006 device operate either as multiple register sections of 4, 5, 8 or 9 stages, or, as a single register of 4, 5, 8, 9, 10, 12, 13, 14, 16, 17 or 18 stages. Longer sizes are possible by cascading two or more 4006 packages. Pins 8 and 11 are not used. The 4006 will operate to frequencies of 5 MHz if the V+ voltage is 12 volts, and 2.5 MHz in +5 volt circuits.

Device Number: 4007

Type of Device: Dual Complementary MOS Pair Plus Spare Inverter

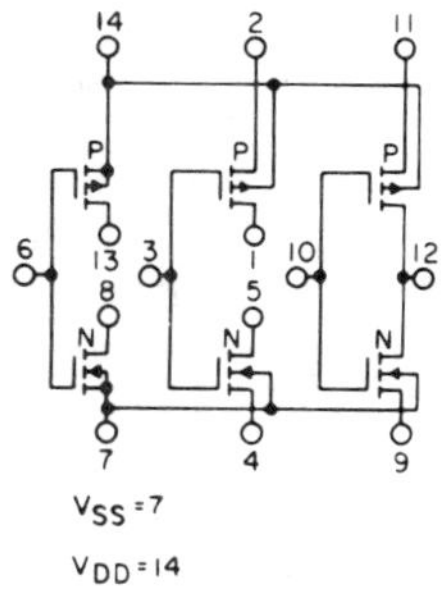

Dual Complementary Pair
Plus Inverter

CD4007A
CD4007UB

Operation & Application: This device is used to form various logic elements and other circuit functions in which MOS capability and/or interfacing are needed. Each complementary pair consists of one N-channel and one P-channel MOSFET with the gate terminals connected together, but with the drain and source terminals separated.

Device Number: 4008

Type of Device: CMOS Four-Bit Full Adder

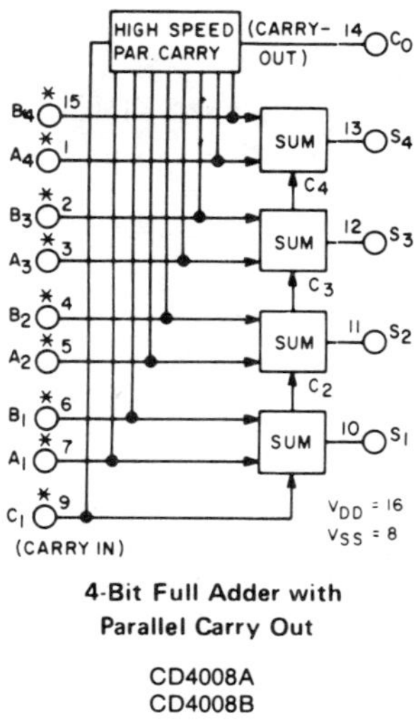

4-Bit Full Adder with
Parallel Carry Out

CD4008A
CD4008B

Operation & Application: The 4008 device contains four full-adder stages with fast look-ahead carry. Four-bit parallel carry-out is provided, and the devices can be connected together to permit high-speed multibyte addition operations. The sum outputs form a four-bit word that is the positive-logic sum of two four-bit input words designated A and B. The usual binary coding is assumed. A carry output is also provided. Average +5-volt addition time is 0.9 milliseconds.

Device Number: 4009

Type of Device: CMOS Hex Buffer/Converters (Inverting) Hex buffer and logic level converter designed to interface CMOS circuitry with either TTL or DTL logic elements. This is an obsolete device, and has been replaced with the 4049 device.

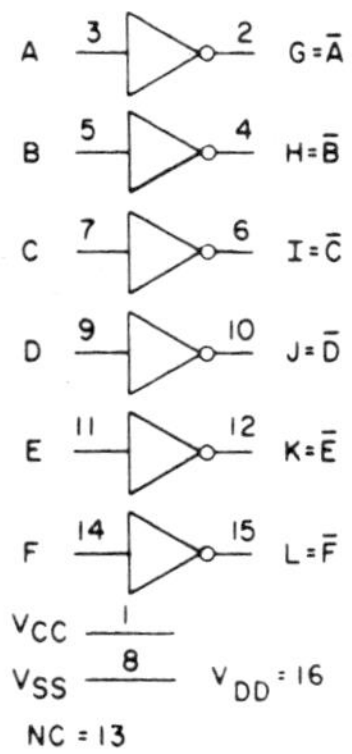

Hex Buffer/Converter
Inverting Type

CD4009A
CD4009UB

Device Number: 4010

Type of Device: CMOS Hex Buffer/Converters (Noninverting)

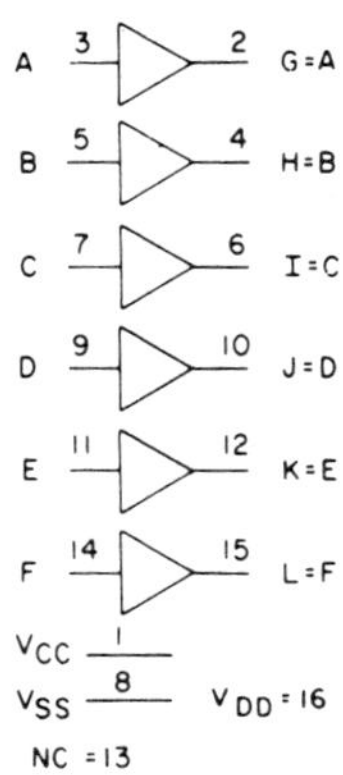

Hex Buffer/Converter
Non-Inverting Type

CD4010A
CD4010B

Operation & Application: Hex buffer and logic level converter used to interface CMOS with either TTL or DTL logic elements. This device is obsolete and has been replaced with the 4050 device.

Device Number: 4011

Type of Device: Quad Two-Input NAND Gate

Quad 2-Input NAND Gate
CD4011A
CD4011B
CD4011UB

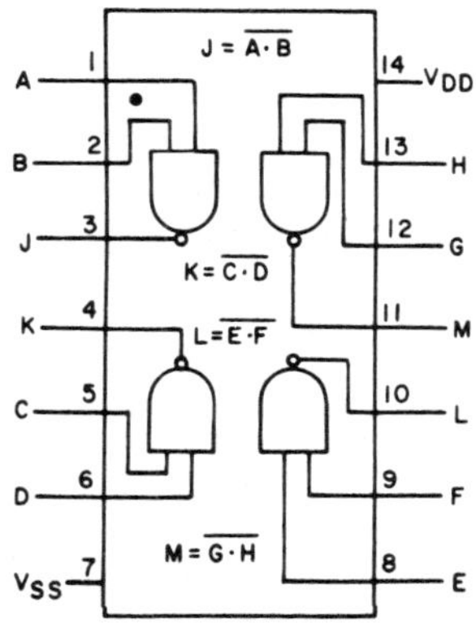

Operation & Application: This device is functionally similar to the 7400 Quad Two-Input NAND Gate from the TTL line. The 4011 contains four two-input NAND gates that are totally independent of each other, except for power-supply connections. A LOW on either input of any section will cause the output of that section to be HIGH. Both inputs must be HIGH for the output to be LOW.

Device Number: 4012

Type of Device: Dual Four-Input NAND Gate

Dual 4-Input NAND Gate
CD4012A
CD4012B
CD4012UB

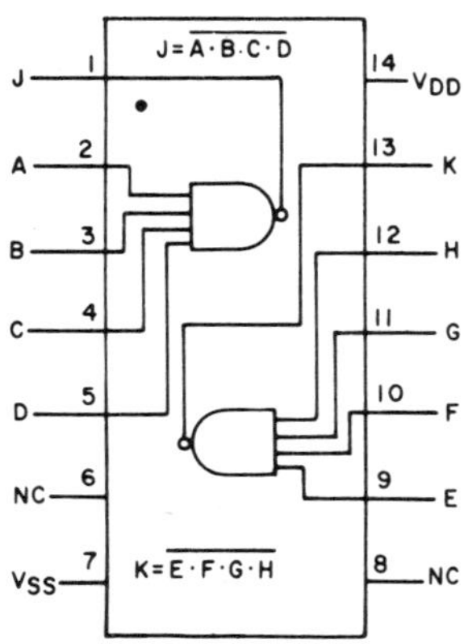

Operation & Application: The 4012 package contains a pair of four-input NAND gates. Each section may be operated independently of the other. A LOW on any one input of either section causes the output of that section to be HIGH. All four inputs must be HIGH for the output to be LOW.

Device Number: 4013

Type of Device: Dual Type-D Flip-Flop with Set/Reset

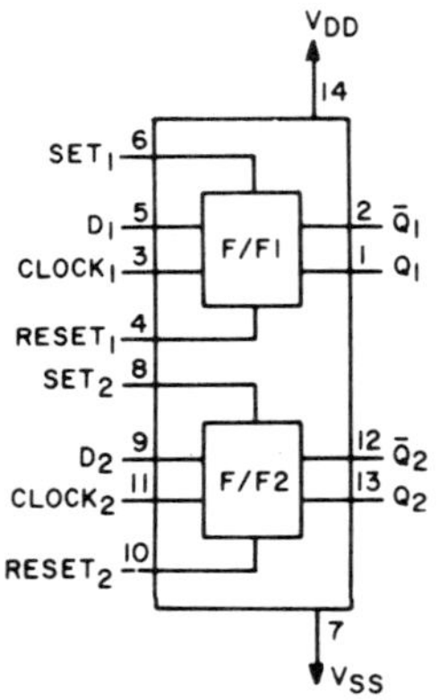

Dual "D" Flip-Flop with Set/Reset Capability
CD4013A
CD4013B

Operation & Application: This package contains two type-D flip-flops, each of which has its own *data, clock, set, reset, Q-output* and *$\overline{Q}$-outputs.* Clocking occurs on the positive-going transition of the clock input. When the clock line snaps HIGH, the data applied to the *data* input is transferred to the Q-output; the complement of the data appears on the not-Q output. In other words, when *data*(D) is HIGH during clock transition, the output goes HIGH; if the D-input is LOW during clock transition then the output goes LOW. *Set* and *reset* are active-HIGH inputs. Making *set* HIGH will force the output to go to the Q = HIGH condition. Making *reset* HIGH forces the output to go to the Q = LOW condition. Set/reset are independent of the clock line. Both type-D flip-flops in the 4013 package may be operated independently. The only common point between the two is the power supply and ground terminals.

Device Number: 4014

Type of Device: Eight-Stage Static Shift Register (PISO or SISO)

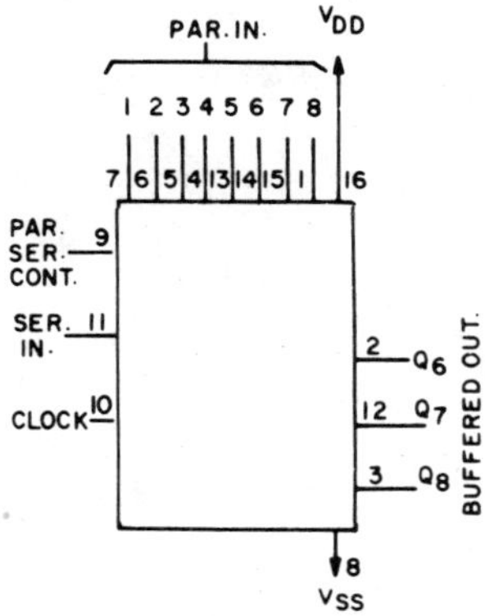

8-Stage Synchronous Shift Register with Parallel or Serial Input/Serial Output
CD4014A
CD4014B

Operation & Application: This chip contains an eight-state static shift register with output taps at the sixth and seventh stages. As a result, it will operate as either 6, 7, or 8 stage shift registers. Both *parallel-in-serial-out* and *serial-in-serial-out* modes of operation are possible.

Device Number: 4015

Type of Device: Dual Four-Stage Static Shift Register

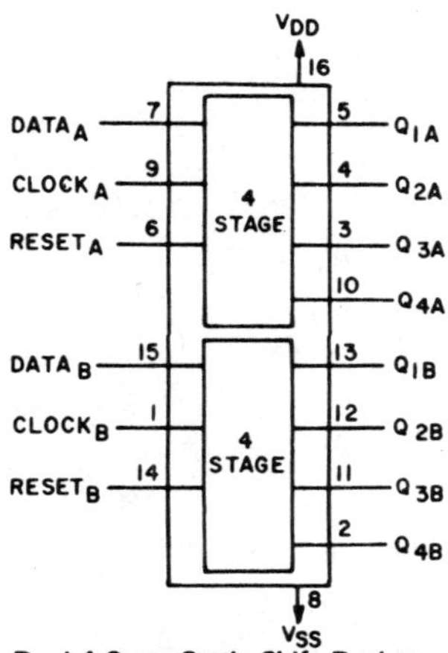

Dual 4-Stage Static Shift Register with Serial Input/Parallel Output
CD4015A
CD4015B

Operation & Application: The 4015 device contains two identical four-bit shift registers, each of which has independent *clock* and *reset* inputs, as well as independent serial data input lines. Each stage of the two flip-flops consist of type-D flip-flops. Data applied to the input will be shifted one state to the right on every positive-going transition of the clock line. The reset line is an active-HIGH input. Bringing reset HIGH will force the shift register to clear.

Device Number: 4016

Type of Device: Quad Bilateral CMOS Electronic Switch

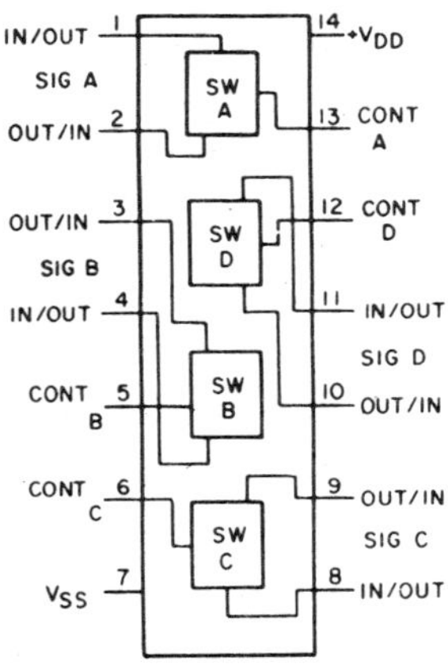

Quad Bilateral Switch

CD4016A
CD4016B

Operation & Application: The 4016 device contains four independent electronic switches that allow the series transmission or multiplexing of either analog or digital signals. Each switch has its own control terminal. Bringing the control terminal to a potential equal to VDD (i.e., HIGH) will make the switch act like a very high series impedance; this is the off condition. If the control line is set to a potential equal to Vss, then the switch is turned on and behaves like a 280 ohm resistor in series with the line. The 4016 is obsolete, and has been largely replaced with the 4066 device.

Device Number: 4017

Type of Device: Decade (Divide-by-10) Synchronous Counter with Decoded 1-of-10 Outputs

Decade Counter/Divider with 10 Decoded Decimal Outputs
CD4017A
CD4017B

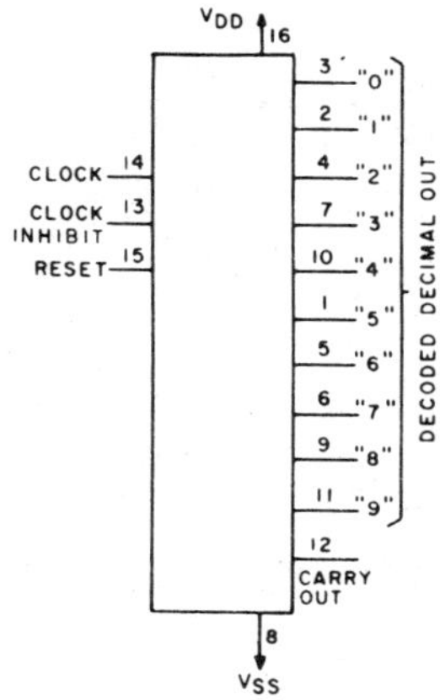

Operation & Application: The 4017 contains a five-stage Johnson counter, and features *clock, reset* and *clock inhibit* inputs. Pulse shaping is provided because of the Schmitt trigger stage in the clock input line. The *clock inhibit* is an active-LOW input that will disable counting without resetting when brought LOW. The counter advances on positive-going transitions of the clock line, provided that *clock inhibit* is LOW. *Reset* is an active-HIGH input and will clear the counter to zero when HIGH. The carry out line completes one cycle for every 10 input cycles. A related device is the 4022 octal counter.

Device Number: 4018

Type of Device: Presettable Divide-by-N Counter

Presettable Divide-by-"N" Counter Fixed or Programmable
CD4018A
CD4018B

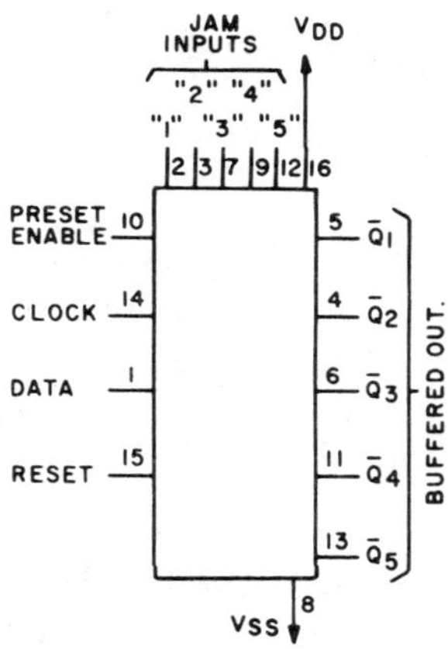

Operation & Application: This chip contains a five-stage Johnson counter, and features buffered Q-outputs, and the following inputs: *clock, reset, data, preset* and *enable*. Alone, the 4018 can offer division ratios of 2, 4, 6, 8, and 10; adding a 4011 device to feedback the appropriate *Q* terminal back to the *data* line we can also derive division ratios of 3, 5, 7, or 9. The counter advances one state for every positive-going transition of the clock line. Schmitt trigger action on the input line provides pulse shaping, and thereby accommodates a large number of signal waveforms. Reset is active-HIGH, so will clear the counter to zero when made HIGH. The preset and enable terminals allow the data applied to the jam inputs to be entered.

Device Number: 4019

Type of Device: Quad AND/OR Select Gate

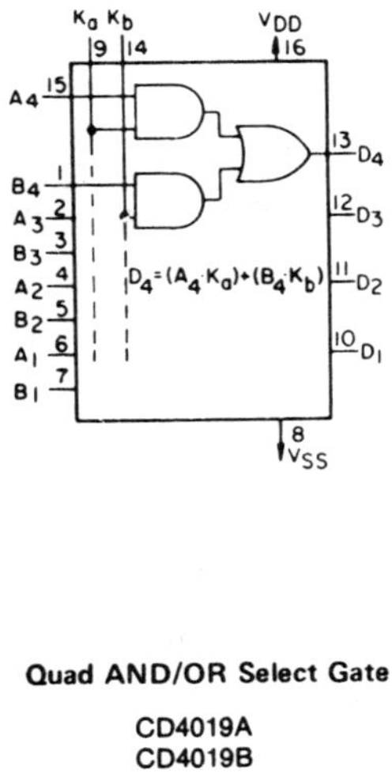

Quad AND/OR Select Gate

CD4019A
CD4019B

Operation & Application: The 4019 device contains a pair of two-input AND gates, each of which feeds from its output one input of a two-input OR gate. Control bits K_a and K_b perform the logic selection that determines Ch. A or Ch. B information is transmitted.

Device Number: 4020

Type of Device: Ripple-Carry Binary Counter/Divider

14-Stage Binary Ripple Counter

CD4020
CD4020B

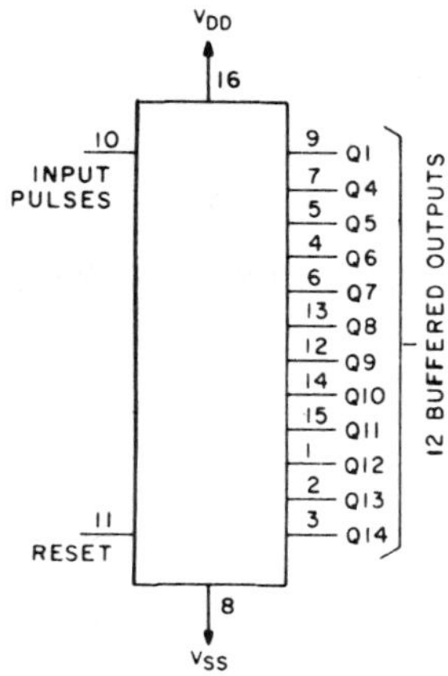

Operation & Application: All counter stages in the 4020 device are master-slave flip-flops. The counter state advances one count for each negative transition of the input pulse. All input and output lines are buffered. Like other CMOS counters, the clock input stage consists of a Schmitt trigger, so will accommodate a wide range of input pulses. The reset line is active-HIGH, so will cause the outputs to go to zero when made HIGH. This device is similar to the 4060 device, except that there is no built-in oscillator circuit. The 4020 is a fourteen-stage counter implying 2^{14} (16,384) division ratio. Every stage except the 2nd and 3rd have output lines, so the device will accommodate division ratios of 2^n, where n = 1, 4, 5, 6, 7, 8, 9, 10, 11, 12, 13, or 14.

Device Number: 4021

Type of Device: Eight-Stage Static Shift Register

8-Stage Static Shift Register Asynchronous Parallel or Synchronous Serial Input/ Serial Output

CD4021A
CD4021B

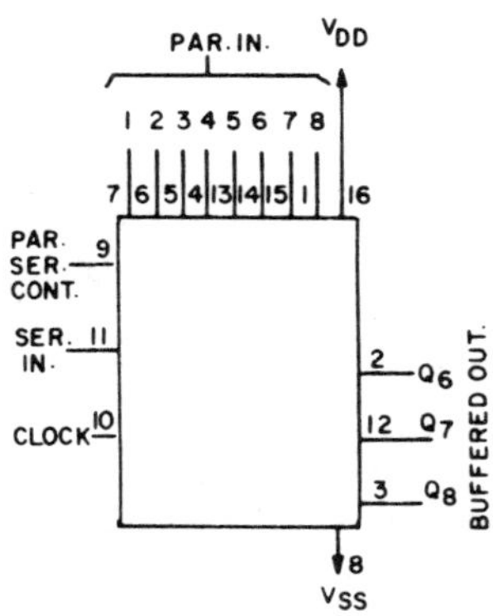

Operation & Application: This device is very similar to the 4014 device, except that it is an asynchronous PISO, or synchronous SISO shift register. The 4014 was synchronous for both PISO and SISO modes.

Device Number: 4022

Type of Device: Octal Counter/Divider

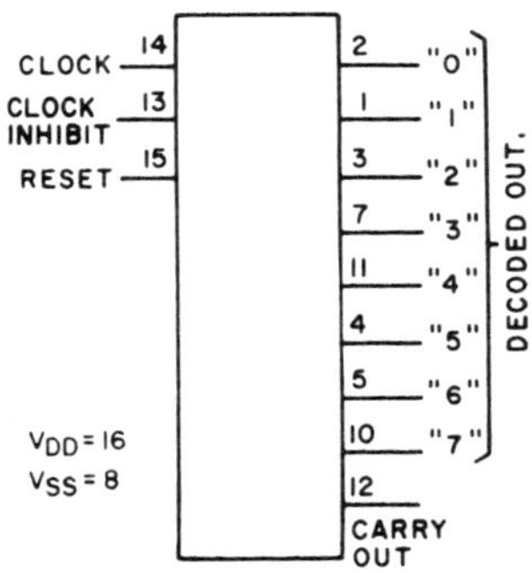

Divide-by-8 Counter/Divider with 8 Decoded Decimal Outputs

CD4022A
CD4022B

Operation & Application: This device is similar to the 4017, except that it has a three-bit octal format rather than four-bit binary coded decimal format.

Device Number: 4023

Type of Device: Triple Three-Input NAND Gate

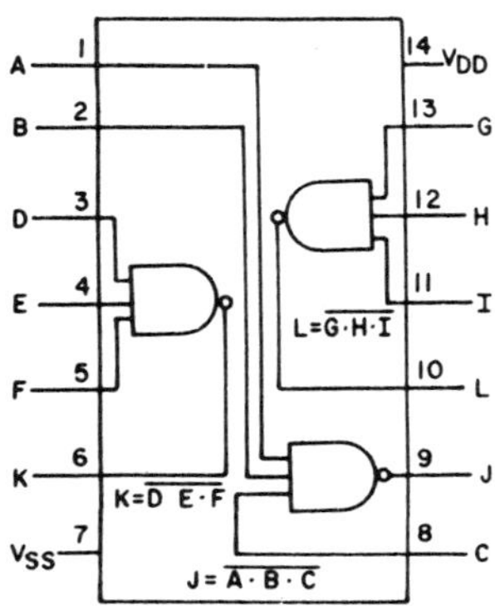

Triple 3-Input NAND Gate

CD4023A
CD4023B
CD4023UB

Operation & Application: The 4023 device contains three independent three-input NAND gates. The usual NAND gate rules apply: a LOW on any one input forces the output HIGH, all three inputs must be HIGH for the output to be LOW.

Device Number: 4024

Type of Device: Seven-Stage Ripple-Carry Binary Counter/Divider

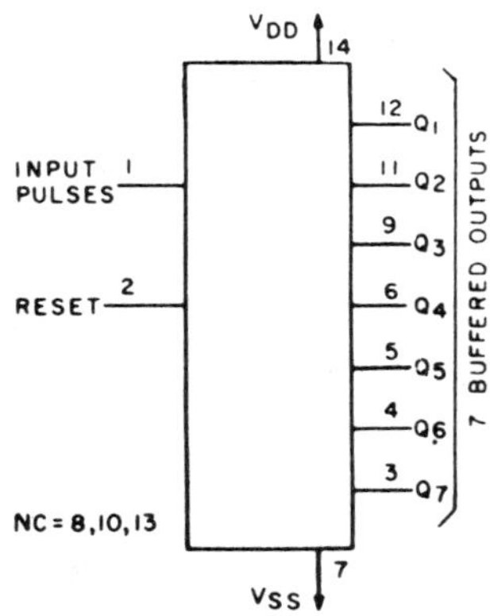

7-Stage Ripple-Carry Binary Counter/Divider
CD4024A
CD4024B

Operation & Application: A binary counter consisting of seven master-slave flip-flops in cascade, with an output at each stage. The state of the counter advances one count on the negative-transition of the clock line. An active-HIGH reset is provided on the 4024 device. Bringing reset HIGH momentarily will reset all stages to zero. There is a Schmitt trigger circuit on the input, so the device will accommodate unusual waveforms with less than optimum rise and/or fall times. At a power-supply voltage of +5 volts dc, the maximum clock frequency is 2.5 MHz, rising to almost 12 MHz at 15 Vdc. Note that the B-version will clock faster than the A-version, the limits being 3.5 MHz@5 Vdc and 24 MHz@ 15 Vdc.

Device Number: 4025

Type of Device: Triple Three-Input NOR Gate

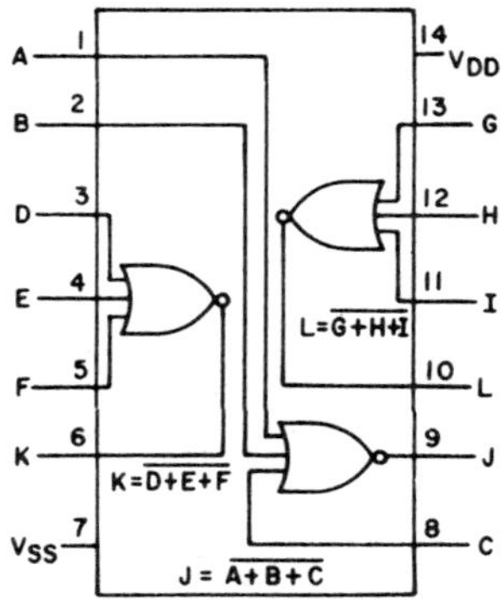

Triple 3-Input NOR Gate

CD4025A
CD4025B
CD4025UB

Operation & Application: The 4025 package contains three independent (except for power-supply connections) three-input NOR gates. The usual NOR gate rules apply: A HIGH on any one input will cause the output to be LOW; all three inputs must be LOW for the output to be HIGH.

Device Number: 4026

Type of Device: Decade (Decimal or Base-10) Counter/Divider with Decoded Seven-Segment Display Outputs and Display Enable

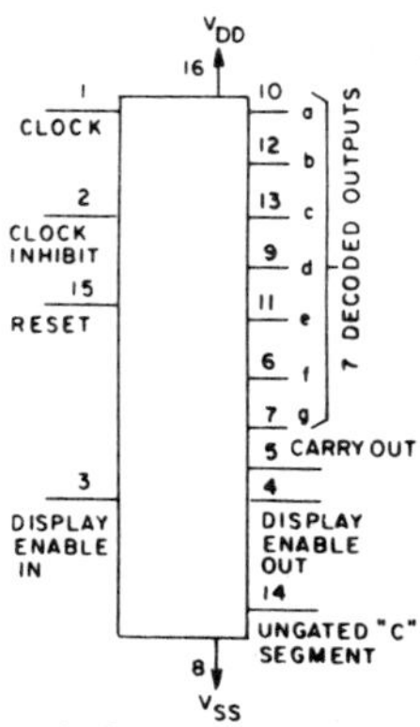

Decade Counter/Divider with 7-Segment Display Outputs and Display Enable

CD4026A
CD4026B

Operation & Application: This counter chip contains a standard five-stage Johnson counter. The chip also contains the decoder circuitry to produce the seven-segment output code for the display device. Inputs present on this chip are *clock, reset, clock inhibit* and *display enable*. The clock input is as on any counter, and will advance the counter on the positive-going transition (i.e., it is positive edge triggered). Reset is active-HIGH, and will cause all stages of the counter to reset to zero when brought HIGH. The clock inhibit line is active-LOW, and will allow the clock signal to affect the output count only when clock inhibit is LOW. Note that the clock inhibit line may be pressed into service as a negative-edge triggered clock if the main clock line is held HIGH. The display enable line is an active-HIGH input that will turn on the display when HIGH. When display enable is LOW, all outputs of the seven-segment display decoder will be LOW (i.e., inactive). Note that the *a, b, c, d, e, f* and *g* display segment outputs are active-HIGH, i.e., they will turn on a display segment when HIGH.

Device Number: 4027

Type of Device: Dual J-K Master-Slave Flip-Flop

Dual J-K Master-Slave Flip-Flop with Set-Reset Capability
CD4027A
CD4027B

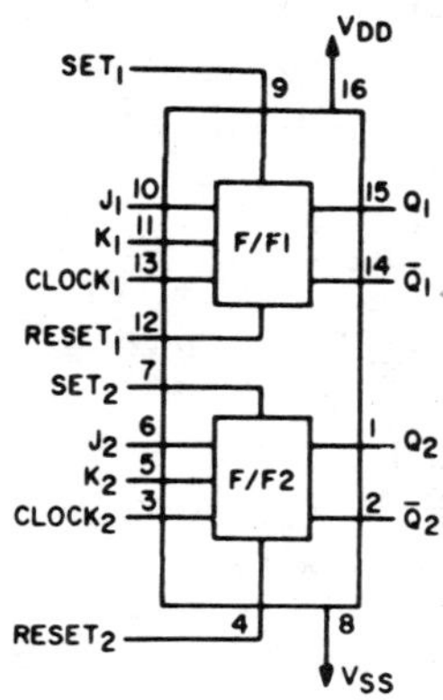

Operation & Application: Package contains two independent J-K flip-flops, both of which have reset and set inputs, in addition to the usual J-K flip-flop inputs. The usual rules for operation of J-K flip-flops in direct and clocked modes apply. See text on J-K flip-flops. The flip-flops will toggle to 3.5 MHz at a supply voltage of 5 Vdc, 8 MHz at 10 Vdc and 12 MHz at 15 Vdc.

Device Number: 4028

Type of Device: BCD-to-Decimal Decoder/Octal Decoder

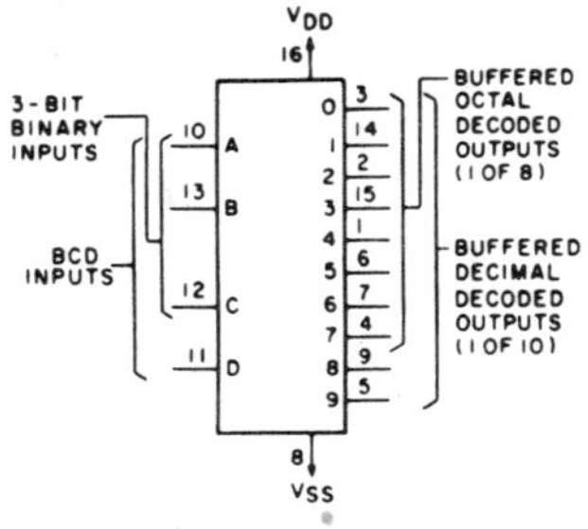

BCD-to-Decimal Decoder

CD4028A
CD4028B

Operation & Application: The 4028 can be used as a four-bit Binary Coded Decimal (BCD)-to-one-of-ten, or three-bit Binary Coded Octal (BCO)-to-one-of-seven decoder. In the decimal application, the four inputs ABCD (weighted 1-2-4-8) are used, where in octal applications the ABC (1-2-4) inputs are used. An output will be HIGH when active. In decimal operation, therefore, the output selected by the four-bit BCD code will be HIGH, while all others are LOW. Similarly, in octal operation, the input selected by the BCO input code will cause the correct one of seven outputs to go HIGH. The output drive capability is approximately 1 mA.

Device Number: 4029

Type of Device: Presettable Up/Down Four-Bit Binary (Base-26) or BCD (Base-10) Counter

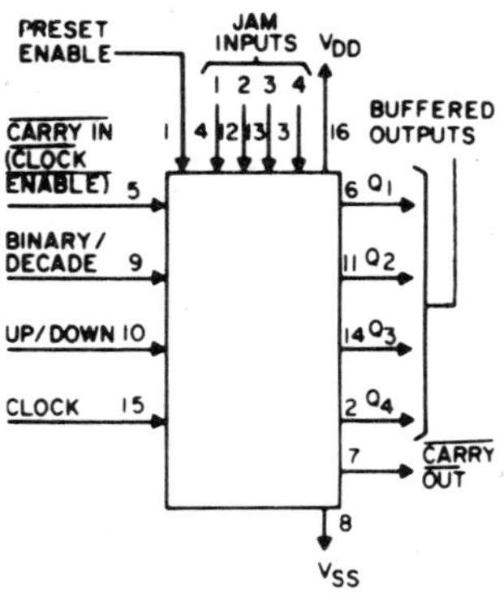

Presettable Up/Down Counter, Binary or BCD-Decade

CD4029A
CD4029B

Operation & Application: Package contains a four-stage counter with look-ahead carry capability, and has the ability to operate in either binary coded decimal (base-10) or straight binary (base-16) modes. A binary/decade input selects the base-mode (binary counting is performed when B/D is HIGH, and BCD when

B/D is LOW. Hence, it is sometimes labeled B/$\overline{D}$. Four jam inputs are provided to permit presetting of the counter.

Device Number: 4030

Type of Device: Quad Exclusive-OR (XOR) Gate

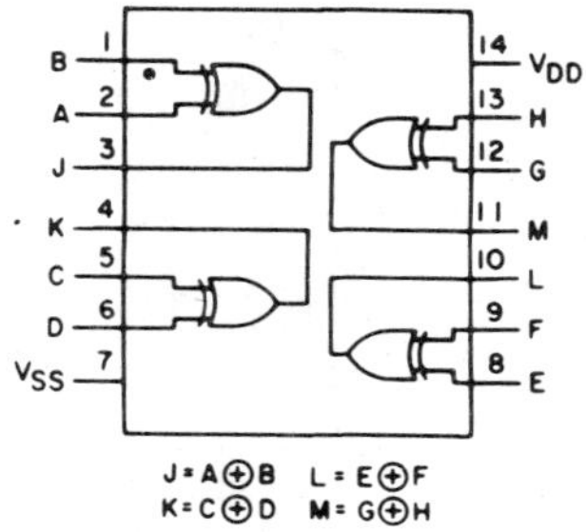

Quad Exclusive-OR Gate

CD4030A
CD4030B

Operation & Application: This package contains four independent Exclusive-OR gates. The usual XOR rules apply: when any *one* input is HIGH, then the output will be HIGH. If both inputs are HIGH, then the output is LOW; if both inputs are LOW then the output is LOW. The general rule is that the output will be HIGH if and only if one input is also HIGH. Anytime both inputs are at the same state, then the output is LOW. Some authorities consider the 4030 obsolete, and prefer to use the 4070 device instead.

Device Number: 4031

Type of Device: Sixty-Four Stage Static Shift Register

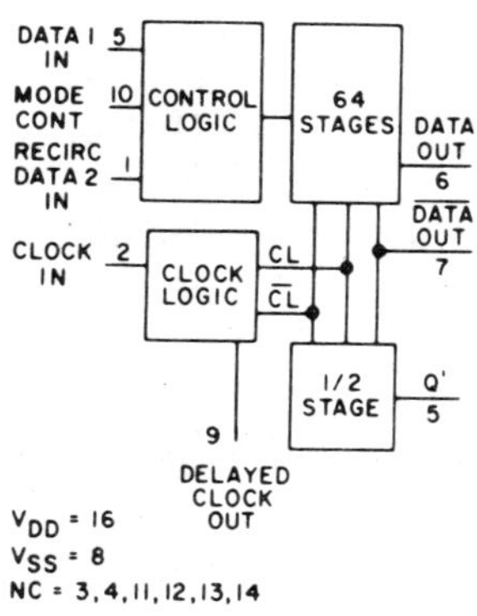

64-Stage Static Shift Register

CD4031A
CD4031B

Operation & Application: This package contains sixty-four type-D master-slave flip-flops in cascade. The general format of the 4031 device is *Serial-In-Serial-Out* with or without recirculation. This device can be used in memory oscilloscopes to store the pattern.

Device Number: 4032

Type of Device: Triple Serial Adder

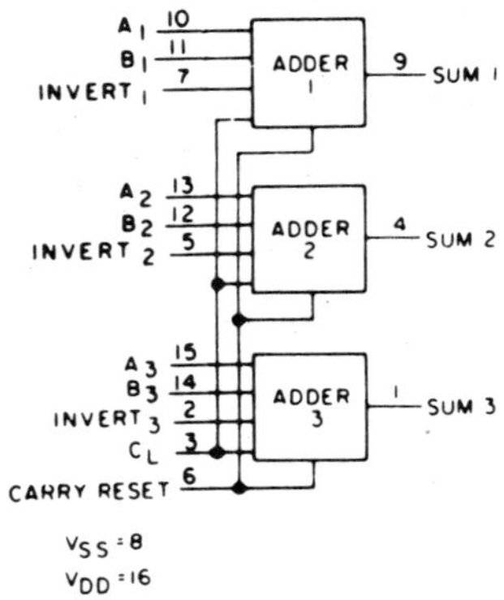

Triple Serial Adder
Positive Logic

CD4032A
CD4032B

Operation & Application: This package contains three independent positive-logic adder circuits. The output of each stage is the sum of the A and B binary inputs plus the carry bit from the previous stage. An active-HIGH *invert* signal is provided that will, when HIGH, cause the output sum to be complemented. A *carry reset* input is also provided.

Device Number: 4033

Type of Device: Decade (i.e., Base-10) Counter/Divider with Seven-Segment Decoded Output and Ripple Blanking

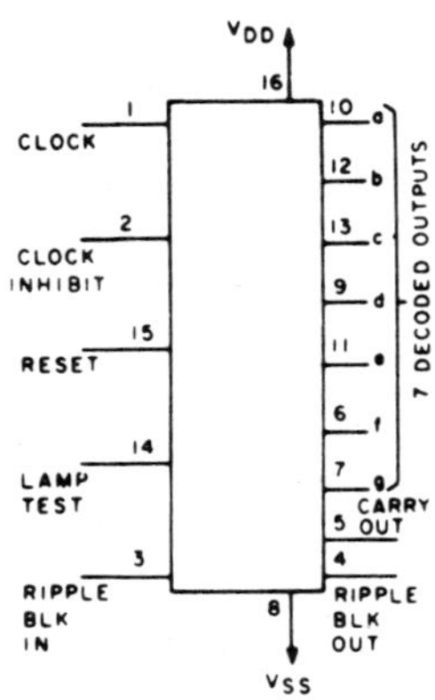

Decade Counter/Divider
with 7-Segment Display
Outputs and Ripple Blanking

CD4033A
CD4033B

Operation & Application: This device contains a five-stage Johnson counter and an output decoder that is compatible with seven-segment digital readout display devices. This device is intended only for low-current LED and fluorescent displays. Inputs on the counter are clock, reset and clock inhibit (see 4026). Signals found on the 4033 and not on its companion device (4026) are ripple-blanking input and output and lamp test input. A HIGH on the reset input will cause the counter to be reset to zero. The counter will advance one step on each positive-going transition of the clock signal. Ripple blanking is used to turn off the displays for leading zeros. Thus, a number such as 00567, which is counted in five cascaded 4033 devices, will display "567," because the leading two zeros were suppressed. Each RBO terminal is connected to the RBI input of the 4033 that is next lower in significance position. The lamp test input is active-HIGH, and will cause the decoder to turn on all segments of the display and thereby display an "8." This feature allows the operator to test the display for burned out segments.

Device Number: 4034

Type of Device: Eight-Stage Static Bidirectional Parallel/Serial Input/Output (I/O) Bus Register

8-Stage Static Bidirectional Parallel/Serial Input/Output Bus Register
CD4034A
CD4034B

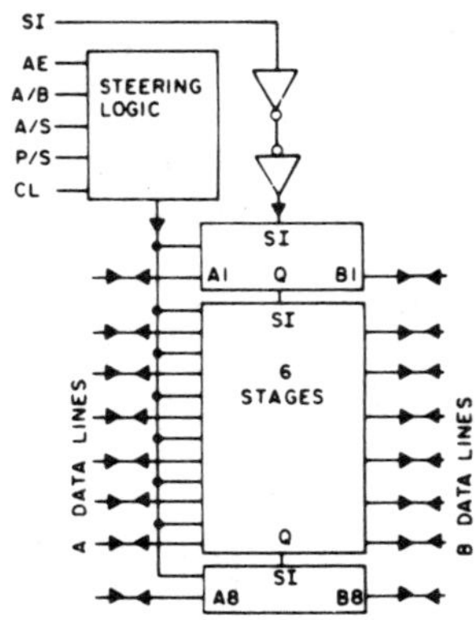

Operation & Application: The 4034 device is used to permit the transfer of parallel-format 8-bit data words between two bus systems (or between computers), it will convert serial data streams to parallel format and then shunt the data word to either of two parallel busses, it will store/recirculate parallel data and it will perform parallel-to-serial conversion on data from either of two eight-bit busses. The eight stages in the 4034 are type-D flip-flops and will operate in the master-slave manner.

Device Number: 4035

Type of Device: Four-Stage Parallel-In/Parallel-Out Shift Register (with J-$\overline{K}$) Serial Inputs and Q/$\overline{Q}$ Outputs

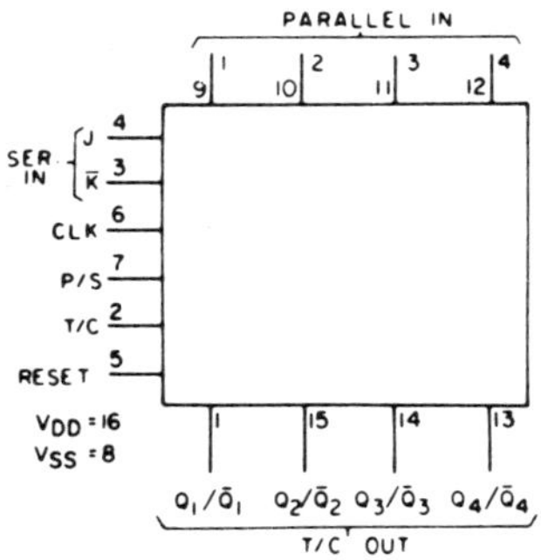

4-Stage Parallel In/Parallel Out Shift Register with J-K Serial Inputs and True/ Complement Outputs

CD4035A
CD4035B

Operation & Application: The 4035 device is a four-stage PIPO shift register with J-$\overline{K}$ input logic, and complementary output logic (both Q and not-Q). The device will accept parallel input data to all four stages, and serial data to the first stage. Asynchronous common reset is provided. A *Parallel/Serial* line permits the selection of input form (HIGH for parallel entry).

Device Number: 4038

Type of Device: Triple Negative-Logic Serial Adders

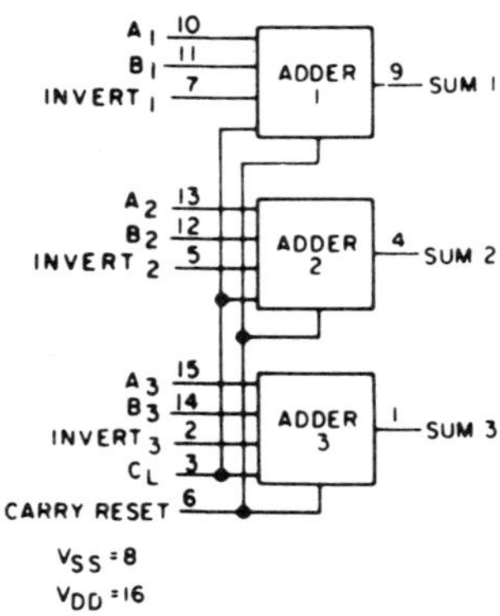

CD4038A
CD4038B

Triple Serial Adder Negative Logic

Operation & Application: This device is similar to 4032, except that the 4038 is a negative-logic adder. See article for 4032 device.

Device Number: 4040

Type of Device: Twelve-Stage Ripple-Carry Binary Counter/Divider

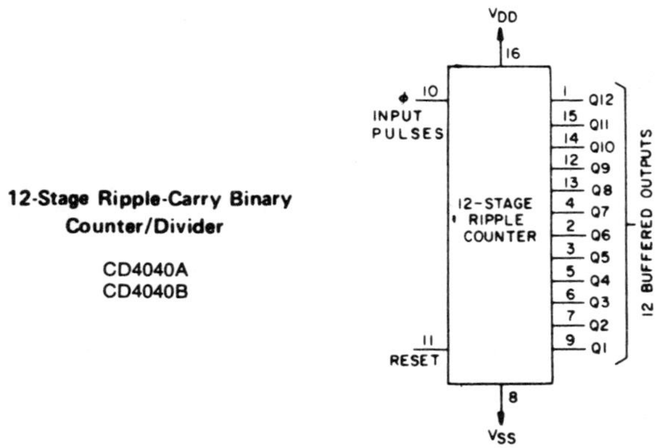

Operation & Application: This device is a 12-bit counter/divider similar to the 14-bit 4020 device and the 7-bit 4024 device.

Device Number: 4041

Type of Device: Quad True/Complement Buffer

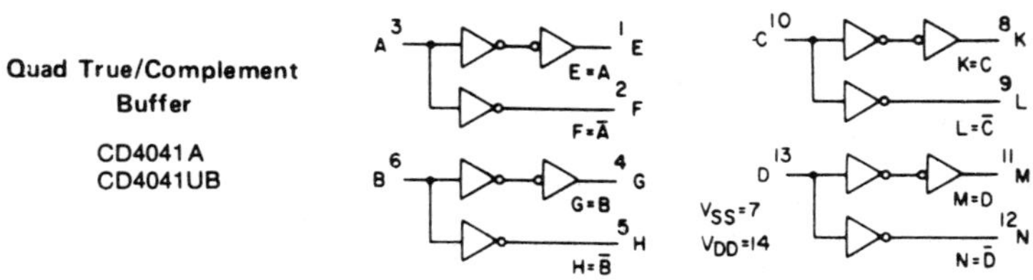

Operation & Application: The 4041 package contains four independent true/complement buffers that will permit CMOS/TTL interfacing, or, it will act as a line buffer for CMOS systems. The outputs will each drive two TTL loads (i.e., fan-out of 2) if the power-supply voltages applied to the package are +5 volts dc to pin no. 14 and ground to pin no. 7.

Device Number: 4042

Type of Device: Quad Clocked "D" Latch

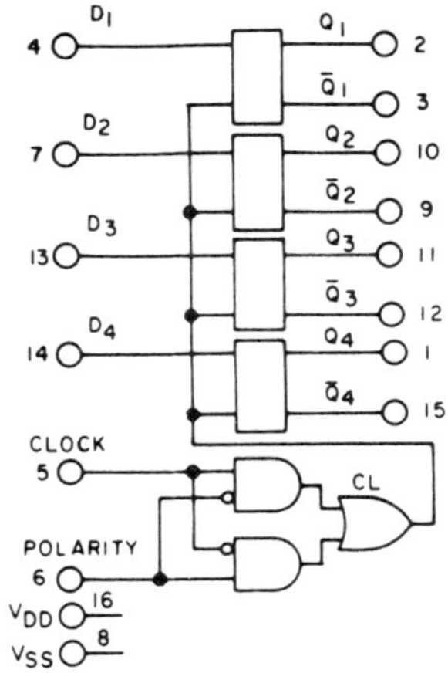

Quad Clocked "D" Latch
CD4042A
CD4042B

Operation & Application: This package contains four type-D flip-flops intended to act as data latches. Each flip-flop is clocked by the common clock or "strobe" line. The clock is programmable on this chip. If the *Polarity* +/– terminal is made HIGH, then the data transfer from input to output occurs on the HIGH clock condition. If the *Polarity* +/– is LOW, on the other hand, the clock is active-LOW so the data transfer takes place on the LOW clock condition.

Device Number: 4043

Type of Device: Quad Tri-State R/S Latches (NOR logic)

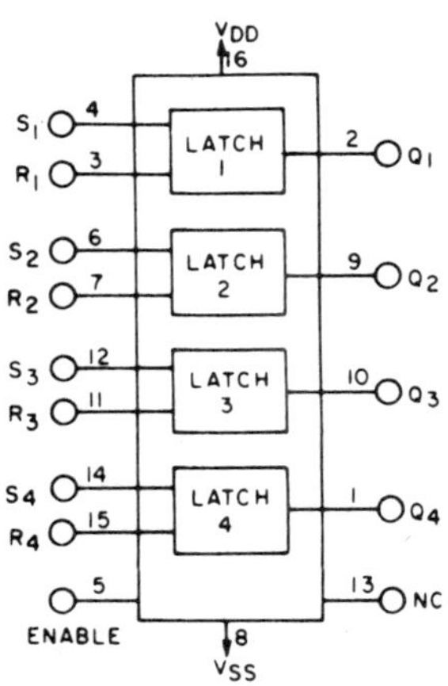

Quad 3-State NOR R/S
Latch
CD4043A
CD4043B

Operation & Application: This package contains four independent R-S flip-flop latches using NOR logic. Each latch has its own Q output and *set* (S)/*reset* (R) inputs. The tri-state Q outputs are controlled by a common chip-enable terminal. When this active-HIGH enable line is HIGH, the Q outputs will be connected to the package outputs; when the enable is LOW then the Q outputs are disconnected from the package output terminals and the outside world sees a high impedance instead of HIGH/LOW states.

Device Number: 4044

Type of Device: Quad Tri-State R/S Latches (NAND Logic)

Quad 3-State NAND R/S Latch
CD4044A
CD4044B

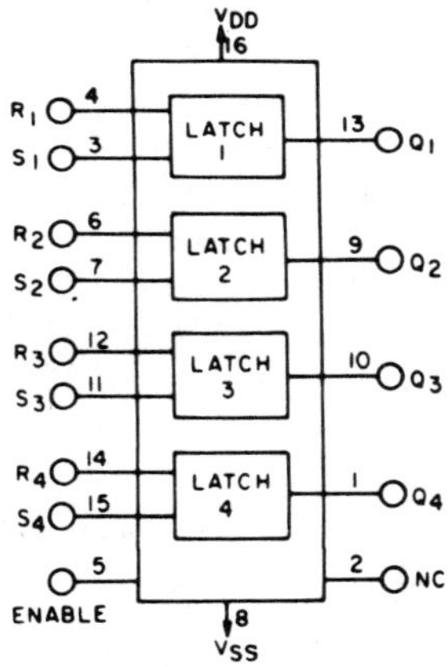

Operation & Application: This device is like the 4043, except that NAND logic is used.

Device Number: 4046

Type of Device: Micropower Phase-Locked Loop (PLL)

Micropower Phase-Locked Loop
CD4046A
CD4046B

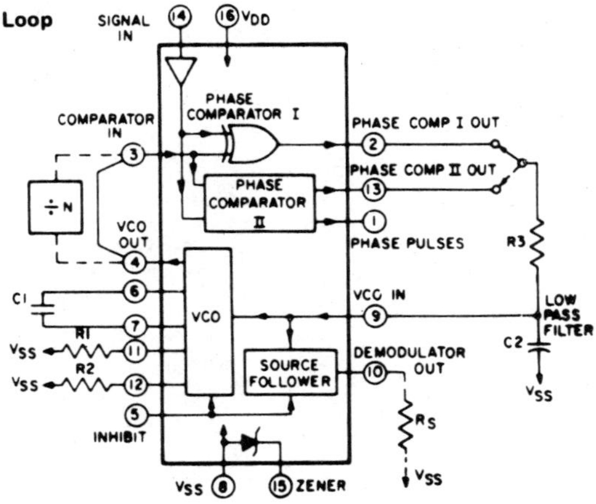

Operation & Application: This device contains a low-powered phase-locked loop (PLL), and consists of a voltage controlled oscillator (VCO) and two different phase comparators. An external RC network supplies the low-pass filtering required for integration of the phase detector outputs.

Device Number: 4047

Type of Device: Low-Power Monostable/Astable Multivibrator

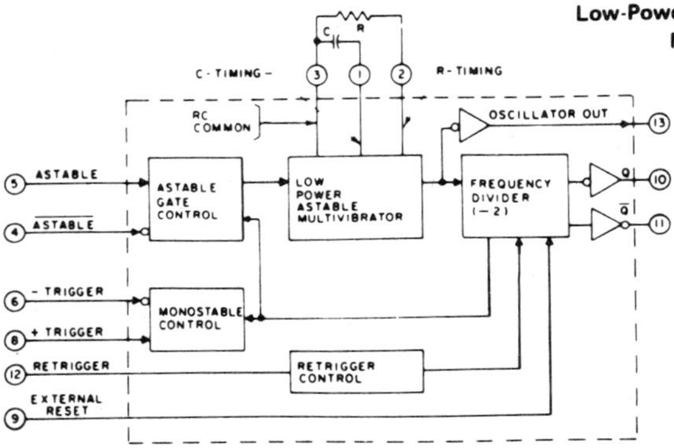

Low-Power Monostable/Astable Multivibrator

CD4047A
CD4047B

Operation & Application: The 4047 device contains circuitry that permits it to operate as either astable or monostable multivibrators. Internal circuitry permits triggering on either positive or negative edge of the trigger input signal. Retriggerability and external counting options are provided.

Device Number: 4049

Type of Device: Hex Inverter with TTL Compatibility

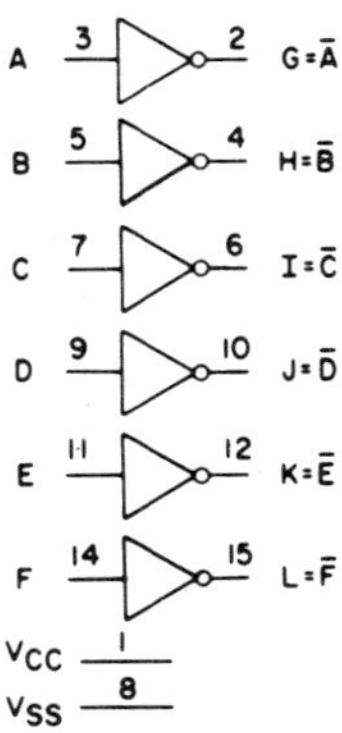

NC = 13
NC = 16
Hex Buffer/Converter Inverting Type

CD4049A
CD4049UB

Operation & Application: The 4049 device is a hex inverter. If the package power supply is +5 Vdc on pin no. 1, and ground on pin no. 8, then the 4049 will drive two TTL loads without need for interfacing. If the supply voltages are any other combination that is legal for CMOS, however, the 4049 will not interface directly to TTL inputs.

Device Number: 4050

Type of Device: Hex Buffer (Noninverting) with TTL Compatibility

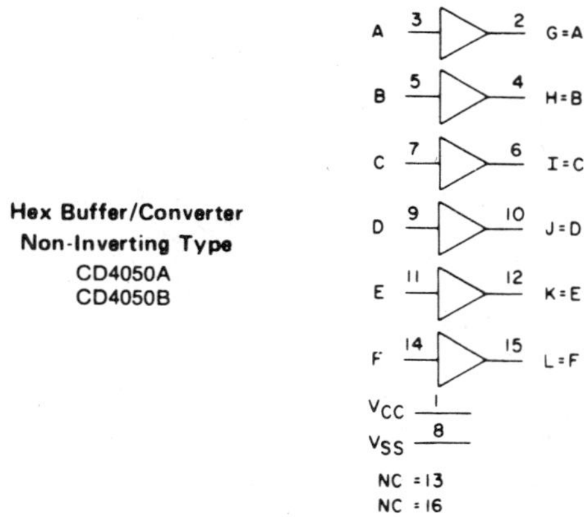

Operation & Application: Noninverting version of the 4050 device.

Device Number: 4051

Type of Device: Single Eight-Channel Multiplexer/Demultiplexer

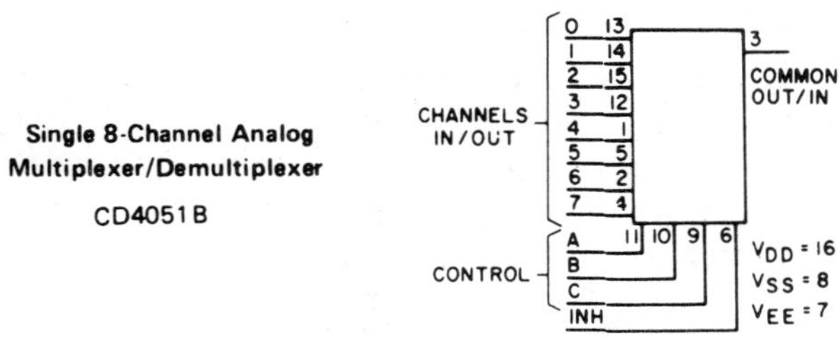

Operation & Application: The 4051 device contains eight bilateral CMOS electronic switches connected such that one terminal of all eight switches are connected to a common line, while the other terminals are connected to unique terminals. There are three Binary Coded Octal (BCO) lines labeled ABC, which are weighted in the familiar 1-2-4 manner. Only the switch that corresponds to the BCO word applied to lines ABC will be turned on. The inhibit line (INH) is an active-LOW input that turns off the electronic switches. The INH line will, therefore, force the eight lines to tri-state condition when HIGH. The switches can be used for either analog or digital signals. If the eight independent switch lines are used as inputs, and the common line is the output, then the 4051 will be a multiplexer, feeding only the selected input to the output. If, however, the common line is the input, the eight independent lines become outputs and the 4051 is a demultiplexer. Related devices are the 4052 and 4053, and distance cousins are the 4016/4066 devices.

Device Number: 4052

Type of Device: Differential Four-Channel Multiplexer/ Demultiplexer

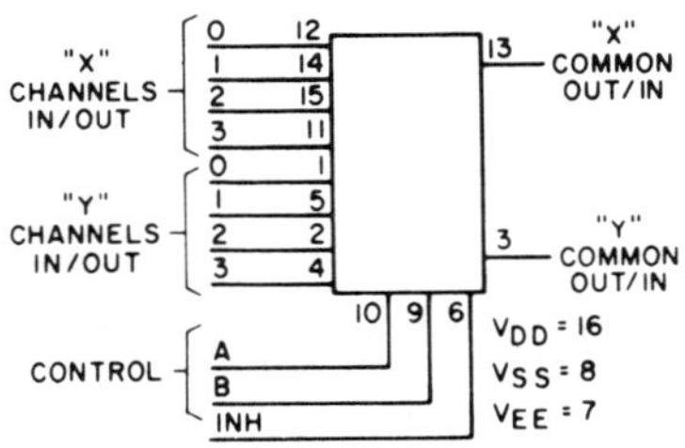

Differential 4-Channel Analog Multiplexer/Demultiplexer

CD4052B

Operation & Application: The 4052 is similar to the 4051 device, except that it contains four pairs of channels, designated X and Y. Like the other devices in this series, the 4053 responds to a binary word applied to inputs in order to determine the switch selected. In the 4052, there are two input selector lines labeled A and B. These lines are weighted such that A = 1 and B = 2. The four permitted states of these lines select channels 0, 1, 2, and 3. For any given state of the input selector lines, there will be two switches turned on. For example, a 01 on AB turns on switches nos. 1 (i.e., X1 and Y1). All X lines terminate on the X-common line, and all Y lines terminate on the Y-common line. If the independent X and Y lines are used as inputs, and the X-common and Y-common are outputs, then the 4052 is a differential multiplexer; if the situation is reversed such that the common lines are inputs, then the 4052 becomes a differential demultiplexer.

Device Number: 4053

Type of Device: Triple Two-Channel Multiplexer/Demultiplexer

Triple 2-Channel Multiplexer/Demultiplexer

CD4053B

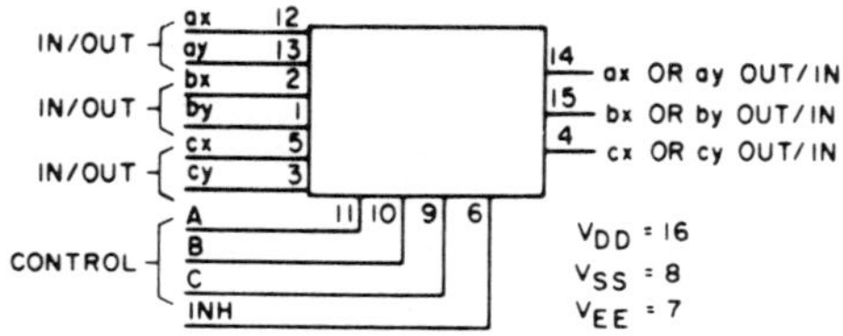

Operation & Application: The 4053 device contains three two-channel CMOS electronic switches. Three control lines are supplied that will each turn on one switch according to the following

Input	HIGH	LOW
A	ay	ax
B	by	bx
C	cy	cx

* INH must be LOW, otherwise all switches are off

Device Number: 4060

Type of Device: Fourteen-Stage Ripple/Carry Counter/Divider and Oscillator

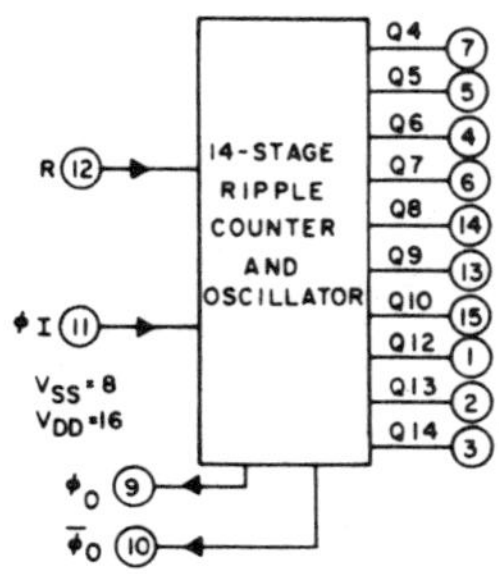

14-Stage Ripple-Carry Binary Counter/Divider and Oscillator

CD4060A
CD4060B

Operation & Application: This device is related to the 4020, and contains the same type of counter. The difference is that the 4060 device has an on-board clock oscillator that is controlled by an RC network.

Device Number: 4063

Type of Device: Four-Bit Magnitude Comparator

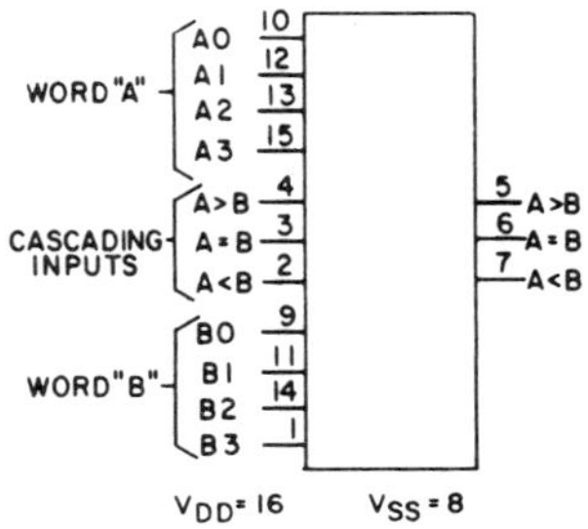

4-Bit Magnitude Comparator

CD4063B

Operation & Application: This device compares two independent four-bit binary words, and issues an output that indicates whether word-A is equal to word-B, word-A is greater than word-B, or word-A is less than word-A. The 4063 is pin-for-pin compatible with the TTL device 7485.

Device Number: 4066

Type of Device: Quad Bilateral CMOS Electronic Switch

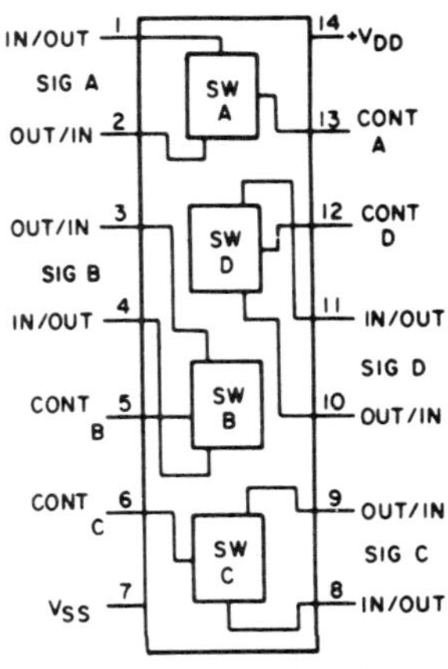

Quad Bilateral Switch

CD4066A
CD4066B

Operation & Application: The 4066 is pin-for-pin compatible with the 4016 device, and is usually preferred over the 4066 for new design. There are four independent CMOS electronic switches inside the 4066 device, and all may be used for either analog or digital signals. Each switch has its own active-LOW control line that will turn the switch ON when LOW, and OFF when HIGH.

Device Number: 4067

Type of Device: Single Sixteen-Channel Multiplexer/Demultiplexer

16-Channel Multiplexer/Demultiplexer

CD4067B

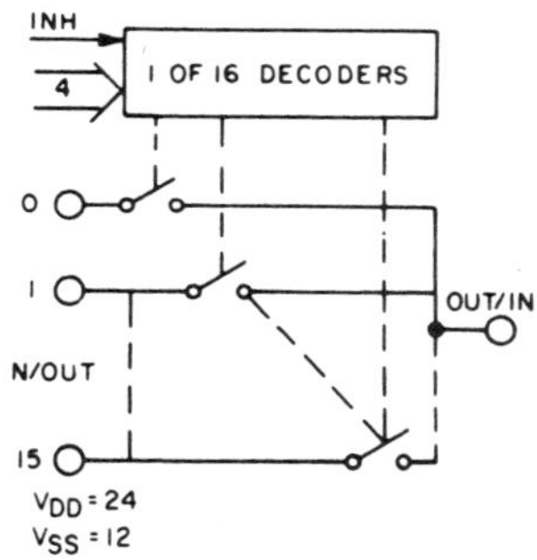

Operation & Application: The 4067 is an array of electronic switches that will multiplex or demultiplex analog or digital signals. Sixteen independent lines connect to one common line according to four binary selector input lines designated ABCD (standard 1-2-4-8 weighting). If the independent lines are used as inputs, then the 4067 is a multiplexer; if the common line is used as the input, and the sixteen independent lines are outputs, then the 4067 is a demultiplexer. An active-LOW inhibit (INH) line turns off the channels when HIGH. A related device is the 4097.

Device Number: 4068

Type of Device: Eight-Input NAND Gate

8-Input NAND/AND Gate

CD4068B

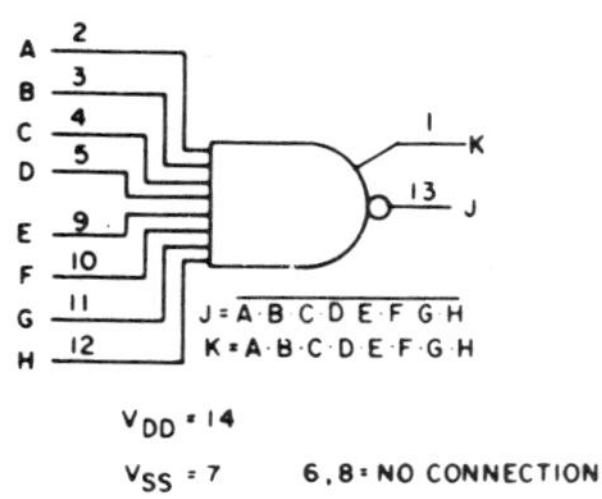

Operation & Application: Device contains a single eight-input NAND gate. A LOW on any one input will cause the output to be HIGH; all eight inputs must be HIGH for the output to be LOW.

Device Number: 4069

Type of Device: Hex Inverter

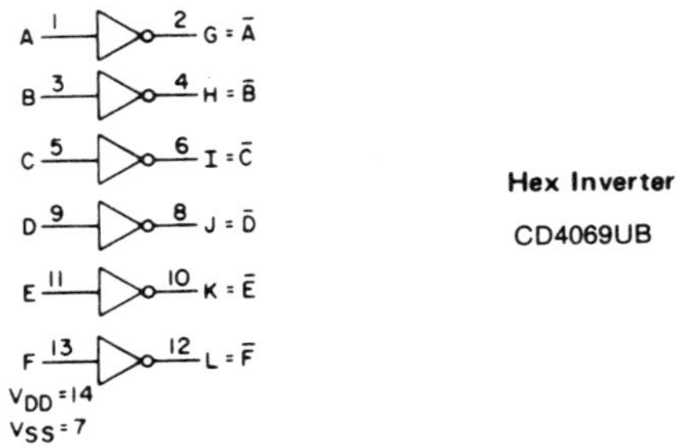

Operation & Application: Package contains six independent inverters. A LOW on any one input will cause the corresponding output to be HIGH; a HIGH on the input will cause the corresponding output to be LOW.

Device Number: 4070

Type of Device: Quad Exclusive-OR (XOR) Gate

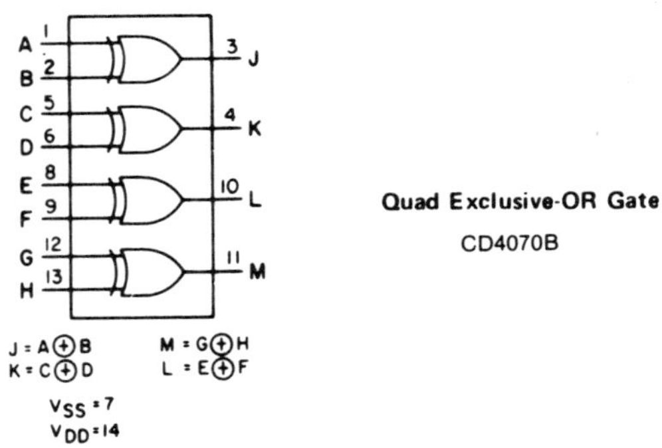

Operation & Application: Each device contains four independent Exclusive-OR (XOR) gates. On any one gate, the output will be HIGH if *one* and only one input is also HIGH. If both inputs are LOW, or, if both inputs are HIGH, then the output will be LOW. Anytime the same level is applied simultaneously to both inputs the output is LOW. Related device is the 4077 (Quad Exclusive-NOR Gate).

Device Number: 4071

Type of Device: Quad Two-Input OR Gate

Quad 2-Input OR Gate
CD4071B

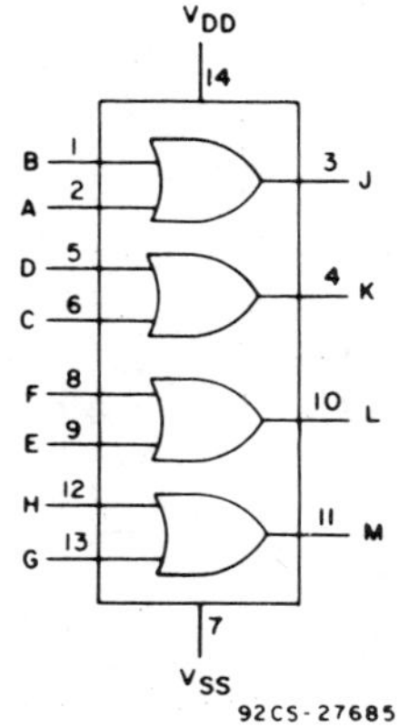

Operation & Application: A HIGH on either input of any one gate will cause the output to be HIGH; a HIGH on both inputs will cause the output to be HIGH; both inputs must be LOW for the output to be LOW. All four gates are independent.

Device Number: 4072

Type of Device: Dual Four-Input OR Gate

Dual 4-Input OR Gate
CD4072B

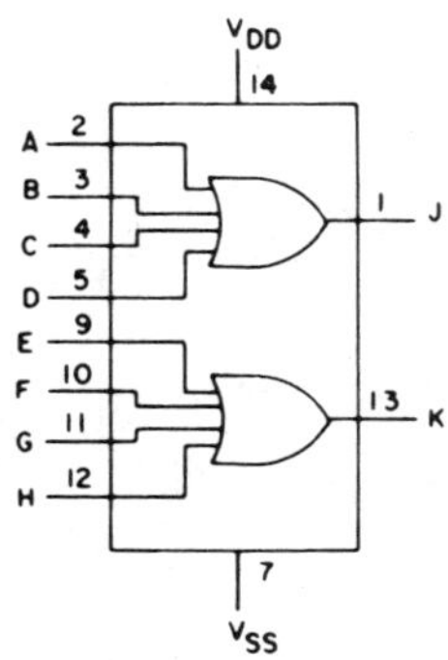

Operation & Application: Package contains two independent four-input OR gates. On either gate, a HIGH on any one terminal, or any combination of the four terminals, will cause the output to be HIGH also. All four inputs must be LOW for the output to be LOW.

Device Number: 4073

Type of Device: Triple Three-Input AND Gates

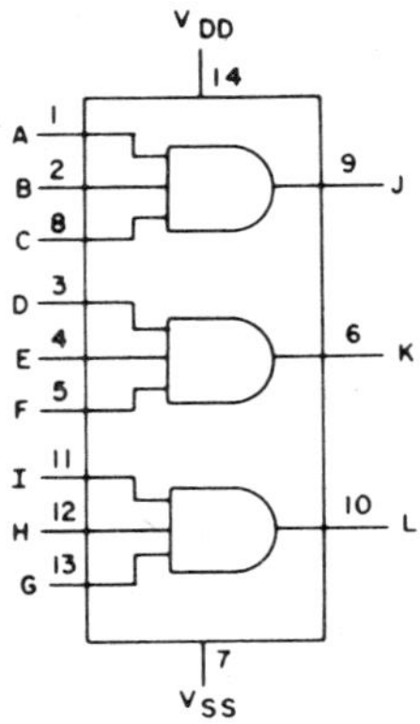

Triple 3-Input AND Gate

CD4073B

Operation & Application: Package contains three indendent three-input AND gates. On any one gate, a HIGH on all three inputs will cause the output to be HIGH. If there is a LOW on any one input, or any combination of inputs, then the output will be LOW also.

Device Number: 4075

Type of Device: Triple Three Input OR Gate

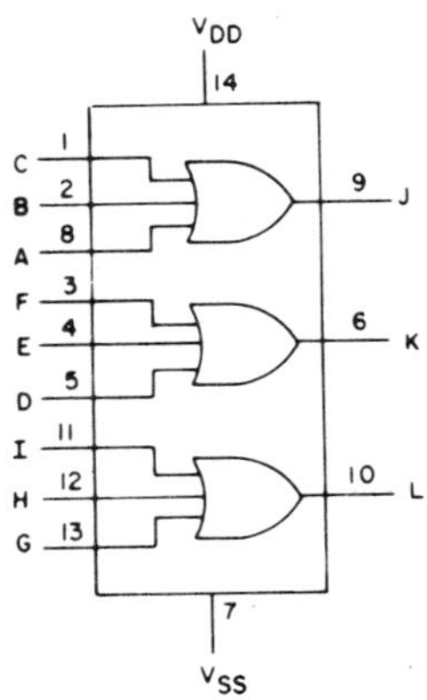

Triple 3-Input OR Gate

CD4075B

Operation & Application: Package contains three independent three-input OR gates. On any one gate, a HIGH on any input, or any combination of the three inputs, causes the output to be HIGH; all three inputs must be LOW for the output to be LOW.

Device Number: 4076

Type of Device: Four-Bit Type-D Registers

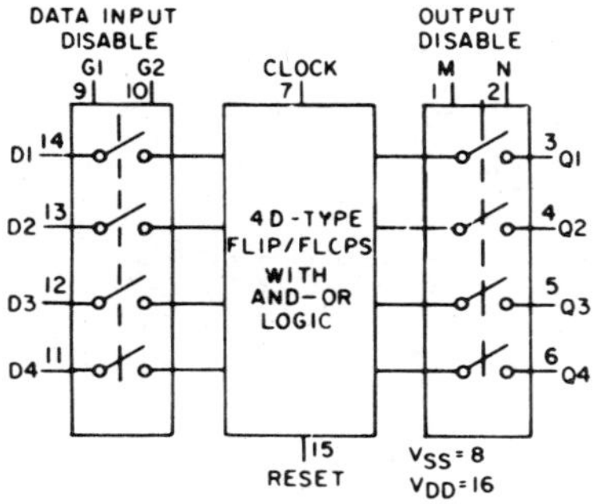

4-Bit D-Type Register

CD4076B

Operation & Application: The package contains four type-D flip-flops arrayed as a data storage register. All four type-D flip-flops are connected to a common clock line, but have independent data and output lines. There are two data disable lines, each of which turns off two flip-flops.

Device Number: 4077

Type of Device: Quad Exclusive-NOR (XNOR) Gate

Quad Exclusive-NOR Gate

CD4077B

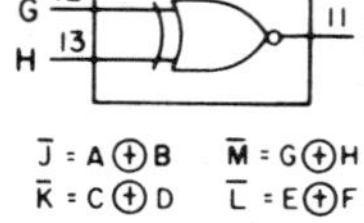

Operation & Application: The package contains four independent Exclusive-NOR gates. On any one gate, the output will be LOW if one and only one input is HIGH. If both inputs are HIGH, or, if both inputs are LOW, then the output will be HIGH. This device is analogous to a 4077 XOR gate with an inverter on the output.

Device Number: 4078

Type of Device: Eight-Input OR/NOR Gate

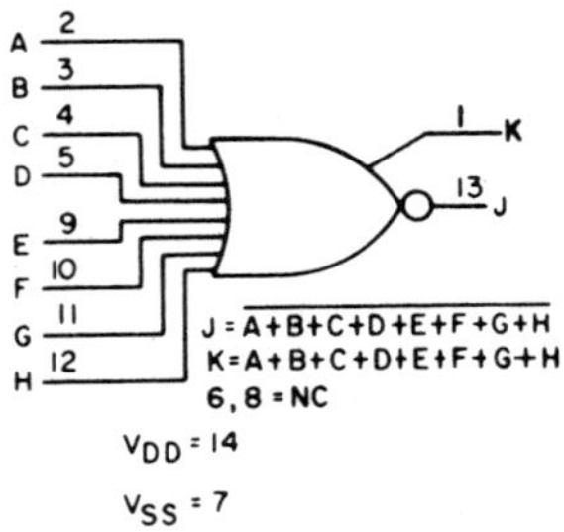

8-Input NOR/OR Gate

CD4078B

Operation & Application: This package is an eight-input gate with complementary outputs. If the inverted output is used, then the gate is a NOR gate, while use of the noninverted output results in an OR gate. In the NOR configuration, a HIGH on any one input, or any combination of inputs, results in a LOW output; all eight inputs must be LOW for the output to be HIGH. In the OR configuration, exactly the opposite occurs.

Device Number: 4081

Type of Device: Quad Two-Input AND Gate

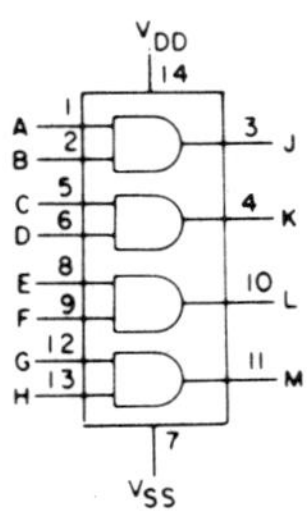

Quad 2-Input AND Gate

CD4081B

Operation & Application: The package contains four independent two-input AND gates. A LOW on any one input, or on any combination of inputs, will cause the output to be LOW. Both inputs must be HIGH for the output to be HIGH also.

Device Number: 4082

Type of Device: Dual Four-Input AND Gate

Dual 4-Input AND Gate

CD4082B

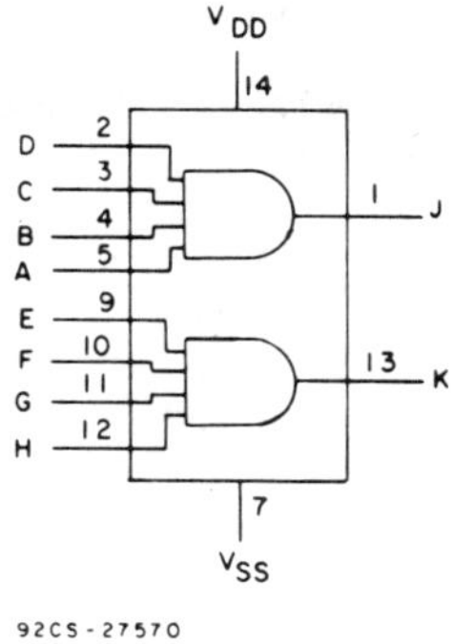

Operation & Application: The package contains a pair of four-input AND gates. On any one gate, a LOW on any input, or on any combination of inputs, will cause the output to be LOW; all four inputs must be HIGH for the output to be HIGH also.

Device Number: 4093

Type of Device: Quad Two-Input NAND Schmitt Triggers

Quad 2-Input NAND Schmitt Trigger

CD4093B

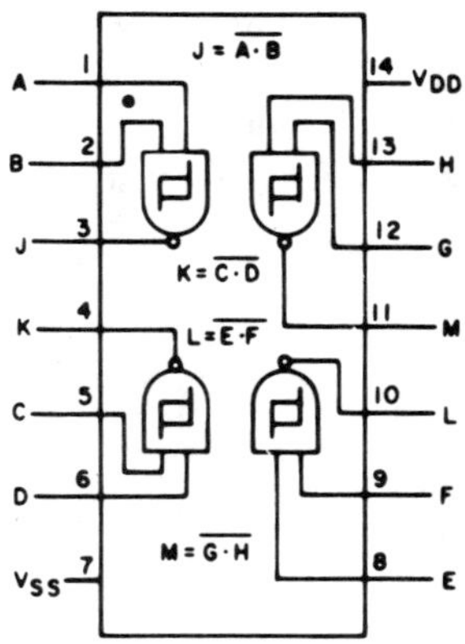

Operation & Application: Device is a quad two-input NAND gate, but with Schmitt trigger inputs. On any one gate, a LOW on either or both inputs causes a HIGH output; both inputs must be HIGH for the output to be LOW. At a power supply potential of +5 volts dc, the positive-going trip voltage threshold is approximately 3.3 volts, while the negative-going trip voltage threshold is 2.3 volts. Other than the Schmitt trigger inputs, this device is essentially a quad two-input NAND gate like any other.

Device Number: 4097

Type of Device: Differential Eight-Channel Analog Multiplexer/Demultiplexer

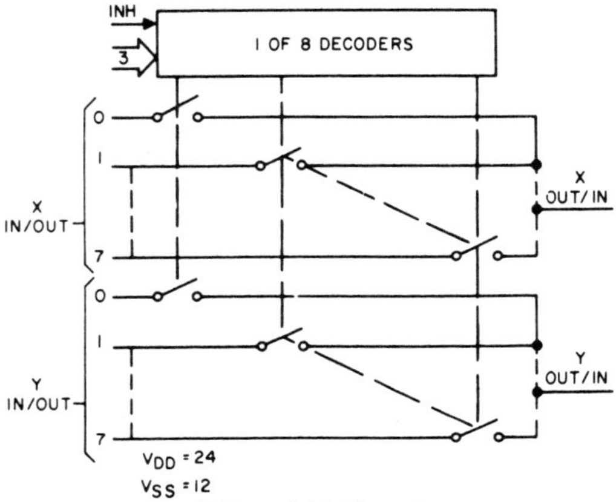

Differential 8-Channel Multiplexer/Demultiplexer

CD4097B

Operation & Application: This device is the same as the 4067, except that it contains eight differential switch pairs rather than sixteen single-ended switches. Only four selector input lines are needed because there are only 8-channels: A, B, and C.

Chapter 5

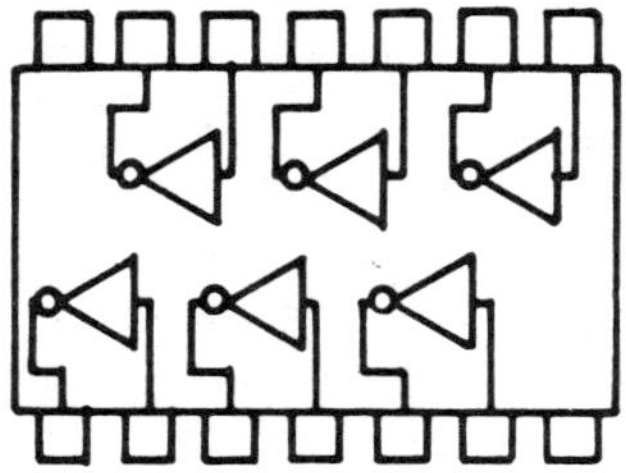

TTL and CMOS Gates

Several elementary types of logic gates form the basic building blocks out of which even the most complex digital circuits are formed. Even the most complicated computers are merely large scale combinations of a few basic types of logic gates. Some of the digital ICs that will become very familiar to you in the near future as you study this text are merely combinations of different gate circuits in monolithic IC form. All complex digital logic functions are made from very elementary gate types. The types of gates that we will discuss are NOT, OR, AND, NAND, NOR and Exclusive-OR (XOR). Various examples of these gates are found in both TTL and CMOS families.

NOT GATES (INVERTERS)

Figure 5-1 shows the circuit symbol and truth table for the most elementary type of gate. This "gate" is called an *inverter, complementor*, or *not-gate*. This type of stage will *invert* the input signal; that is to say the output of the inverter will be the opposite of the input. If the input is HIGH, then the output will be LOW. Similarly, if the input is LOW, then the output will be HIGH. Such gates are sometimes called *complementors* because, in digital circuits, HIGH and LOW levels are considered to be complements of each other.

We use the NOT terminology because of the symbols used denote input and output. If the input is called A, then the output will be called *not-A* because it will always be the logic state that A is not.

The symbol used for not-A is "$\overline{A}$," sometimes called *bar-A*. Not-A, however, is the preferred expression. The truth table for a NOT gate is shown in Fig. 5-1. Whenever A is at logic level 1 (HIGH), then the output $\overline{A}$ will be at logic 0 (LOW). When A goes to level 0, then $\overline{A}$ goes to 1.

The symbol for an inverter, or NOT gate, as shown in Fig. 5-1, is a triangle with a circle at the output. Anytime you see a digital logic element in a schematic that has a circle at the output, the output is inverted. Similarly, some multiple-input devices have a circle on one or more inputs. This symbol means that the logic function *for that input only* is inverted.

You will often see a triangle without a circle on the output. This is the schematic representation for *noninverting buffers*. Such devices will produce an output that is identical to the input but is usually capable of delivering more current than ordinary TTL outputs. Buffers are used to isolate one circuit from another, or to increase the drive capability (fan-out) of a circuit. Whenever a logic element must drive a long transmission line or a circuit that is heavily loaded with many inputs, a buffer might be required. Figures 5-2 and 5-3 show popular examples of digital inverters from the TTL and CMOS logic families.

The TTL 7404 device is shown in Fig. 5-2. This chip is called a hex inverter because it contains six independent TTL inverters. The positive side of the +5 Vdc power supply is applied to pin no. 14, while the ground side is applied to pin No. 7. Although by no

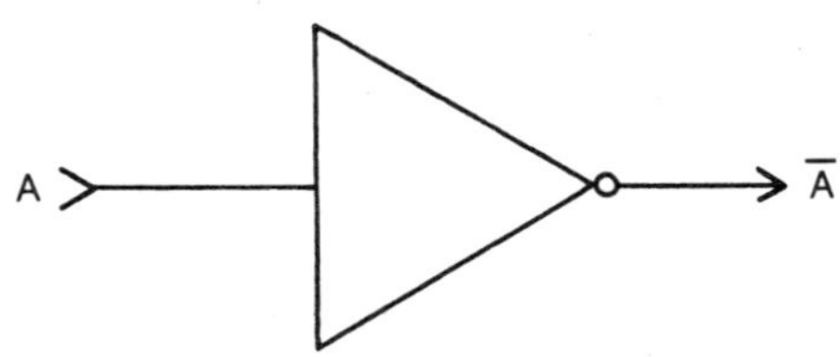

A	$\overline{A}$
HIGH	LOW
LOW	HIGH

Fig. 5-1. Inverter symbol.

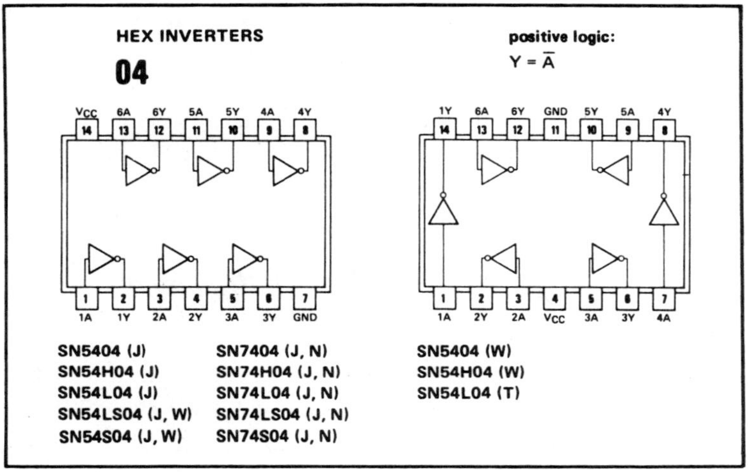

Fig. 5-2. 7404 pinouts.

means universal, this pinout configuration for the power terminals is very common in digital IC devices. Note that there is no interaction between the inverter sections, so all six may be used independently. The average current drain is 12 to 15 mA.

Three related devices, using the same pinouts, are the 7405, 7406, and 7416. All three of these are *open-collector* chips, meaning that the output transistor requires an external pull-up resistor to the V+ power supply. The resistor connects the collector of the transistor, connected internally to the IC output terminal, to its power source. The 7405 is very similar to the 7404, except for the required 2.2 kohm pull-up resistor to the +5 V supply. The 7405 draws 12 to 15 mA.

The 7406 and 7416 are open-collector devices, like the 7405, but are able to withstand 30 V (7406) and 15 V (7416), respectively. Note that the package voltage applied to pin no. 14 must remain at +5 V; only the output transistor collector can operate at the higher potential. The higher collector voltage is intended to make it possible for the IC to drive an external load.

The 7406 device can sink up to 30 mA, while the 7416 can sink up to 40 mA, through the output transistor collector. This specification means that the 7406 load must have a dc resistance not less than R = (30 V)/(0.03 A), or 1000 ohms. The 7416 load must be not less than R = (15 V)/(0.04 A), or 375 ohms. These values refer to the *dc resistance* between the open-collector output and the V+ power supply. Both 7406 and 7416 devices are frequently used as drivers for relay coils, LEDs, and incandescent lamps.

Figure 5-3 shows the pinouts for the 4009, 4049, and 4069 CMOS hex inverter chips. The 4009 is now considered to be obsolete for new designs and has been replaced by the 4049 device. The 4009 device uses two VCC+ terminals and will be destroyed if these voltages are applied in wrong sequence. The voltage applied to pin no. 1 must always be equal to or greater than the voltage applied to pin no. 16. Again, all six inverters may be used independently.

The 4049 device may be used as a level translator between CMOS devices and TTL devices. The voltage applied to pin no. 1 defines the output voltage swing, so if +5 Vdc is used, the device becomes TTL-compatible. The 4049 can handle 3.2 mA, so it will drive two standard TTL loads. The input voltages applied to the

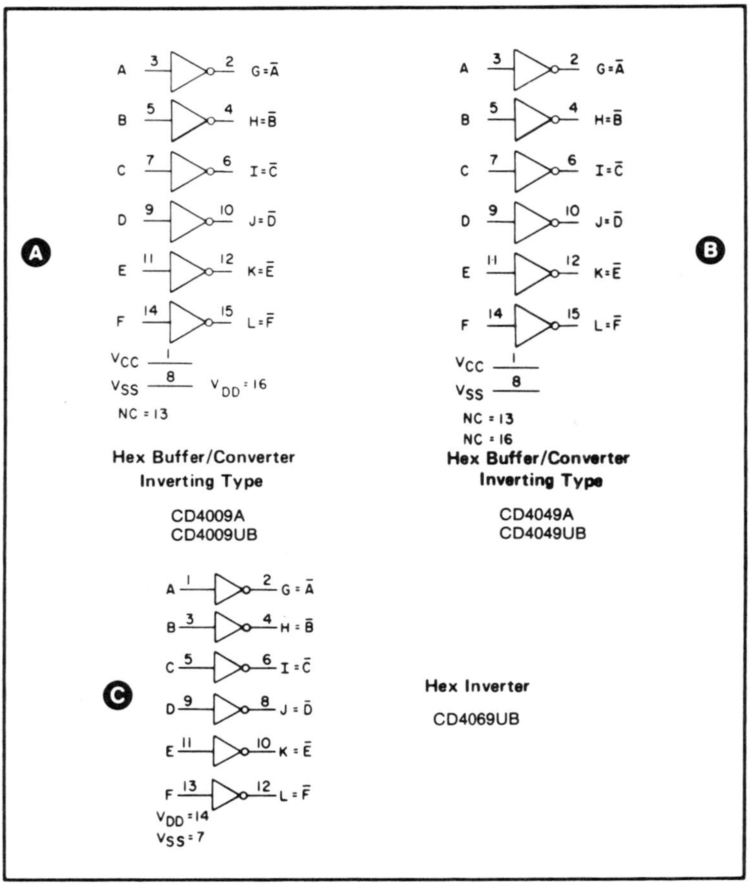

Fig. 5-3. 4009, 4049, and 4069 pinouts.

4049 inverter sections can swing to +15 Vdc, regardless of the Vcc+ voltage applied to pin no. 1.

Although some people call the 4069 device a "low-power" version of the 4049, it is not. The 4069 has different pinouts and is not suitable for either direct interfacing with TTL logic or level translation of any other type. The 4069 will only interface directly with other CMOS devices.

The difference between the current drains of TTL and CMOS devices can be seen by comparing the 7404 and 4069 devices. At +5 Vdc (used by all TTL devices), the 4069 draws 0.5 mA, as opposed to 15 mA for the 7404; this is a difference of 3000 percent! It is little wonder that designers of battery-powered devices prefer CMOS over TTL.

OR GATES

An OR gate will produce a logical-1 or HIGH output if any input is HIGH. Figure 5-4A shows the symbol for an OR gate, Fig. 5-4B shows an equivalent circuit that produces the same action, and Fig. 5-4C shows the truth table for the OR gate. Let us consider Fig. 5-4B in the light of the definition of an OR gate. Let the voltage at point "C" represent the output of the OR gate, and switches A and B the inputs of the OR gate. The logic is as follows:

Switch Condition	Input logic level	Output Condition
open	0	LOW
closed	1	HIGH

By the definition of an OR gate, we would expect C to be HIGH when either input is HIGH also (switch A *or* switch B is closed), and LOW only when both inputs are also LOW (switch A and switch B are open). This is exactly the action found in the circuit of Fig. 5-4B. Voltage C goes to +5 Vdc if either switch is closed and will be zero only when both switches are open. This condition is reflected in the truth table shown in Fig. 5-4C: C is HIGH if A *or* B is also HIGH.

Examples of TTL and CMOS OR gates are shown in Figs. 5-5 and 5-6, respectively. The type 7432 TTL device is a quad, two-input OR gate. Note that the Vcc+ and ground terminals are pins 14 and 7, respectively. This is the same power-supply configuration as found on the 7404 (discussed in the previous section). The 7432 requires approximately 20 mA of current per package.

A CMOS example, the 4071 device, is shown in Fig. 5-6. This

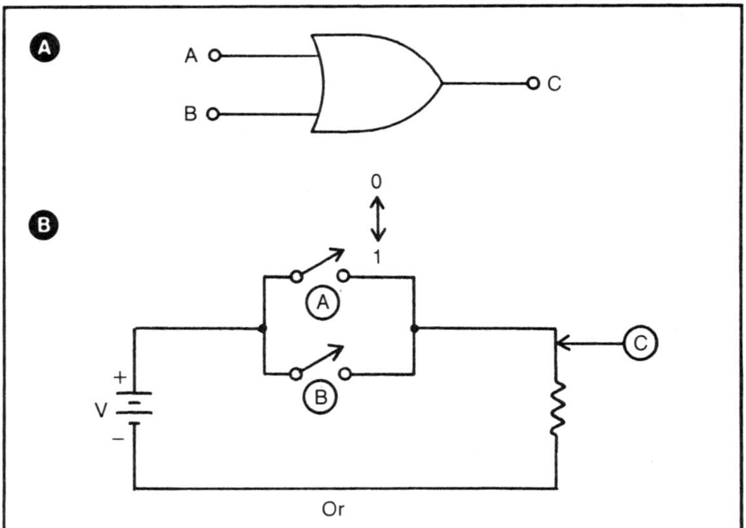

C

A	B	C
0	0	0
0	1	1
1	0	1
1	1	1

Fig. 5-4. NOR gate (A) symbol, (B) equivalent circuit, (C) truth table.

IC is also a quad, two-input, OR gate. The current requirement is also 0.5 mA, and the difference between TTL and CMOS is seen even more clearly. But power consumption also brings speed. The propagation time—the time between input and the output response that it causes—is 12 ns for TTL, and 80 ns for CMOS.

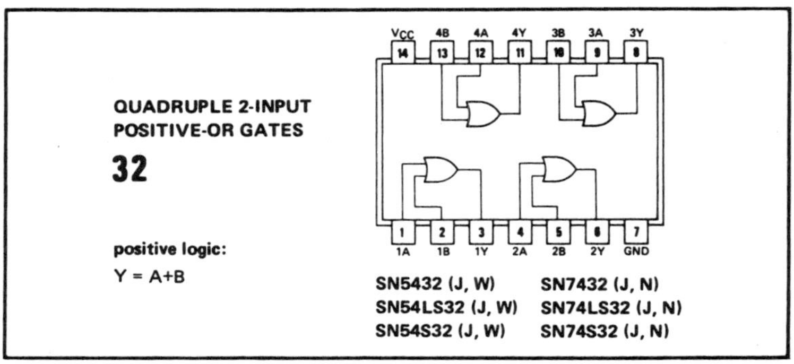

Fig. 5-5. 7432 pinouts.

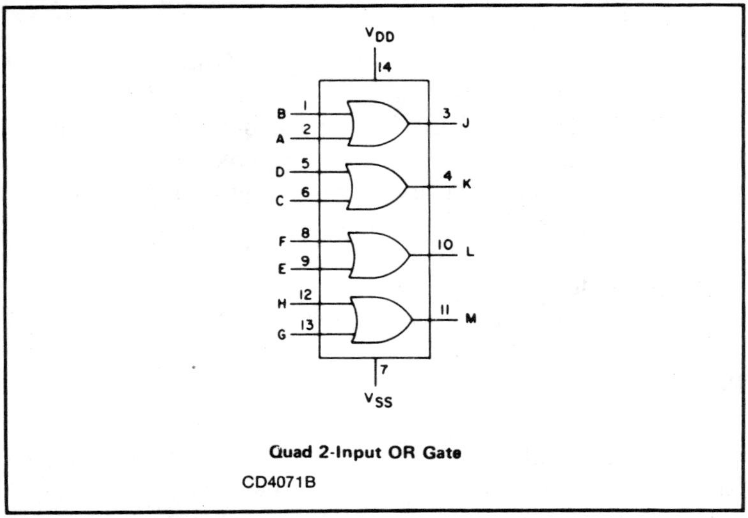

Fig. 5-6. 4071 pinouts.

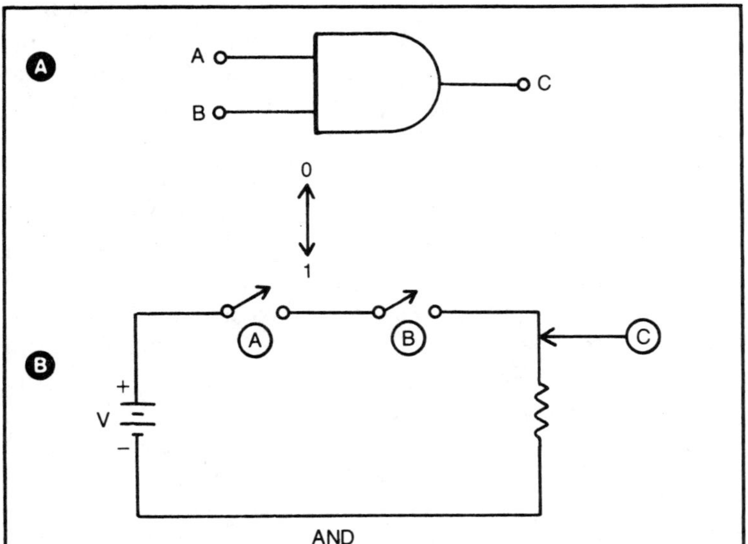

A	B	C
0	0	0
1	0	0
0	1	0
1	1	1

Fig. 5-7. AND gate (A) circuit symbol, (B) equivalent circuit, (C) truth table.

AND GATES

The AND gate produces a HIGH output if *both* inputs are HIGH and a LOW input if *either* input is LOW. Figure 5-7A shows the circuit symbol for an AND gate, while Fig. 5-7B shows an equivalent circuit that does the same job, and Fig. 5-7C shows the AND gate truth table. In this type of circuit we obtain a HIGH output only if A *and* B are also HIGH. There are few example of TTL AND gates, but in the CMOS line we find the 4073, 4081, and 4082 devices. The 4073 is a triple, three-input AND gate. The 4081 is a quad, two-input AND gate, and the 4082 a dual, four-input device.

NAND GATES

A NAND gate can be viewed as an AND gate with an inverted output; in fact, you can simulate the NAND gate by connecting the input of an inverter (NOT gate) to the output of an AND gate. The designation "NAND" is derived from NOT-AND. We sometimes see the NAND expression written in the form $\overline{\text{AND}}$, $\overline{\text{AXB}}$, or A•B.

Figure 5-8 shows the NAND gate. Figure 5-8A shows the circuit symbol used in schematic diagrams. Figure 5-8B is an equivalent circuit that performs the same job, and Fig. 5-8C is the NAND gate truth table. The rules of operation for the NAND gate are as follows:

- ☐ A LOW on either input creates a HIGH output.
- ☐ A HIGH on both inputs is required for a LOW output.

We can see this same action in the equivalent circuit shown in Fig. 5-8B. Recall our switch logic protocol: Switch open is a LOW on the input, and a switch closed is a HIGH on the input. The voltage at point "C" will be HIGH if either switch A or switch B is open (either input is LOW). But if both switches are closed (both inputs are HIGH), the load is shorted out and point "C" is LOW.

The NAND gate is one of the most popular gates used in digital equipment. An example of a quad, two-input, TTL NAND gate is shown in Fig. 5-9, the popular 7400 device. It is possible to construct circuits to replace any of the other gates, using only 7400 devices, although such would be unnecessary and uneconomic course of action.

Recall that TTL is a current-sinking logic family, meaning that the output will sink, or pass to ground, current. The inputs of a TTL device form a 1.6-mA current source. Any TTL input is considered HIGH when greater than +2.4 V is applied, and must be held to less than 0.8 V when LOW. This requirement means that any ground

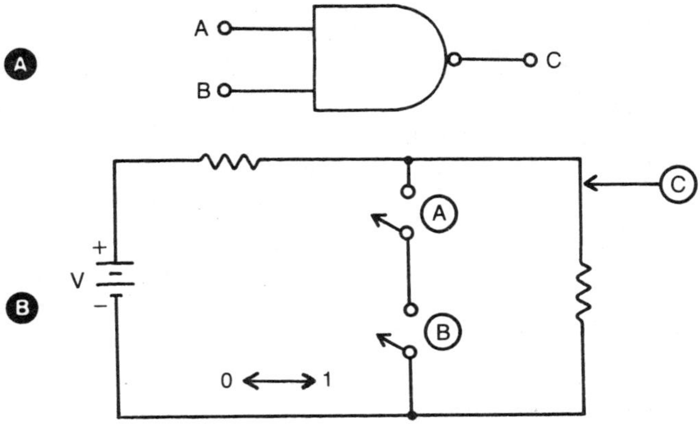

NAND

A	B	C
0	0	1
0	1	1
1	0	1
1	1	0

Fig. 5-8. NAND gate (A) Circuit symbol, (B) equivalent circuit, (C) truth table.

QUADRUPLE 2-INPUT POSITIVE-NAND GATES

00

positive logic:
$Y = \overline{AB}$

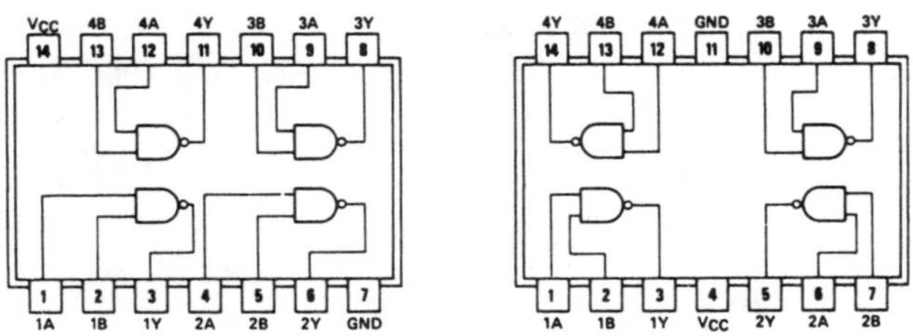

SN5400 (J)
SN54H00 (J)
SN54L00 (J)
SN54LS00 (J, W)
SN54S00 (J, W)

SN7400 (J, N)
SN74H00 (J, N)
SN74L00 (J, N)
SN74LS00 (J, N)
SN74S00 (J, N)

SN5400 (W)
SN54H00 (W)
SN54L00 (T)

Fig. 5-9. 7400 pinouts.

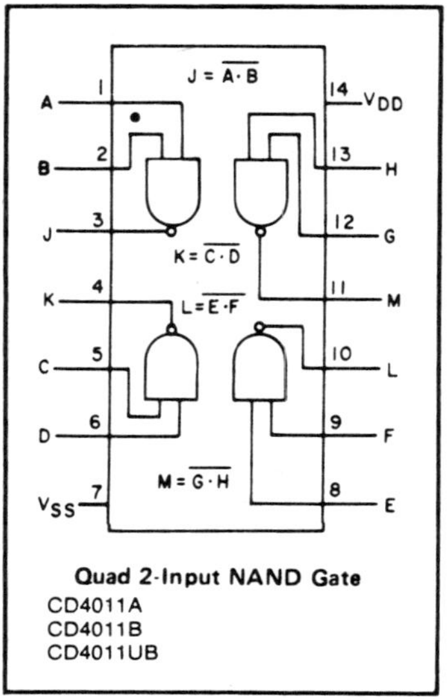

Fig. 5-10. 4011 pinouts.

connection must have a resistance less than R = (0.8 V)/(0.0018 A), or 440 ohms. If a higher ground resistance is encountered, or if a voltage between 0.8 V and 2.4 V is applied, then operation is undefined—and all bets as to operation are off! The 7400 devices requires approximately 12 mA of current per package.

A CMOS NAND gate IC, the 4011, is shown in Fig. 5-10. This device is also a quad, two-input, NAND gate. The 4011 requires only 0.4 mA per package, again a substantial savings over TTL. Various combination circuits using NAND gates will be discussed in subsequent chapters.

NOR GATES

A NOR gate is the same as an OR gate with an inverted output. The NOR designation is derived from "NOT-OR", and the NOR functions can be symbolized by $\overline{A + B} = C$. The circuit diagram symbol for the NOR gate is shown in Fig. 5-11A, while the equivalent circuit is shown in Fig. 5-11B. A truth table for the NOR gate function is shown in Fig. 5-11C. The rules governing the operation of the NOR gate are:

- ☐ A HIGH on either input produces a LOW output.
- ☐ If both inputs are LOW, then the output is HIGH.

We can see this action in the circuit of Fig. 5-11B. If either switch A or switch B is closed, then output C will be LOW. But when both switches are open, then output C is HIGH.

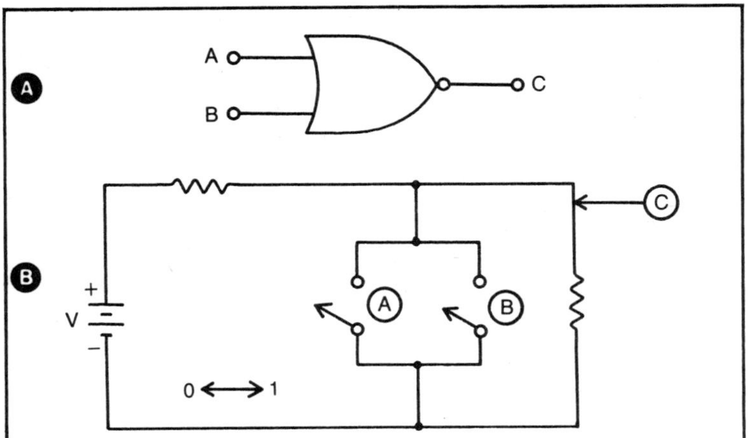

A	B	C
0	0	1
0	1	0
1	0	0
1	1	0

Fig. 5-11. NOR gate (A) symbol, (B) equivalent circuit, (C) truth table.

Figure 5-12 shows an example of a popular TTL 7402 NOR gate. This device contains four two-input gates that may be used independently of each other. The package draws about 12 mA.

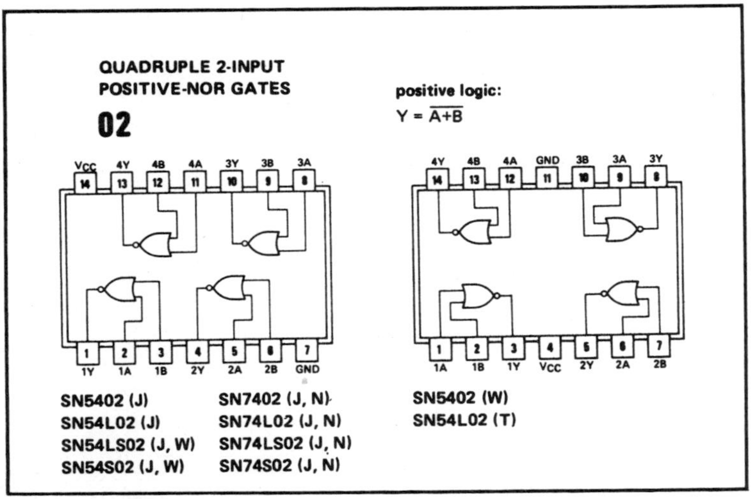

Fig. 5-12. 7402 pinouts.

EXCLUSIVE-OR (XOR) GATES

The Exclusive-OR, or XOR as it is designated, is shown in Fig. 5-13. Examine the truth table for the XOR gate in Fig. 5-13C. From this truth table we may infer the rules for operation governing the Exclusive-OR gate:

☐ The input is LOW if both inputs are LOW or if both inputs are HIGH.

☐ The output is HIGH if either input is HIGH.

In other words, a HIGH on either input alone produces a HIGH output, but a HIGH on both inputs simultaneously produces a LOW output. Any time we find all inputs of an XOR gate in the same condition (HIGH or LOW), then the output will be LOW.

SUMMARY

In this chapter we have discussed five basic gates (AND, OR, NAND, NOR, and XOR), and the inverter (sometimes called a NOT

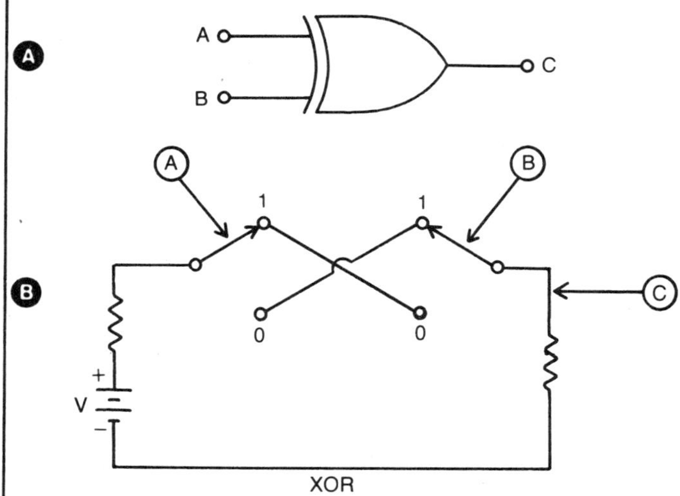

C

A	B	C
0	0	0
0	1	1
1	0	1
1	1	0

Fig. 5-13. Exclusive-OR (XOR) gate (A) circuit symbol, (B) equivalent circuit, (C) truth table.

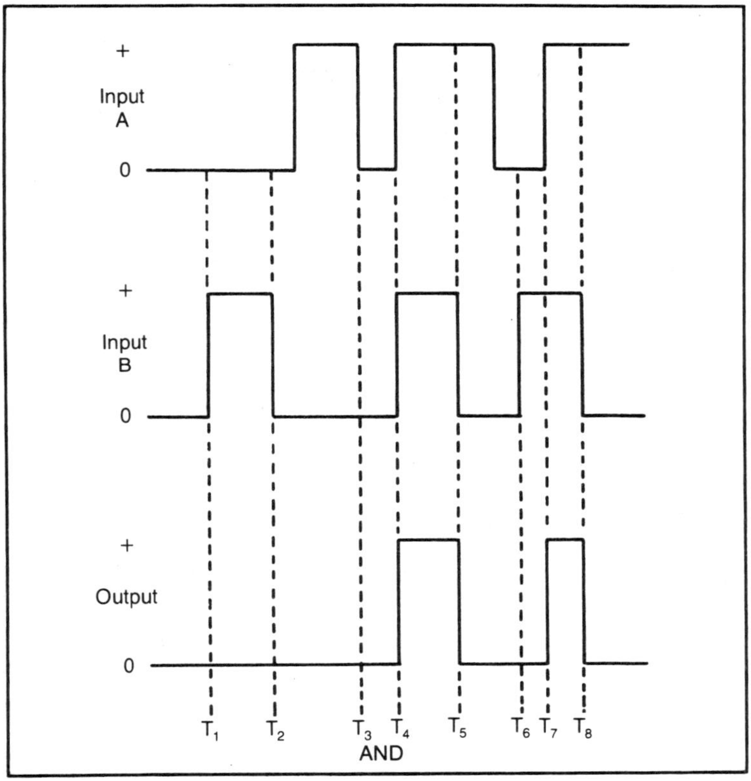

Fig. 5-14. AND gate timing waveform.

gate). From these basic forms of gate, all digital functions can be constructed; in fact, even the largest special function LSI (large-scale integration) digital ICs contain little more than a complex interconnection of these six basic functions: AND, OR, NAND, NOR, XOR, and NOT. If we want to split hairs even further, we could say that only AND, OR, XOR, and NOT are needed, because the NOR and NAND functions are NOT-OR and NOT-AND respectively.

Figures 5-14 through 5-19 shows a series of waveform timing diagrams that summarize the actions of these basic gates. Such diagrams are frequently used in digital electronics as a graphic means of showing dynamic circuit action. In fact, in many cases the timing diagram is the only reliable way to visualize circuit action.

In the timing diagrams, we have followed the same denotation as before: A and B are inputs, while C is the output. Note that these diagrams are for two-input devices (except, of course, for the

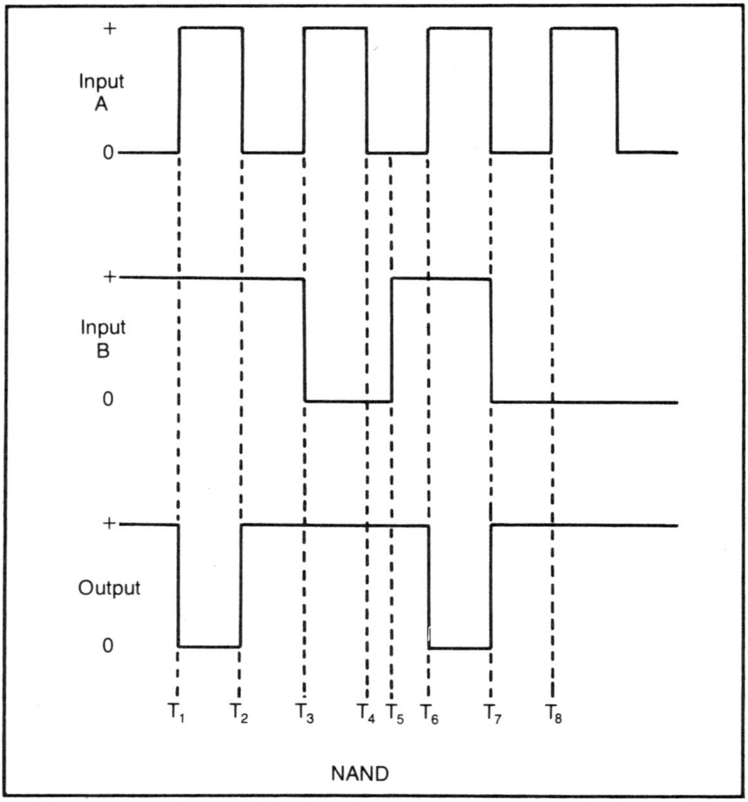

Fig. 5-15. NAND gate timing waveforms.

inverter), but the basic concepts can be extended to devices with three or more inputs as well.

In Fig. 5-14, we see the operation of the AND gate. In this case, a train of square waves is applied to input A, while a single pulse is applied to input B. Note that output C remains LOW until B goes HIGH. This is due to the requirement that both inputs be HIGH before the output can be HIGH. Once input B has gone HIGH, because of the same rule, the output will follow input A. To emphasize again, the rules for AND gates are:

- ☐ The output will be LOW if either input is LOW.
- ☐ The output is HIGH only if both inputs are HIGH.

A timing diagram for the NAND gate is shown in Fig. 5-15. Recall that a NAND gate is an AND gate with an inverted output. The NAND gate circuit symbol shows this relationship because it is an AND gate symbol with a small circle at the output terminal.

The NAND gate timing diagram uses input condition similar to the previous case for the purpose of comparing their operations. A square wave is applied to input A, while a single pulse is applied to input B. If either input is LOW, then the output will be HIGH. Therefore, the output is HIGH until both A and B are HIGH. When B goes HIGH, the output will follow A, but is inverted, i.e., 180 degrees out of phase. To reemphasize, the rules for NAND gates are:

- ☐ The output is HIGH if either input is LOW.
- ☐ The output will be LOW only if both inputs are HIGH.

The timing diagram for an OR gate is shown in Fig. 5-16. Recall that an OR gate will produce an output if either input is HIGH. This action is shown in the output waveform of Fig. 5-16. The output waveform would also be HIGH if both inputs were HIGH. To reemphasize the action of an OR gate, the rules are as follows:

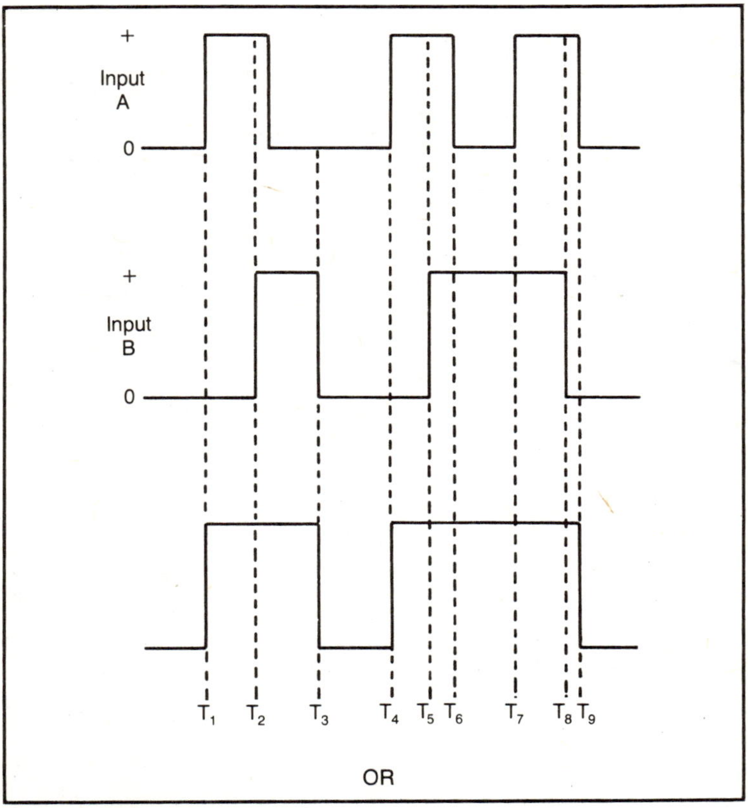

Fig. 5-16. OR gate timing waveforms.

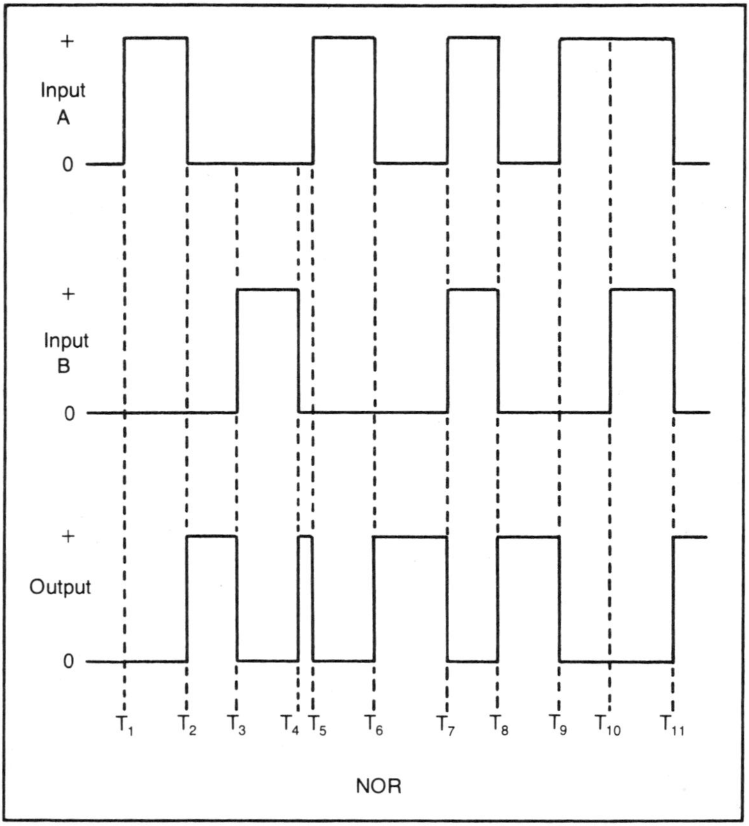

Fig. 5-17. NOR gate timing waveforms.

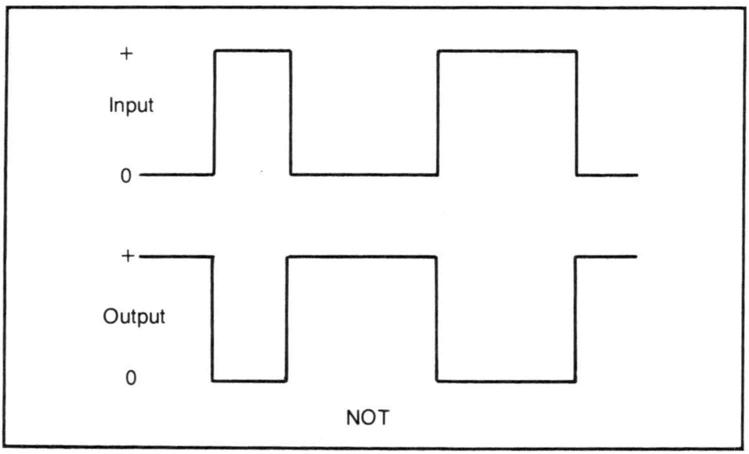

Fig. 5-18. NOT gate (inverter) waveforms.

☐ The output is HIGH if either input, or both inputs, are HIGH.

☐ The output is LOW only when both inputs are also LOW.

The timing diagram for the NOR gate is shown in Fig. 5-17. Recall that the output of a NOR gate will be LOW if either input is HIGH. To reemphasize, the rules for the operation of a NOR gate are:

☐ The output will be LOW if either or both inputs are HIGH.

☐ The output will be HIGH only if both inputs are LOW.

The timing diagram for an inverter is shown in Fig. 5-18. Notice that the output $\overline{A}$ is merely the inverse of the input; that is, it is exactly 180 degrees out of phase with the input. Although this seems like a trivial case, it is nonetheless important.

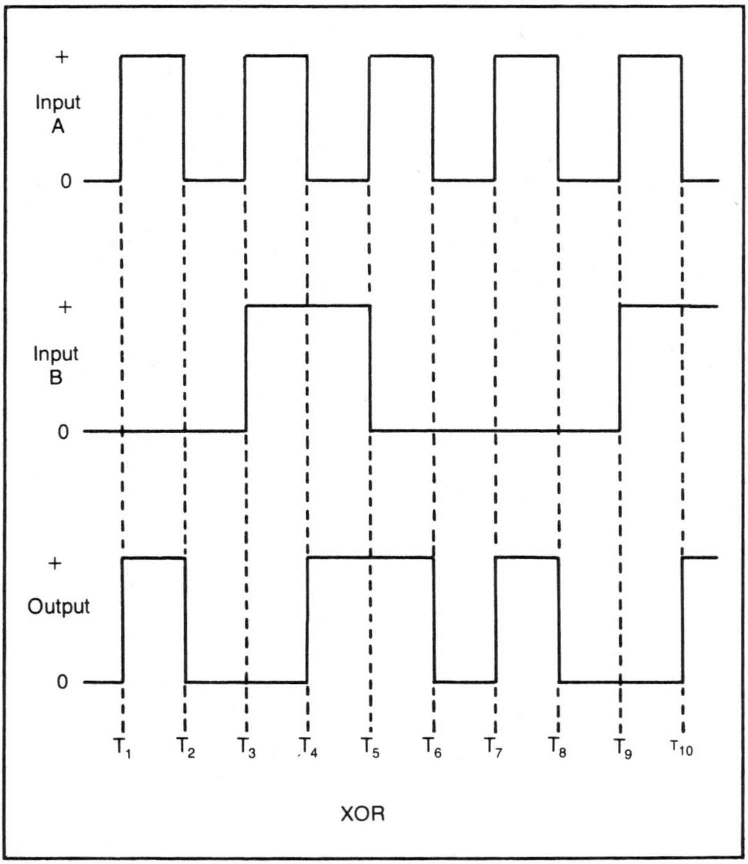

Fig. 5-19. Exclusive-OR (XOR) timing waveforms.

The Exclusive-OR gate timing diagram is shown in Fig. 5-19. Recall that the XOR gate will produce a HIGH if either input is HIGH, but not if both inputs are LOW, then the output will be LOW. This action can be seen in Fig. 5-19. Input A sees a train of square waves, and we have labeled the different sections 1, 2, 3 . . . etc. to more closely study the effects. At time 1, both inputs are LOW, so the output C is also LOW. But at times 2, input A is HIGH and input B is LOW. This will make output C HIGH until the beginning of time 3. At that time, both inputs are again LOW. Note in time period 4 that both inputs A and B are HIGH, so the output is LOW. Similarly, the output is LOW during the time period 8. To reemphasize, the rules for the Exclusive-OR (XOR) gate are as follows:

☐ The output is LOW any time that both inputs are in the same logic condition—HIGH or LOW.

☐ The output is HIGH if either input is HIGH.

You may have noted in some of the descriptions in this chapter that one must be careful in using the terms AND/OR and NAND/NOR. In the NAND gate description, for example, the output is HIGH if either A or B is LOW. But isn't this NOR gate talk? No, not if we are talking about positive logic. Most of the designations are based on positive logic. But when you talk about negative logic, in which logical-1 is zero volts, and logical-0 is V+, the definitions reverse. A positive logic NAND gate is a negative logic NOR gate. Similarly, a positive logic AND gate is a negative logic OR gate. This is why some older catalogs, published before the "standard" of positive logic was firmly established, listed the 7400 as a NAND/NOR gate; it all depends upon your point of view.

Chapter 6

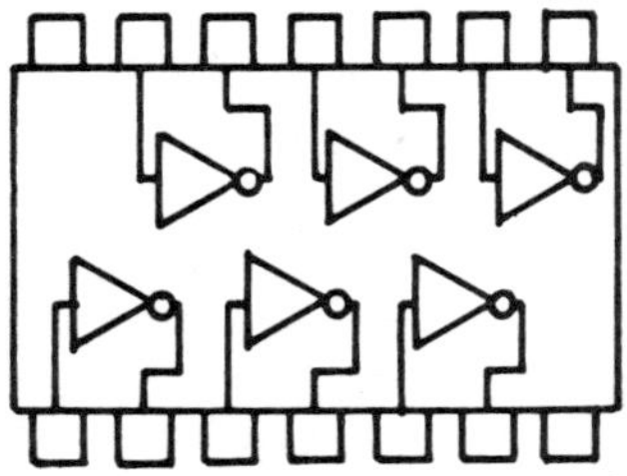

Unclocked Flip-Flops

The objective of this chapter is for you to learn the operation of clocked and unclocked flip-flops. You will also become familiar with the various TTL/CMOS flip-flops available.

All computer and digital logic functions can be made from combinations of AND, OR, NAND, NOR, and NOT gates; in fact, even the NAND and NOR functions are merely combinations of NOT gates with AND and OR gates, respectively. All of the gate functions are transient circuits; that is, they are incapable of *storing* information, even for a short period of time.

Flip-flops are circuits that will store single-bits of information. Most solid-state computer memories are arrays of flip-flops organized in a manner that allows storage of the digital words of a computer.

Most flip-flops are *bistable* circuits; that is, there are *two* stable output states. The flip-flop is not particularly concerned with which state is in existence at any given time; it is happy in either state. If the output is labeled Q, we find that Q can be either HIGH or LOW and will be stable when it is HIGH or LOW.

R-S FLIP-FLOPS

One of the simplest flip-flops is the *reset-set* (R-S) flip-flop. These circuits can be made from either NAND or NOR gates, although the performance characteristics are different for the two different types. The NOR gate implementation is shown in Fig. 6-1, while an occasionally used circuit symbol is shown in Fig. 6-2.

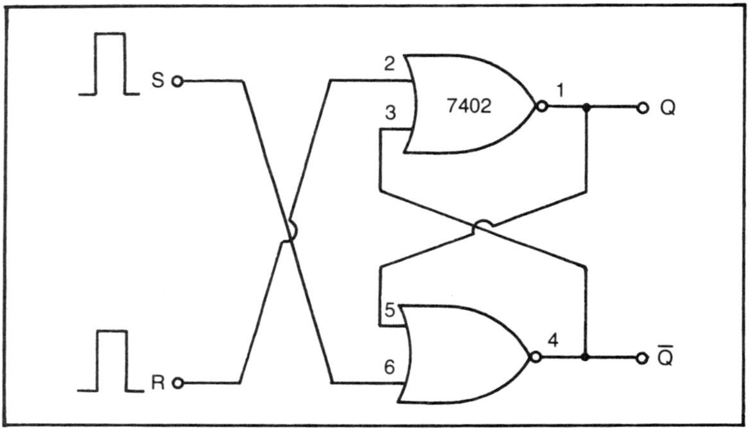

Fig. 6-1. NOR-logic RS flip-flop.

Note that the circuit has two outputs, labeled Q and $\overline{Q}$ (read "not-Q"). These outputs are complementary, meaning that one will be HIGH while the other will be LOW. If Q is HIGH, then not-Q must be LOW. Similarly, when not-Q is HIGH, Q must be LOW.

In the NOR gate R-S flip-flop, the output changes state when an appropriate input is momentarily brought HIGH. Only a brief pulse at the input is needed to effect the change. The rules governing the operation of the NOR-gate R-S flip-flop are summarized in the truth table in Table 6-1. These rules follow:

☐ When both S and R input are LOW, no change occurs in the output.

☐ If both S and R are simultaneously brought HIGH, this is a *disallowed state*. This input condition is to be avoided.

☐ Bringing the R input momentarily HIGH will set the output, i.e., make Q HIGH and not-Q Low.

☐ Bringing the R input momentarily HIGH forces the flip-flop to reset; i.e., make Q LOW and not-Q HIGH.

The basic NOR gate R-S flip-flop can be constructed from either TTL or CMOS NOR gate ICs. In the case shown in the example, TTL type 7402 NOR gates have been used to make the flip-flop.

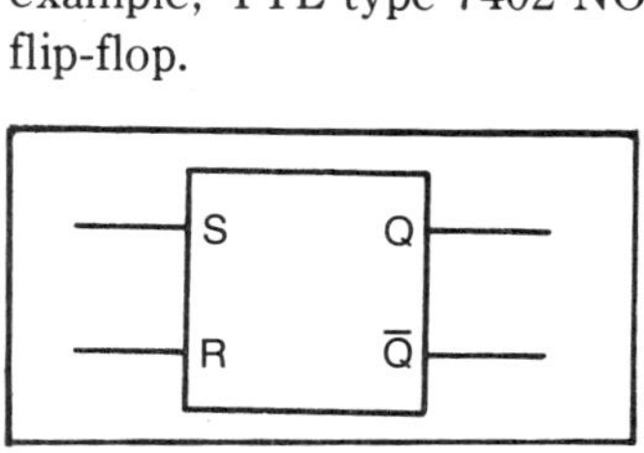

Fig. 6-2. RS symbol.

S	R	Q	$\overline{Q}$
0	0	NO CHANGE	
0	1	0	1
1	0	1	0
1	1	DISALLOWED	

NOR-IMPLEMENTATION

Table 6-1. NOR/Logic Truth Table (RS FF).

The NAND gate R-S flip-flop uses inverted logic; i.e., the output state changes are caused by bringing the appropriate input LOW. An example of a NAND gate R-S flip-flop is shown in Figs. 6-3 and 6-4.

Because the inputs are active-LOW, this circuit is sometimes called the $\overline{\text{reset-set}}$ (or $\overline{\text{R-S}}$) flip-flop. In this text, however, it is more convenient to simply refer to both types as R-S flip-flops, and then be sure that the NAND/NOR distinction is made in the text. The example of Fig. 6-3 is implemented using the type 7400 TTL NAND gate.

The output conditions are governed by the following rules and summarized in the truth table in Table 6-2.

☐ If the $\overline{S}$ and $\overline{R}$ inputs are both simultaneously LOW, we have a disallowed state. Such states must be avoided.

☐ If both S and R inputs are made simultaneously HIGH, then no change in the output state will occur.

☐ If the S input is momentarily brought LOW while the R remains HIGH, then the Q output goes to the HIGH state and not-Q is LOW.

☐ If the R input is momentarily brought LOW and the S input

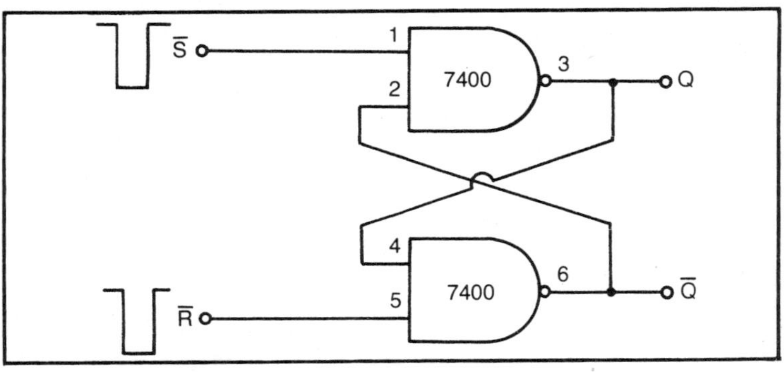

Fig. 6-3. NAND-logic RS flip-flop.

S	R	Q	$\overline{Q}$
0	0	DISALLOWED	
0	1	1	0
1	0	0	1
1	1	NO CHANGE	

Table 6-2. NAND-Logic RS FF Truth Table.

remains HIGH, then the Q output goes LOW and the not-Q goes HIGH.

The R-S flip-flop is used frequently in applications where a pulse to the S or R input sets a condition and then a subsequent pulse is used to reset the circuit to its initial conditions.

Although it is common practice to construct R-S flip-flops using discrete gates (in contrast to other forms of flip-flop which are purchased already built in integrated circuit form), there are at least two IC R-S flip-flops on the market. Both of the devices are CMOS, and they are numbered 4043 and 4044. These devices are CMOS quad three-state R-S "latches" (i.e., flip-flops). This means that there are four R-S flip-flops inside of each 16-pin DIP IC package (see Figs. 6-5 and 6-6). Each flip-flop is independent from the others with the excepton of power-supply connections and the *enable* terminal (pin no. 5 in both cases). The difference between the two devices is that the 4043 is a NOR-logic R-S flip-flop, while the 4044 is a NAND-logic R-S flip-flop. Note that only Q outputs are provided, there are no not-Q outputs.

"Three state" (i.e., tri-state) devices have the property of allowing the output to be disconnected from the output pin in order to float the output across a bus or other such structure. When the *enable* input is active (i.e., HIGH), then the outputs of the four flip-flops are connected to their respective output terminals. When the *enable* input is inactive, on the other hand, that is to say LOW, then the outputs are disconnected from their respective terminals and the "outside" world looking back into those terminals sees a

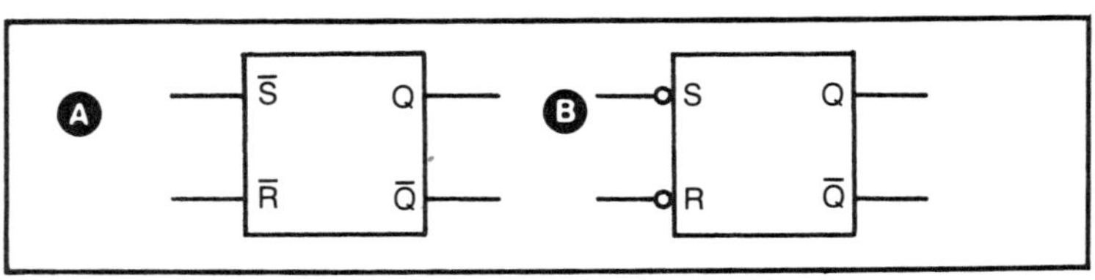

Fig. 6-4. (A) Usual circuit symbol, (B) alternate symbol.

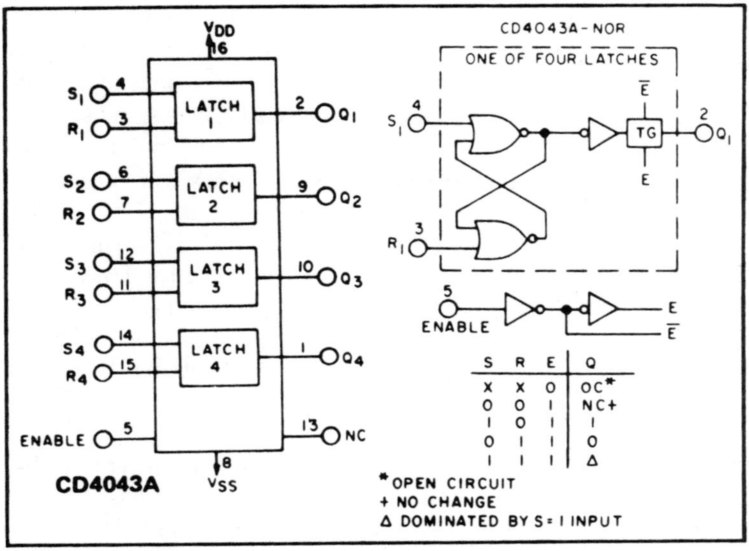

Fig. 6-5. 4043 symbol.

high impedance. The main purpose of the tri-state outputs is to allow multiple bussing of the devices on a common line. The circuit will then command which is used by placing a HIGH on the *enable* input.

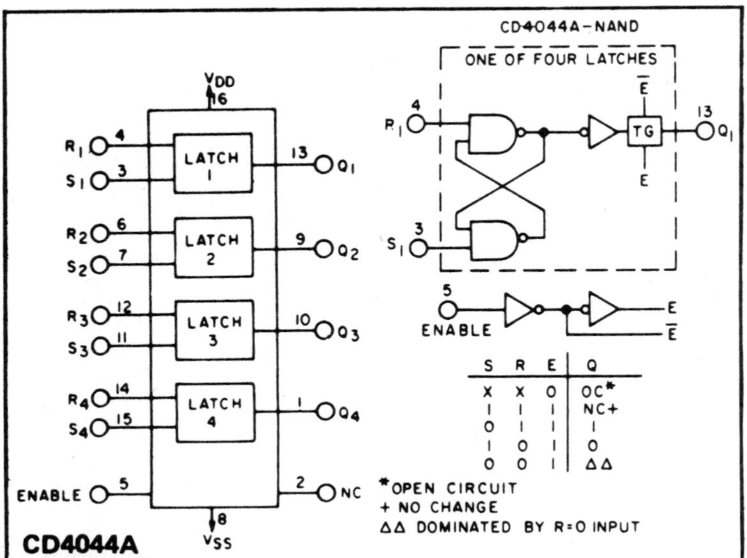

Fig. 6-6. 4044 symbol.

The R-S flip-flop is also sometimes called (especially in older texts) a *bistable multivibrator.* We will discuss multivibrators in general in a later chapter but the bistable is represented by the R-S flip-flop. The designation of the multivibrator is always a function of the number of stable output states. The bistable multivibrator, therefore, has two stable output states. In other words, when a trigger pulse is received on either input, the outputs will go to the state indicated by the operating rules (i.e., Q = HIGH when *set* and Q = LOW when reset, with not-Q being opposite Q) and *stay there.* The bistable multivibrator will remain happy in either state.

The R-S flip-flop operates in an asynchronous manner. While this form of operation is needed in many applications, there are times when it is thoroughly inappropriate. For those operations we will need a clocked flip-flop. In the next chapter we will consider some of the different forms of clocked flip-flops including the clocked R-S, master-slave, J-K, type-D and type-T designs. Most of those flip-flops are readily available in both TTL and CMOS versions, even though it is also possible for you to construct them from scratch using NAND and NOR gates.

Chapter 7

Clocked Flip-Flops

The operation of the R-S flip-flop is unconditional. The output state changes immediately when an appropriate input pulse is received. Such circuits are only able to operate *asynchronously*. The clocked R-S flip-flop of Fig. 7-1 is able to operate in a synchronous manner; i.e., the output will change state only when the input pulse coincides with a *clock* pulse. The behavior is obtained by adding a pair of NAND gates to the NAND-type R-S flip-flop circuit. A commonly used circuit symbol for the clocked R-S flip-flop is shown in Fig. 7-2. This circuit is sometimes called an RST flip-flop.

An example of a timing diagram for an RST flip-flop is shown in Fig. 7-3. The clock produces a fixed-frequency chain of square waves at the C input. Note that the Q output does not change state immediately when the S input pulse goes HIGH. The output waits until the clock input is also HIGH. Similarly, the output is not reset until the clock pulse and the R input pulse are coincident.

MASTER-SLAVE FLIP-FLOPS

It is often difficult, or impossible, to transfer data through a circuit in an orderly manner. We often get into high-speed electronic analogies of the old-fashioned "relay race" problem. Recall that type of problem in circuits where two relays are supposed to close simultaneously? If one relay is a little sluggish or the other is a little faster than usual, then there will be a brief instant where one is open and the other is still closed. This condition often produces unpredictable results. The same action exists in digital circuits, and

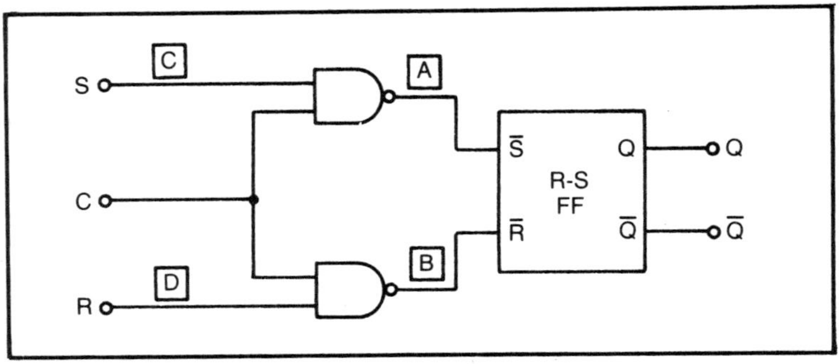

Fig. 7-1. Clocked RS FF.

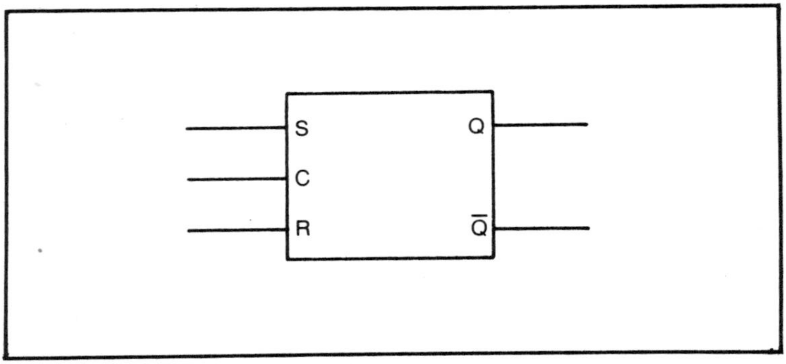

Fig. 7-2. Clocked RS FF circuit symbol.

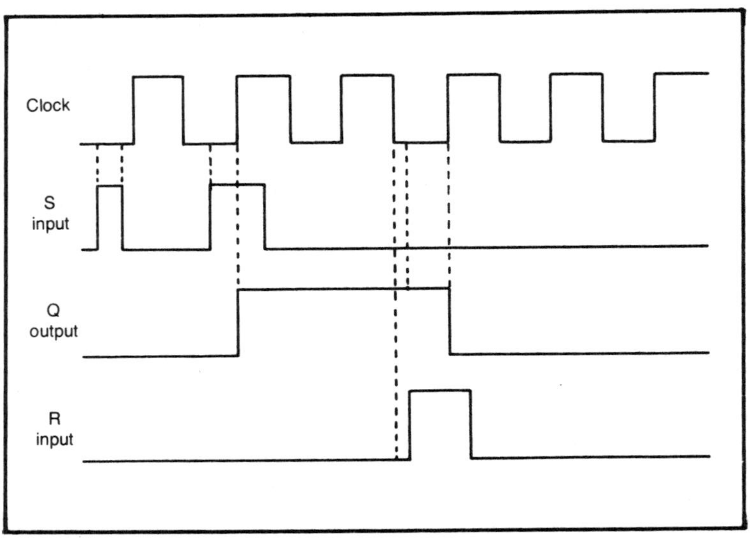

Fig. 7-3. Clocked RS FF truth table.

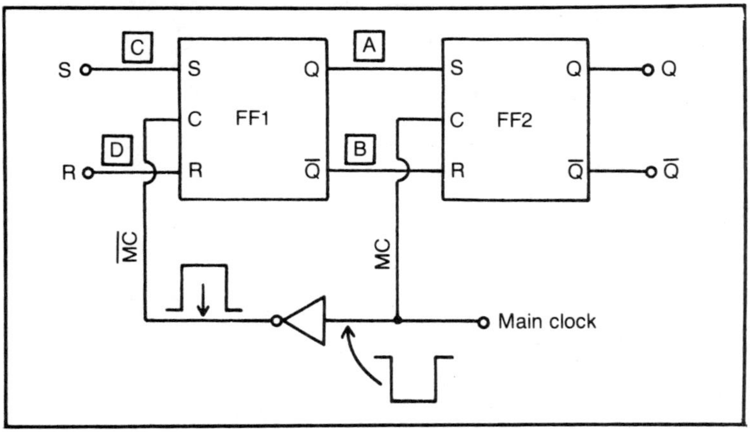

Fig. 7-4. Master-slave FF.

is caused by device propagation time. Note that two RS flip-flops and an inverter can be used to synchronize a data transfer.

The circuit, shown in Fig. 7-4 is called a *master-slave flip-flop*. The inputs to FF1 are the inputs to the circuit as a whole. The outputs for the overall circuit are the outputs of FF2. Also the clock affects FF2 directly but must pass through an inverter before it can affect FF1. Both FFs only become active when their respective clock inputs are HIGH.

A timing diagram for this circuit is shown in Fig. 7-5. Recall from above that FF1 is enabled when the clock pulse is LOW, and FF2 is enabled when the clock pulse is HIGH. When the pulse is applied to the S input, nothing happens at Q1 (the Q output of FF1) until the clock pulse drops LOW. At that time Q1 snaps HIGH, thereby making the S input of FF2 HIGH. But at this time the clock pulse is LOW, so no change occurs at the FF2 output terminal. When the next clock pulse arrives, however, the FF2 clock input goes HIGH (Q1 is still HIGH), so the Q output of FF2 goes HIGH.

Similarly, the reset pulse arrives when the clock is LOW, so the Q1 output immediately drops LOW. The Q output of FF2, however, remains in the HIGH condition until the HIGH transition of the next clock pulse. FF1 is considered the master flip-flop, while FF2 is the slave. The action at FF1 is given time to settle before the changes can be reflected at the output of FF2. This provides an orderly transfer of data between input and output.

TYPE-D FLIP-FLOPS

A type-D flip-flop is a modified RST flip-flop that has only one

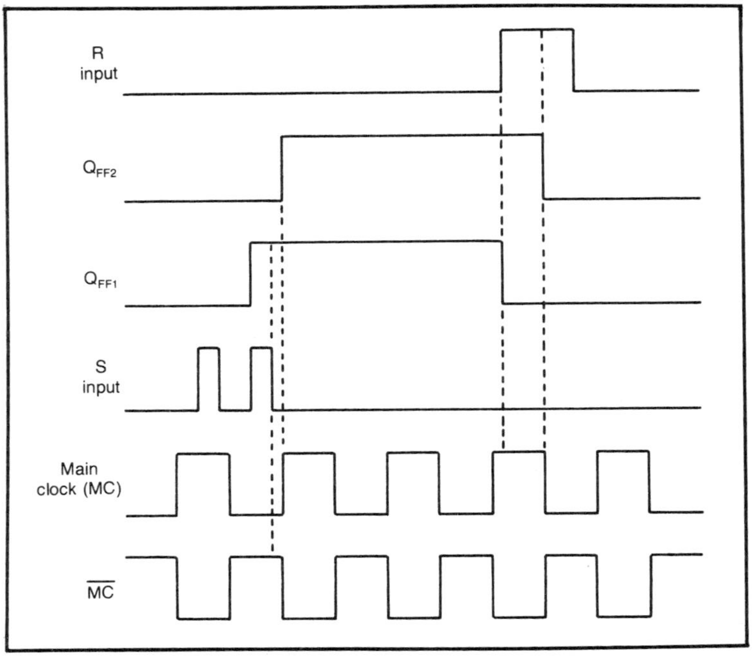

Fig. 7-5. Timing waveform.

input (labeled the D or data input). See Fig. 7-6. The type-D flip-flop will transfer data from the D input to the Q output only when the clock terminal is HIGH. The following are the rules governing the operation of the type-D flip-flops:

□ When the clock input goes HIGH, the data present on the D input is transferred to the Q output.

□ If the clock input remains HIGH, the Q output will follow changes in the data present at the input.

□ If the clock remains LOW, then the Q output will retain the data that was present on the D input at the instant the clock dropped LOW.

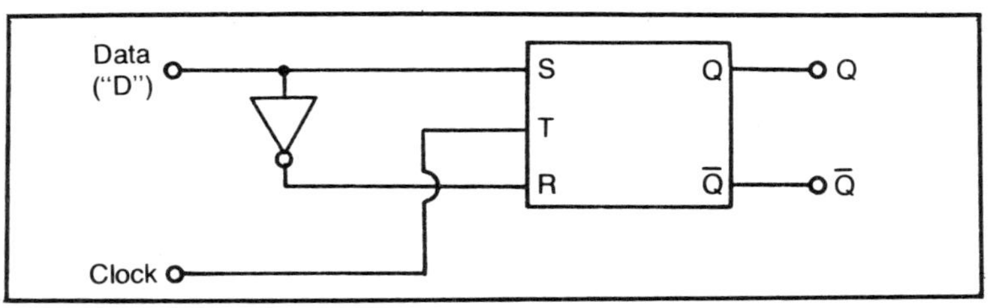

Fig. 7-6. RS FF implementation of type-D FF.

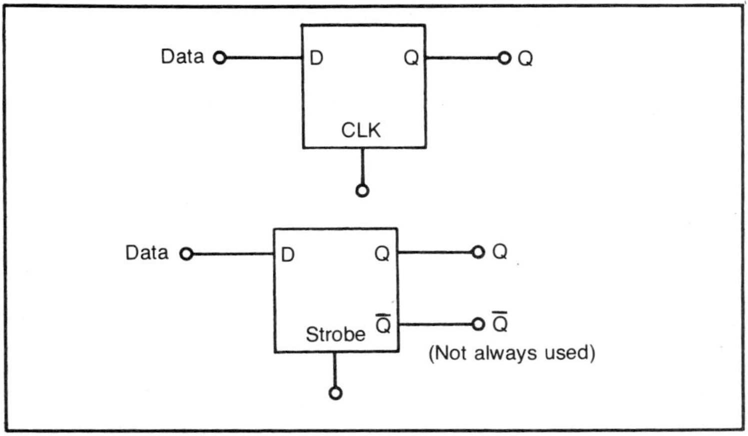

Fig. 7-7. Type-D flip-flop circuit symbols.

Because of the behavior presented in the above rules, the type-D flip-flop is sometimes called a *data latch*, or simply *latch*. The circuit symbol is shown in Fig. 7-7.

We can see these rules more graphically in Figs. 7-8 through 7-10. Consider Fig. 7-8 first. Recall that the data on the D input will be transferred to the Q output only when the clock terminal is HIGH. At time t_0 in Fig. 7-8, the clock goes HIGH, the D input is LOW, though, so the Q output is LOW also.

At time t_1, the D input goes HIGH but cannot affect the Q output because the clock is LOW. The clock goes HIGH at t_2, so the output will also go HIGH. The output pulse exists only for the interval t_3 - t_2, because the HIGH conditions on the D input only coincide in that interval.

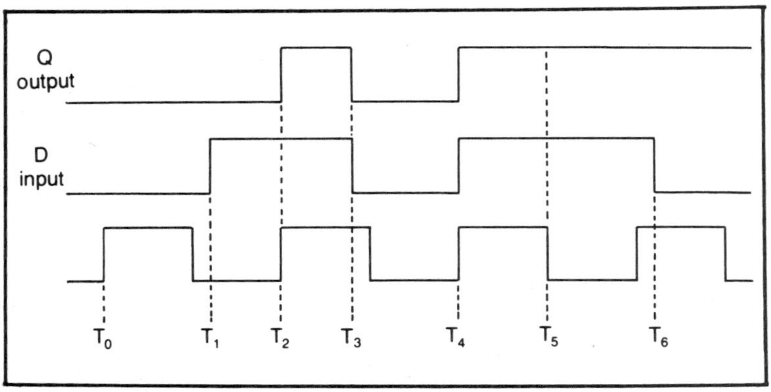

Fig. 7-8. Timing waveforms.

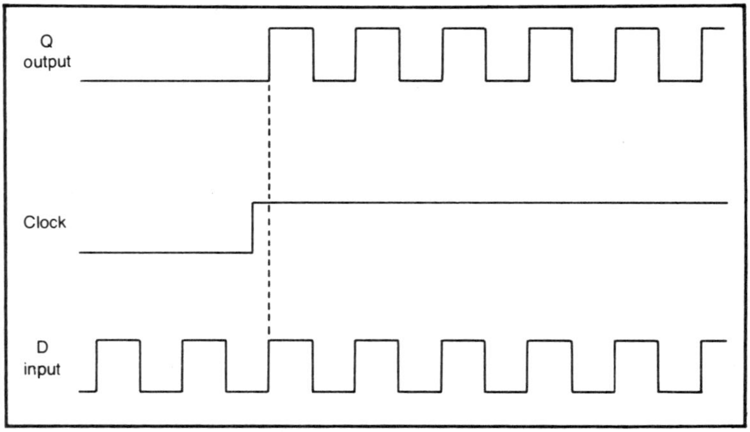

Fig. 7-9. Timing waveforms.

At time t_4 both D and the clock inputs go HIGH, but note that the D line remains HIGH even after the clock pulse disappears. By the third rule, then, the output must remain HIGH after the clock pulse passes.

Another situation is shown in Fig. 7-9. In this case, the clock line goes HIGH and remains HIGH. The Q output, therefore, is in an unlatched condition, so will follow the data at the D input. Because the D input data is a square wave, the output data will also be a square wave.

Still another condition is shown in Fig. 7-10. Again the D input data is a square wave, but the clock is not permanently HIGH in this example. At time t_1, both the clock and the D input are HIGH, so the Q output is also HIGH. At time t_2, the clock drops LOW, but since D

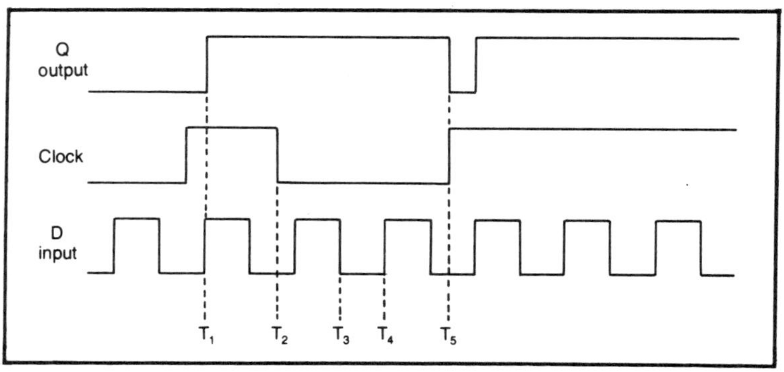

Fig. 7-10. Timing waveforms.

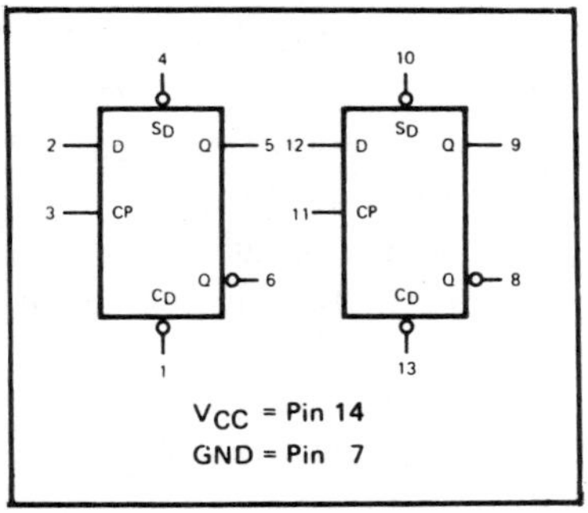

Fig. 7-11. Circuit symbols for 4013.

is HIGH at that instant, the output will remain in the HIGH condition. Note that the D input can change at will, without affecting the output, as long as the clock input remains LOW. But at the time t_5, the clock goes HIGH, and since the D line is LOW, the output goes LOW also. At time t_6, however, the D line goes HIGH, while the clock is still HIGH, so the output goes HIGH.

EXAMPLES OF TYPE-D FLIP-FLOPS

Only rarely will modern circuit designers use individual logic gates to make a flip-flop of any variety. The only common example is the R-S flip-flop. Even those, however, are commonly available in the form of standard CMOS chips, where they had not been in TTL. There are too many good integrated circuit flip-flops on the market for anyone to seriously consider building their own. Figures 7-11 through 7-14 show two TTL and one CMOS version.

A TTL dual type-D flip-flop is shown in Fig. 7-11. The logic diagram is shown in Fig. 7-12. This particular type-D flip-flop has

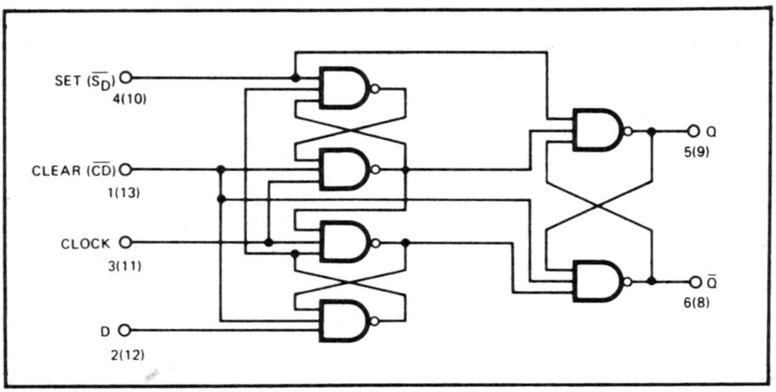

Fig. 7-12. Clocked RS FF.

both *set* and *clear* inputs, in addition to the normal clock and D inputs. The 7474 device will obey the following rules of operation:

☐ Data on the D input is transferred to the Q output only on positive-going transitions of the clock pulse.

☐ No changes in the data are reflected as output changes unless the clock is going HIGH. The device is, therefore, an edge-triggered flip-flop.

☐ If the *set* input is grounded, then Q immediately goes HIGH and not-Q goes LOW.

☐ If the *clear* input is grounded, then Q goes LOW and not-Q goes HIGH.

☐ *Set* and *clear* inputs must not be simultaneously grounded. If these inputs are not being used, then they should be tied HIGH to +5 Vdc.

☐ The two flip-flops in the 7474 IC are completely independent of each other, except for sharing common power supply and ground terminals.

The TTL 7475 device shown in Fig. 7-13 is a special case of the type-D flip-flop called a *quad-latch*. This IC device is level sensitive, so the output will follow changes as long as the clock terminal is HIGH. This is more like the behavior expected of traditional type-D flip-flops, as given earlier in this section.

The 7475 is used primarily to hold four bits of data; i.e., one bit in each flip-flop. The pinouts for the 7475 are shown in Fig. 7-13. The clock inputs are labeled *enable* inputs. When the enable inputs are HIGH, then, data on the D inputs are transferred to the Q outputs. The 7475 is arranged in a 2 × 2 format, meaning that the enable inputs of two flip-flops are tied together and are brought out to a package pin. The enable inputs to the other two remaining

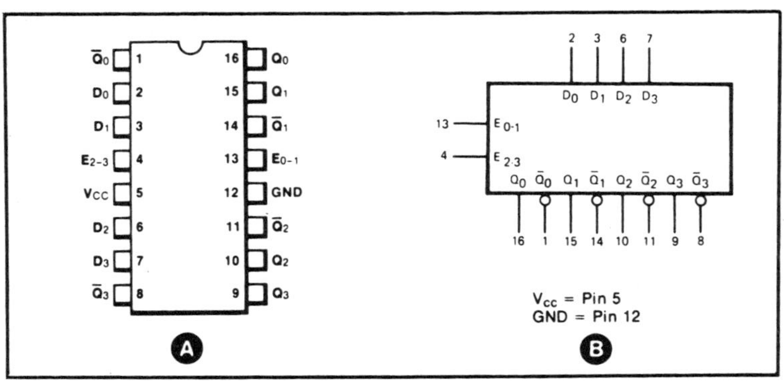

Fig. 7-13. Package pinouts (A) and logic diagram (B).

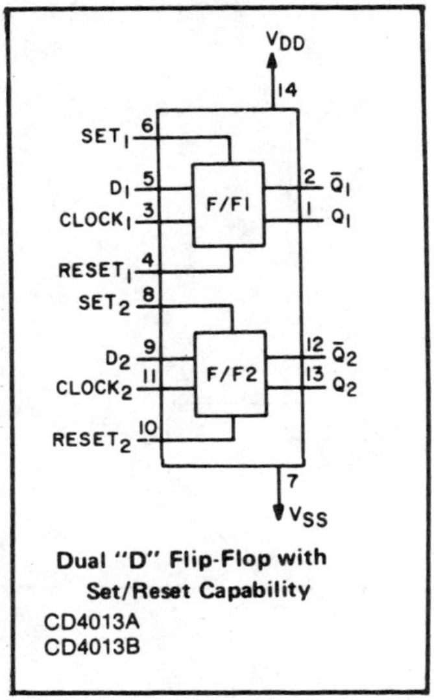

Fig. 7-14. 4013 pinouts.

flip-flops are also tied together and are brought out to another package terminal.

One common place to find 7475s is in digital counter applications, where they are used to hold the BCD data being displayed while the counter chip, which is often a 7490 device, updates the data. The display can be updated when the count is completed by momentarily bringing the enable terminals HIGH. A single 7475 will store all four bits required for a single BCD display decoder.

A device similar to the 7475 is the TTL IC type 74100. This 24-pin IC contains a pair of four-bit latches and therefore has the ability to store up to eight bits of digital data, a fact that is not lost on microcomputer circuit designers.

A CMOS type-D flip-flop is the CD4013 or, simply 4013, device shown in Fig. 7-14. Like the 7474 device from the TTL line,

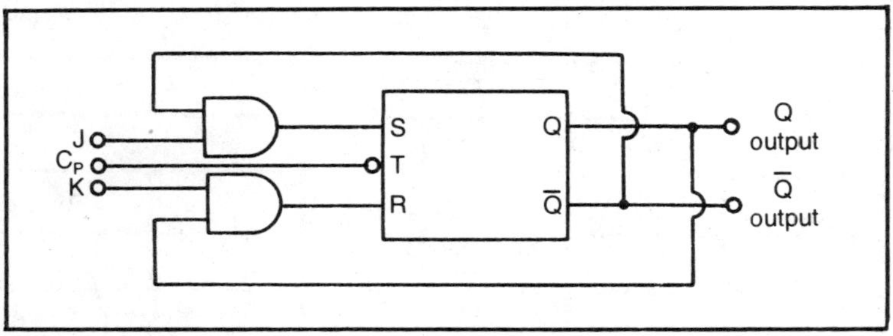

Fig. 7-15. JK FF.

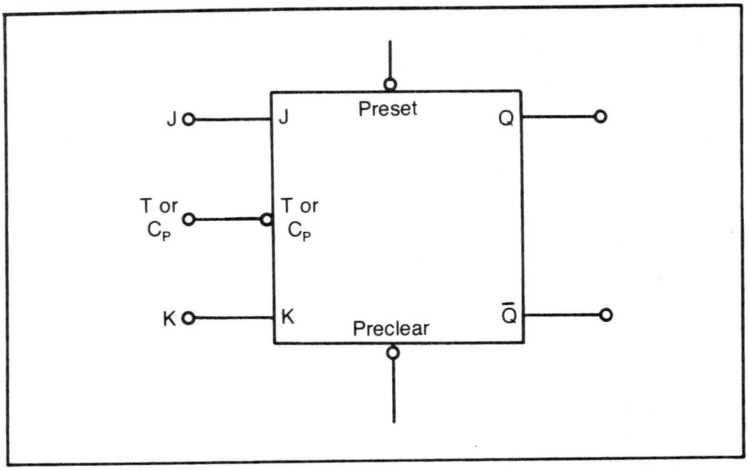

Fig. 7-16. JK FF.

the CMOS CD4013 is a dual type-D flip-flop. It can be used in either direct or clock modes, i.e., it has set and clear direct inputs as well as D and clock inputs. The 4013 differs from the 7474, however, in that direct inputs become active when HIGH. If unused, therefore, these inputs should be tied to ground, instead of +5 volts, as in the case of the TTL version. This is exactly the opposite protocol from the TTL.

In the clocked mode, the 4013 behaves much like the 7474. It is a positive-edge triggered device, so data is transferred to the Q output when the clock line goes HIGH. The action occurs on the positive-going transition of the clock pulse. In this respect, the 4013 and 7474 follow the same rules.

In the direct mode, the operation is similar to the 7474 but requires pulses of the opposite polarity. The rules for the set and clear inputs of the 4013 are as follows:

Table 7-1. Truth Table for Clocked J-K Operation.

J	K	CLOCK		
0	0		(NO CHANGE)	
0	1		0	1
1	0		1	0
1	1		BINARY DIVISION	

☐ If the set input is made HIGH, then the Q output goes HIGH and not-Q goes LOW.

☐ If the clear input is made HIGH, then the Q output goes LOW and not-Q goes HIGH.

J-K FLIP-FLOPS

The J-K flip-flop is very similar to the other clocked flip-flops, although certain differences exist that make the J-K special. Figure 7-15 shows a gate logic diagram for a J-K flip-flop, while the usual schematic symbol is shown in Fig. 7-16. Table 7-1 shows the truth table for clocked operation. Like the TTL and CMOS type-D flip-flops discussed earlier, the J-K flip-flop is capable of operating in either direct or clocked modes.

Direct operation uses the set and clear inputs to force the Q and not-Q outputs into specific states—HIGH or LOW. The truth table for TTL J-K flip-flops (7473, 7476, etc.) is shown in Table 7-2. The rules for these devices are:

☐ LOW conditions on both set and clear inputs results in a forbidden, or disallowed, state.

☐ Making the clear input HIGH and the set input LOW forces Q HIGH.

☐ Making the clear input LOW and the set input HIGH forces Q LOW.

☐ Making both set and clear inputs HIGH causes the flip-flop to operate in the clocked mode.

Again the CMOS version operates in a similar manner, but with opposite polarity pulses. CD4027 is a dual J-K flip-flop, so:

☐ If both set and clear are LOW, then normal clocked operation is obtained.

☐ Making clear LOW, and set HIGH, forces the Q output HIGH.

Table 7-2. Truth Table for Direct-Mode J-K Operation.

J	K	CLOCK	Q	$\overline{Q}$
0	0	(DOESN'T CARE)	DISALLOWED	
0	1		1	0
1	0		0	1
1	1		CLOCKED OPERATION (SEE TABLE 7-1)	

Table 7-3. 74143 Pin Functions.

FUNCTION	PIN NO.	DESCRIPTION
CLEAR INPUT	3	When low, resets and holds counter at 0. Must be high for normal counting.
CLOCK INPUT	2	Each positive-going transition will increment the counter provided that the circuit is in the normal counting mode (serial and parallel count enable inputs low, clear input high).
PARALLEL COUNT ENABLE INPUT (PCEI)	23	Must be low for normal counting mode. When high, counter will be inhibited. Logic level must not be changed when the clock is low.
SERIAL COUNT ENABLE INPUT (SCEI)	1	Must be low for normal counting mode, also must be low to enable maximum count output to go low. When high, counter will be inhibited and maximum count output will be driven high. Logic level must not be changed when the clock is low.
MAXIMUM COUNT OUTPUT	22	Will go low when the counter is at 9 and serial count enable input is low. Will return high when the counter changes to 0 and will remain high during counts 1 through 8. Will remain high (inhibited) as long as serial count enable input is high.
LATCH STROBE INPUT	21	When low, data in latches follow the data in the counter. When high, the data in the latches are held constant, and the counter may be operated independently.
LATCH OUTPUTS (Q_A, Q_B, Q_C, Q_D)	17, 18, 19, 20	The BCD data that drives the decoder can be stored in the 4-bit latch and is available at these outputs for driving other logic and/or processors. The binary weights of the outputs are: $Q_A = 1$, $Q_B = 2$, $Q_C = 4$, $Q_D = 8$.
DECIMAL POINT INPUT	7	Must be high to display decimal point. The decimal point is not displayed when this input is low or when the display is blanked.
BLANKING INPUT (BI)	5	When high, will blank (turn off) the entire display and force RBO low. Must be low for normal display. May be pulsed to implement intensity control of the display.
RIPPLE-BLANKING INPUT (RBI)	4	When the data in the latches is BCD 0, a low input will blank the entire display and force the RBO low. This input has no effect if the data in the latches is other than 0.
RIPPLE-BLANKING OUTPUT (RBO)	6	Supplies ripple blanking information for the ripple blanking input of the next decade. Provides a low if BI is high, or if RBI is low and the data in the latches in BCD 0; otherwise, this output is high. This pin has a resistive pull-up circuit suitable for performing a wire-AND function with any open-collector output. Whenever this pin is low the entire display will be blanked; therefore, this pin may be used as an active-low blanking input.
LED/LAMP DRIVER OUTPUTS (a, b, c, d, e, f, g, dp)	15, 16, 14, 9 11, 10, 13, 8	Outputs for driving seven-segment LED's or lamps and their decimal points. See segment identification and resultant displays on following page.

☐ Making clear HIGH, and set LOW, forces Q LOW.

☐ Making both set and clear HIGH is a disallowed condition.

Clocked operation means that all changes occur synchronously with the clock pulse. The J-K flip-flop operates on the negative-going transition of the clock pulse, in distinct contrast to the behavior of the 7474 (type-D) device discussed earlier.

Table 7-3 shows the operation of TTL J-K flip-flops in the clocked mode. The changes that are shown occur as illustrated on the negative transition of the clock pulse. The rules of operation are as follows:

☐ If both J and K are LOW, no change occurs in the output, regardless of clock pulse transitions.

☐ If J is LOW and K is HIGH, the Q output is forced LOW.

☐ If J is high and K is LOW, Q is forced HIGH.

☐ If both J and K are HIGH, the Q output will go to the

DUAL J-K FLIP-FLOPS WITH CLEAR

73

'73, 'H73, 'L73
FUNCTION TABLE

INPUTS				OUTPUTS	
CLEAR	CLOCK	J	K	Q	$\bar{Q}$
L	X	X	X	L	H
H	⎍	L	L	Q_0	$\bar{Q}_0$
H	⎍	H	L	H	L
H	⎍	L	H	L	H
H	⎍	H	H	TOGGLE	

'LS73A
FUNCTION TABLE

INPUTS				OUTPUTS	
CLEAR	CLOCK	J	K	Q	$\bar{Q}$
L	X	X	X	L	H
H	↓	L	L	Q_0	$\bar{Q}_0$
H	↓	H	L	H	L
H	↓	L	H	L	H
H	↓	H	H	TOGGLE	
H	H	X	X	Q_0	$\bar{Q}_0$

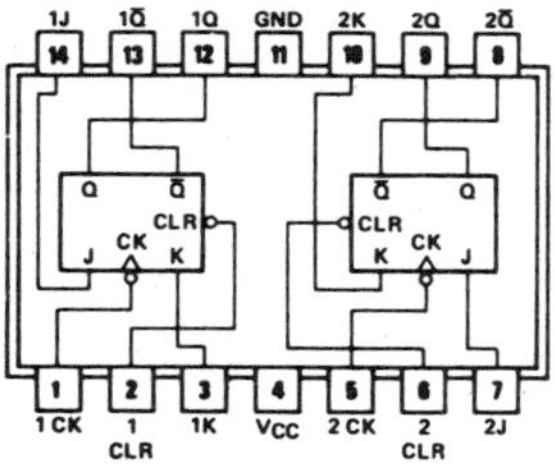

SN5473 (J, W) SN7473 (J, N)
SN54H73 (J, W) SN74H73 (J, N)
SN54L73 (J, T) SN74L73 (J, N)
SN54LS73A (J, W) SN74LS73A (J, N)

Fig. 7-17. 7473 pinouts.

DUAL J-K FLIP-FLOPS WITH PRESET AND CLEAR

76

'76, 'H76
FUNCTION TABLE

INPUTS					OUTPUTS	
PRESET	CLEAR	CLOCK	J	K	Q	$\bar{Q}$
L	H	X	X	X	H	L
H	L	X	X	X	L	H
L	L	X	X	X	H*	H*
H	H	⎍	L	L	Q_0	$\bar{Q}_0$
H	H	⎍	H	L	H	L
H	H	⎍	L	H	L	H
H	H	⎍	H	H	TOGGLE	

'LS76A
FUNCTION TABLE

INPUTS					OUTPUTS	
PRESET	CLEAR	CLOCK	J	K	Q	$\bar{Q}$
L	H	X	X	X	H	L
H	L	X	X	X	L	H
L	L	X	X	X	H*	H*
H	H	↓	L	L	Q_0	$\bar{Q}_0$
H	H	↓	H	L	H	L
H	H	↓	L	H	L	H
H	H	↓	H	H	TOGGLE	
H	H	H	X	X	Q_0	$\bar{Q}_0$

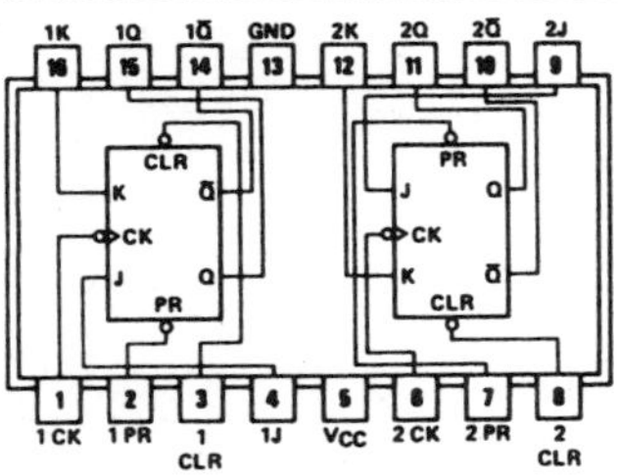

SN5476 (J, W) SN7476 (J, N)
SN54H76 (J, W) SN74H76 (J, N)
SN54LS76A (J, W) SN74LS76A (J, N)

Fig. 7-18. 7476 pinouts.

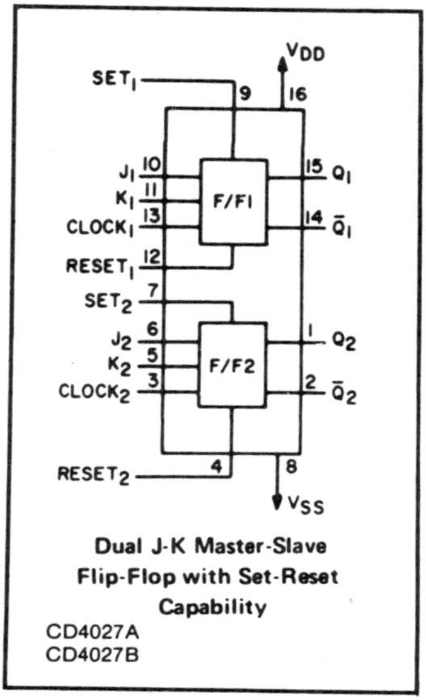

Fig. 7-19. 4027 pinouts.

opposite of its present state; for example, if it is HIGH, then it will go LOW, and vice versa.

In the clocked mode, the CMOS 4027 device follows these same rules, except that the changes occur on the positive-going transition of the clock pulse.

J-K FLIP-FLOP EXAMPLES

We have mentioned three examples of commercial IC J-K flip-flops have been mentioned: 7473 (TTL), 7476 (TTL), and 4027 (CMOS). Because the rules for these devices have already been given, they will not be repeated here. Figures 7-17, 7-18, and 7-19 show the pinouts for the 7473, 7476, and 4027 devices, respectively.

Chapter 8

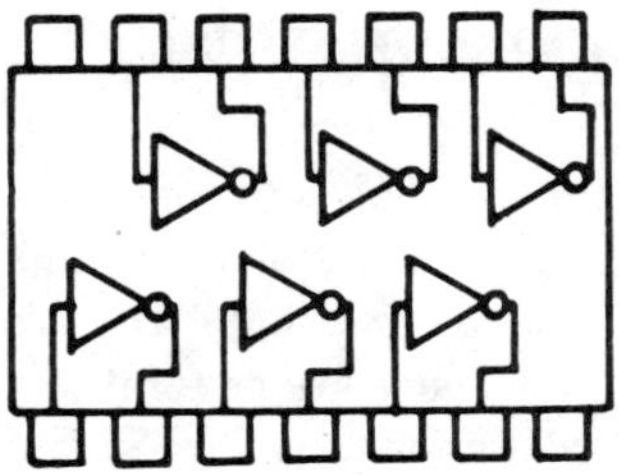

Timer and Clock Circuits

A *timer* is a circuit that is able to produce an output level for a precise length of time. This output level may be normally HIGH and will drop LOW for the specified period of time; or, alternatively, it could be normally LOW, and then snap HIGH for the specified period when active. Devices of both sorts are known.

Many timers are also part of a family of circuits called *multivibrators*. In the case of the timer described above, we are talking about a *monostable multivibrator* or one-shot stage. There are also *bistable* and *astable* multivibrators.

The names of these multivibrators are derived from the number of stable output states that each will allow. The monostable multivibrator, for example, has only one stable state. It normally rests in its stable, or dormant, state. But when a trigger pulse is received, the output snaps to the active state. It will remain in the active state only for a given period of time, after which the output snaps back to the stable state.

The bistable multivibrator is our old friend, the R-S flip-flop. Recall from Chapter 6 that one trigger input on this stage will SET the output and make Q HIGH, while a pulse on the other input will RESET the output to make Q LOW. The R-S flip-flop, however, doesn't much care which state it is in. It would remain happily in either SET or RESET condition; two stable states exist, which is where the name *bistable* multivibrator comes from.

The astable multivibrator has *no* stable states. It will merrily flip and flop back and forth between SET and RESET conditions.

The astable multivibrator, therefore, functions as a *clock generator*, producing a square wave output wave train.

Both astable and monostable multivibrators function as timer circuits. Bistables, however, are not easily used in this capacity.

TIMERS VERSUS CLOCKS

It is usually the practice to distinguish *timers* from *clocks*. A timer circuit is the one-shot described earlier in this chapter. On the other hand, a clock is an *astable multivibrator*. Clocks are used to synchronize digital circuits.

There are several ways to generate a timer pulse. One of the most popular is the 555 integrated circuit timer. There is also a 556 device, which is a dual 555.

The 555 is inside an 8-pin miniDIP integrated circuit package. It has the advantage of being essentially free of period drift caused by variations in power-supply voltages; something that cannot be said of the unijunction transistor (UJT) relaxation oscillators previously used as timers.

The block diagram of the 555 is shown in Fig. 8-1, while the pinouts of the 8-pin miniDIP package is shown in Fig. 8-2. The 555 is one of the most popular IC devices used for timer circuits. Perhaps one of the reasons for this is that it is neither TTL nor CMOS, yet it can be interfaced with either of those logic families. The 555 is of bipolar transistor construction.

One principal difference between the 555 and TTL devices is that the 555 can be operated over a wide range of potentials; i.e., 5

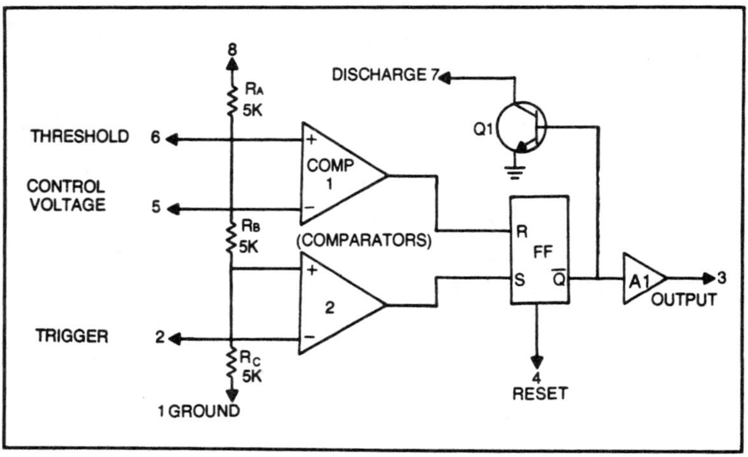

Fig. 8-1. Internal block diagram of the 555.

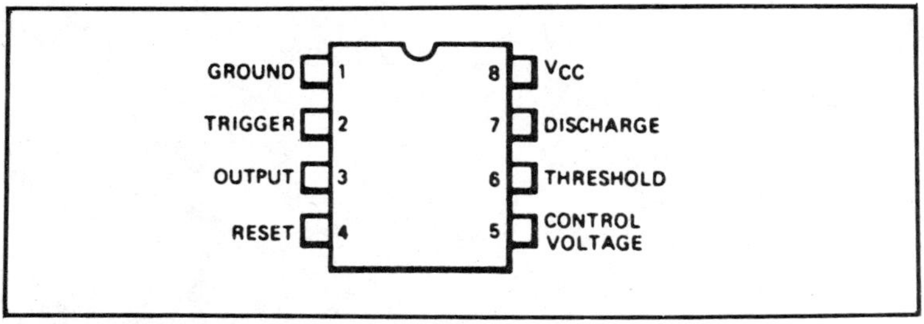

Fig. 8-2. 8-pin miniDIP 555 pinouts.

V to 15 V, although potentials between 9 V and 12 V seems optimum. When the output terminal is LOW, the output terminal (pin no. 3) can *sink* up to 200 mA. When the output is HIGH, on the other hand, it will *source* up to 200 mA.

The 555 is capable of operating in either monostable or astable modes. The major difference between these modes of operation is the use of the output to retrigger the device during astable operation.

The 555 is one of those nice devices that will allow the clever designer to use a lot of imagination. Before you can fully understand the wide range of uses this chip can offer, however, it is necessary to know the inner workings intimately. For this reason, we are going to explain the modes of operation using modified versions of Fig. 8-1 in the illustrations.

Figure 8-3 shows a monostable multivibrator using the 555 IC timer. Figure 8-3 is the block diagram, while Fig. 8-4 is the circuit as it appears in a schematic diagram.

The heart of the 555 timer is an R-S flip-flop (FF) that is controlled by a pair of voltage comparators. An R-S flip-flop, you will recall, is a bistable circuit; that is, it has two stable states. In its initial state the Q output is LOW, and the not-Q is HIGH. If a pulse is applied to the FF-SET (S) input the situation reverses itself, and the not-Q output becomes LOW, while the Q is HIGH.

A comparator is a device that is capable of comparing two voltage levels, and issuing an output that indicates whether they are equal or unequal. A comparator can be made by using an operational amplifier or other high gain linear amplifier without any negative feedback. In the 555, two comparators are designed so that their outputs go HIGH when the two inputs are at the same potential.

Under initial conditions, at time t_0, the not-Q terminal of the R-S flip-flop is HIGH, and this biases transistor Q1 on, placing pin no. 7 of the 555 at ground potential. This keeps capacitor C1

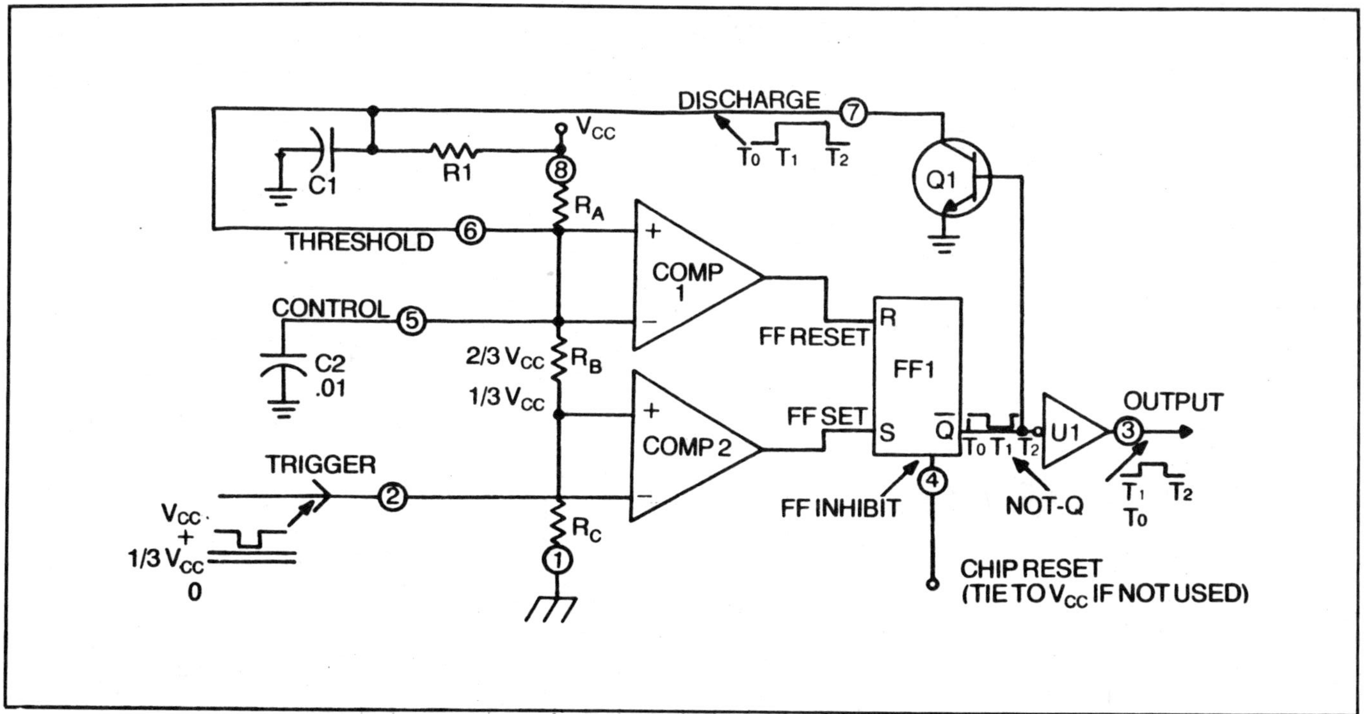

Fig. 8-3. 555 used as a monostable multivibrator.

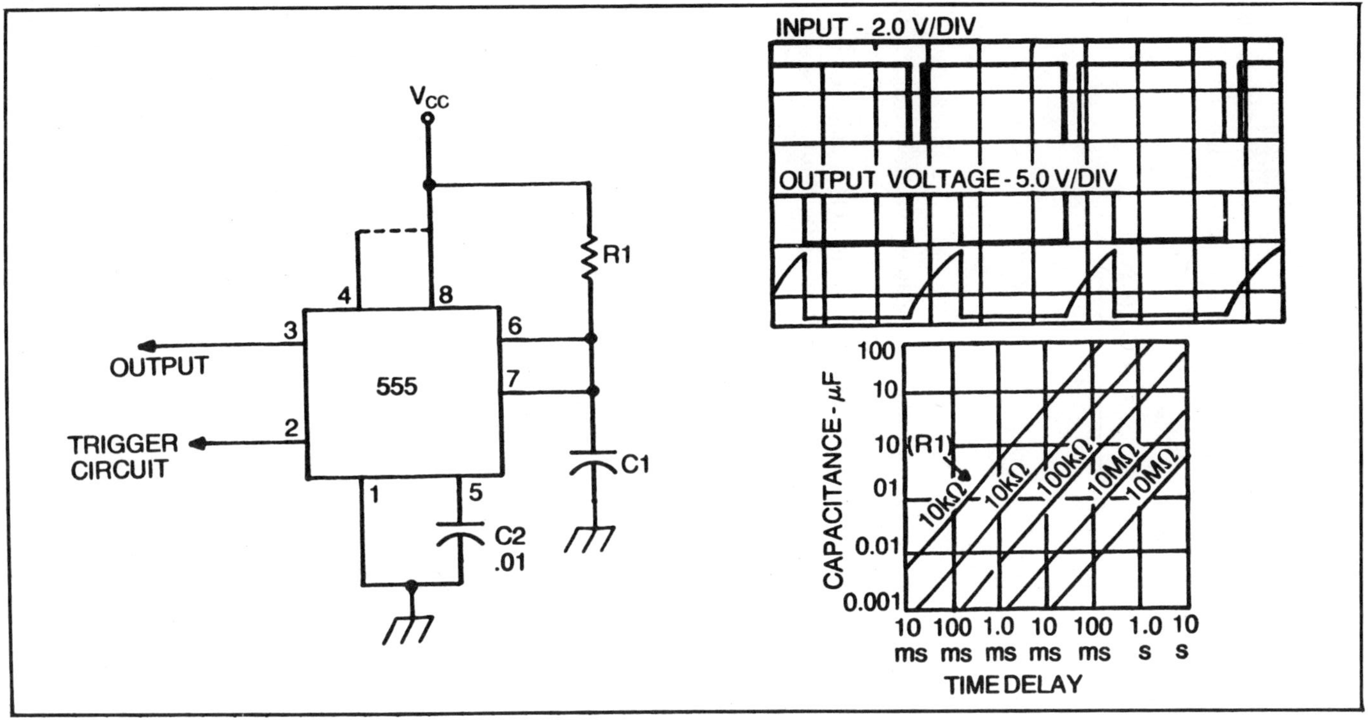

Fig. 8-4. Circuit for 555 one-shot.

discharged. Also, amplifier A1 is an inverter, so the output terminal (pin no. 3) is initially in a LOW condition.

Resistors R_A, R_B, and R_C are inside the 555 IC and are of equal value, nominally 5000 ohms. These form a voltage divider that is used to control the voltage comparators. The inverting (−) input of comparator no. 1 is biased to a potential of:

$$E_1 = \frac{(V_{CC})(R_B + R_C)}{R_A + R_B + R_C}$$

$$E_1 = \frac{2}{3} V_{CC}$$

This means that the output of comparator no. 1 will go HIGH when the control voltage to IC pin no. 5 is equal to 2/3-Vcc. Similarly, the same voltage divider is used to bias comparator No. 2. The voltage applied to the noninverting input of the second comparator is given by:

$$E_Z = R_A + R_B + R_C \times \frac{R_C}{V_{CC}}$$

$$= \frac{1}{3} \times V_{CC}$$

When the voltage applied to the trigger input (IC pin no. 2) drops to 1/3-Vcc, the output of the second comparator goes HIGH.

The control voltage is, in this case, the voltage across capacitor C2. This capacitor charges through resistor R_A and will reach 2/3-Vcc in less that 1 ms after power is applied. If a short, negative-going pulse is applied to the trigger input at time t, in Fig. 8-4, then the output of comparator no. 2 will drop LOW as soon as the trigger pulse voltage drops to a level equal to 1/3-Vcc. This puts the flip-flop in the SET condition and causes the not-Q output to drop to the LOW state.

A drop to the LOW state by the not-Q output at time t_1 causes two things to occur simultaneously: One is to force the output of buffer amplifier A1 HIGH, and the other is to turn off transistor Q1. This allows capacitor C1 to begin charging through resistor R1. The voltage across C1 is applied to the noninverting input of comparator no. 1 through the threshold terminal (pin no. 6). When this voltage rises to 2/3-Vcc, comparator no. 1 will toggle to its HIGH state and will RESET the R-S flip-flop. This occurs at time t_2 and forces the not-Q output again to its HIGH state.

The output of amplifier A1 again goes LOW, and transistor Q1 is turned on again. When Q1 is on, we find that capacitor C1 is discharged. At this point the cycle is complete, and the 555 timer is again in its dormant state. The output terminal will remain LOW until another trigger pulse is received. The approximate length of time that the output terminal remains in the HIGH condition is given by:

$$t = t_2 - t_1$$
$$t = 1.1\ R1C1$$

Where:

t is the time duration in seconds.
R1 is the resistance of *R1* in ohms.
C1 is the capacitance of *C1* in farads.

This function is graphed in Fig. 8-5 for times between 0.01 and 10.0 seconds with values of R1 and C1 that are easily obtainable. The time relationship between the trigger pulse, the output pulse, and the voltage across capacitor C1 is shown in Fig. 8-6.

If the reset terminal is not used, it should be tied to Vcc to prevent noise pulses from jamming the flip-flop. If, however, negative-going pulses are applied simultaneously to the trigger input (pin no. 2) and the reset terminal (pin no. 4), the output pulse

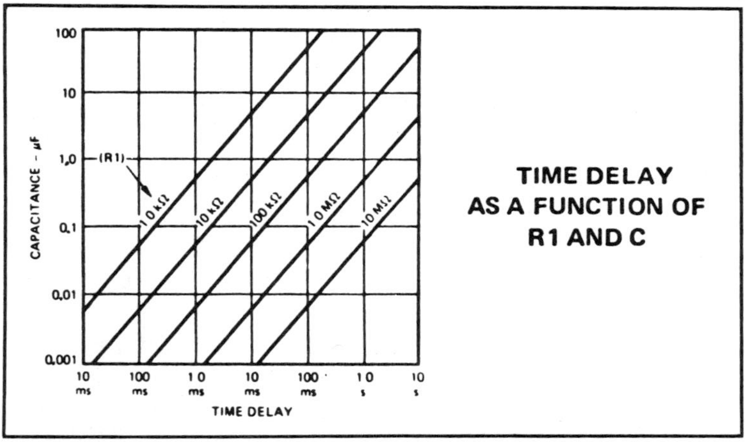

Fig. 8-5. Operating parameters of the 555 in monostable operation.

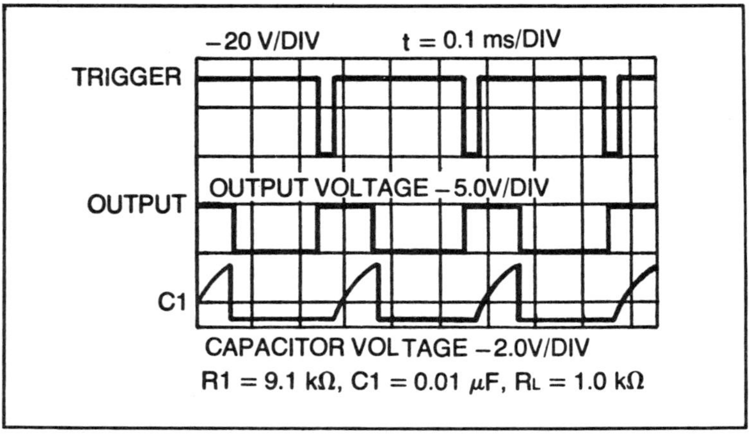

Fig. 8-6. Waveform of the 555.

will terminate. When this occurs, the output terminal drops back to the LOW state, even though the time duration (t) has not yet expired.

An astable multivibrator is similar in many respects to the monostable variety, except that it is self-retriggering. The astable multivibrator has no stable states, so its output will swing back and forth between the HIGH and LOW states. This action produces a wave train of square wave pulses. An example of a 555 astable multivibrator circuit is shown in Figs. 8-7 through 8-9. Again, we have a version of Fig. 8-1 for block diagram analysis in Fig. 8-7, and the circuit as it will appear in schematic diagrams is shown in Fig. 8-8. As in the previous case, the inverting input of comparator no. 1 is biased to 2/3-VCC, and the noninverting input of comparator no. 2 is biased to a level of 2/3-VCC, and the noninverting input of comparator no. 2 is biased to a level of 1/3-VCC through the action of resistor voltage divider network R_A, R_B, and R_C. The remaining two comparator inputs are strapped together and are held at a voltage determined by the time constant C1(R1 + R2). Under initial conditions, the not-Q output of the R-S flip-flop is HIGH. This turns on transistor Q1, keeping the junction of resistors R1 and R2 at ground potential. Capacitor C1 has been charged, but when Q1 turns on, C1 will discharge through resistor R2. When the voltage across capacitor C1 drops to a level of 2/3-VCC, the output of comparator no. 1 goes HIGH and that resets the flip-flop. This action again turns on Q1 and allows C1 to discharge to 1/3-VCC. Capacitor C1, then, alternatively charges to 2/3-VCC and then discharges to 1/3-VCC. Figure 8-9 shows the relationship between HIGH and LOW times.

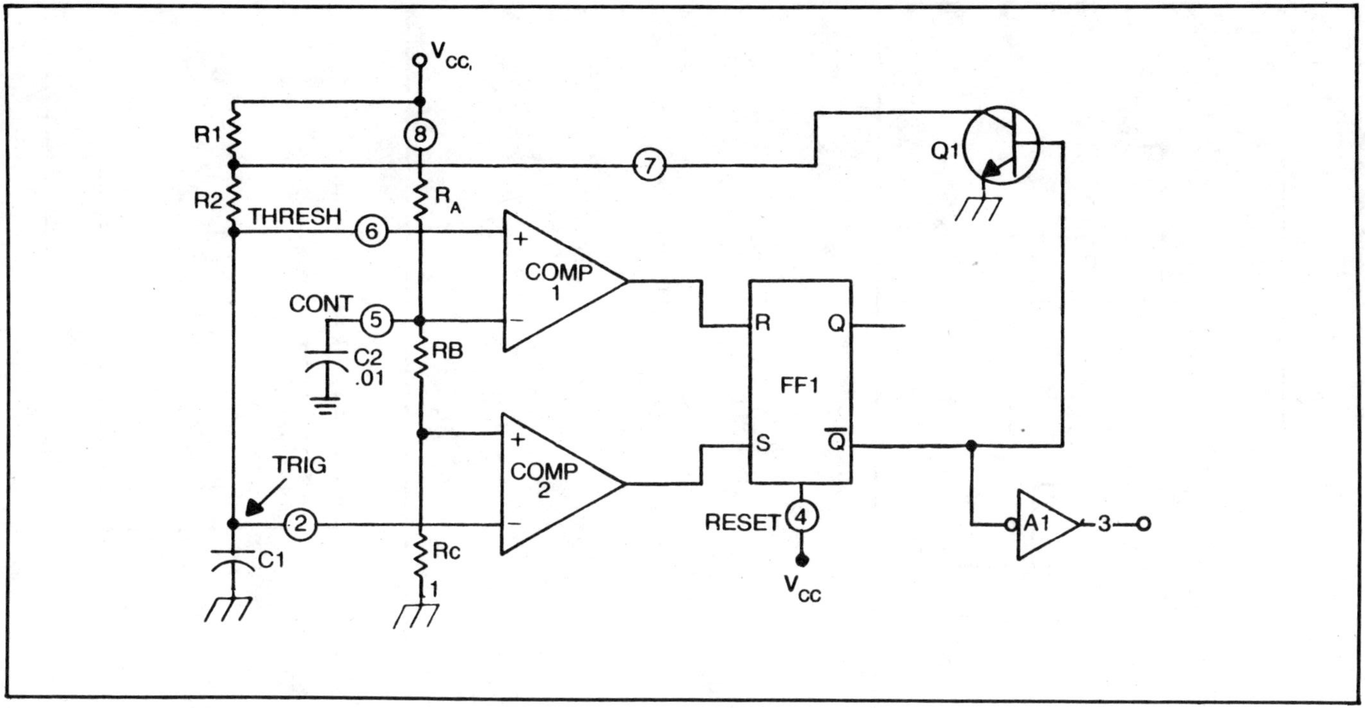

Fig. 8-7. Block diagram of 555 astable multivibrator.

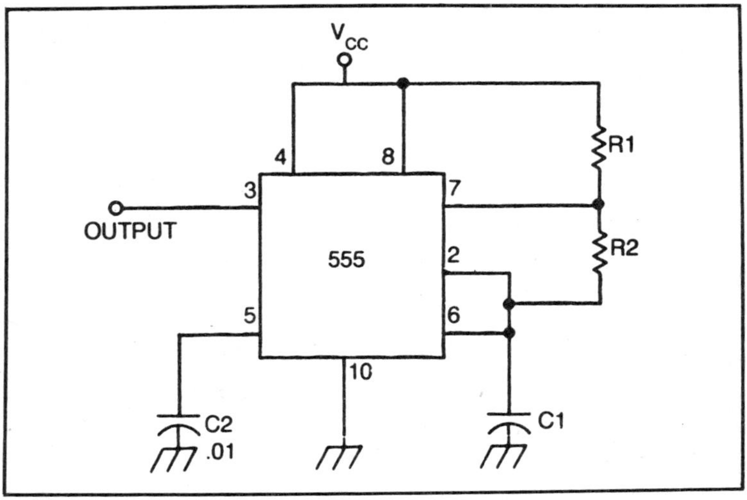

Fig. 8-8. Circuit for 555 astable multivibrator.

The high time, t_1, is given by:

$$t_1 = 0.693\ (R1 + R2)\ C1$$

and the low time by:

$$t_2 = 0.693\ (R2)\ C1$$

The total period of the waveform, t, is the sum t_1 and t_2, and is given by:

$$t = t_1 + t_2$$
$$t = 0.693\ (R1 + 2R2)\ C1$$

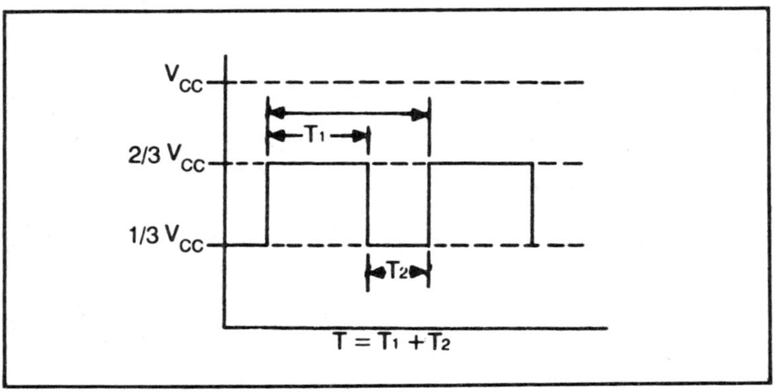

Fig. 8-9. 555 astable duty cycle.

In any electrical circuit, or any other physical system for that matter, the frequency of an oscillation is the reciprocal of the period. In this case, then, the frequency of oscillation is:

$$F_{Hz} = 1/t_{sec}$$

$$F_{Hz} = \frac{1}{0.693\ (R1 + 2R2)\ C1}$$

$$F_{Hz} = \frac{1.44}{(R1 + 2R2)\ C1}$$

The last equation is shown graphically in Fig. 8-10 for frequencies between 0.1 Hz and 100,000 Hz, using easily obtainable component values. The relationship between the C1 voltage and the output state is shown in Fig. 8-11. The *duty cycle*, also called *duty factor*, is the percentage of the total period that the output is HIGH. This is given by:

$$DF = \frac{t_1}{t_1 + t_2}$$

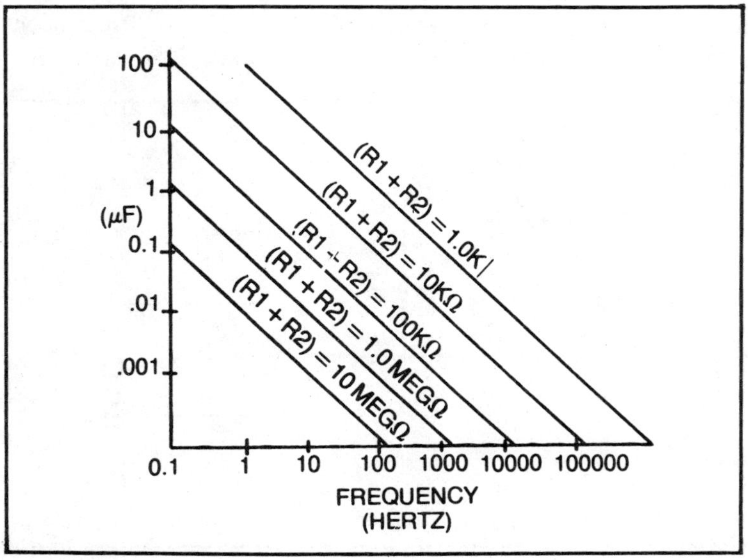

Fig. 8-10. Timing chart.

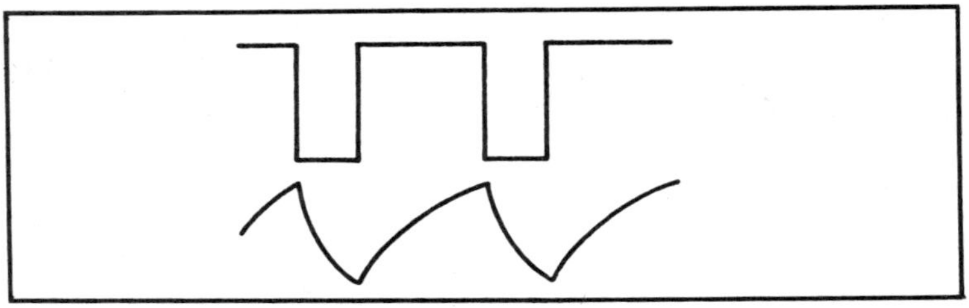

Fig. 8-11. Timing waveform.

$$DF = \frac{R2}{R1 + R2}$$

The trigger input of the 555 timer is held in the HIGH condition in normal dormant operation. To trigger the chip and initiate the output pulse pin no. 2 must be brought down to a level of 1/3-Vcc. Figure 8-12 shows how this can be done manually. Resistor R4 is a pull-up resistor used to keep pin no. 2 HIGH. Capacitor C3 is charged through resistors R3 and R4. When S1 is pressed, the junction of R3 and C3 is shorted to ground, rapidly discharging C3. The sudden decay of the charge on C3 generates a negative-going pulse at pin no. 2 that triggers comparator no. 2, initiating the output pulse sequence.

The 555 timer is also available in a dual version called the 556. This IC is a 14-pin DIP package containing two timers, independent of each other except for the power-supply voltage. It is, then, a dual

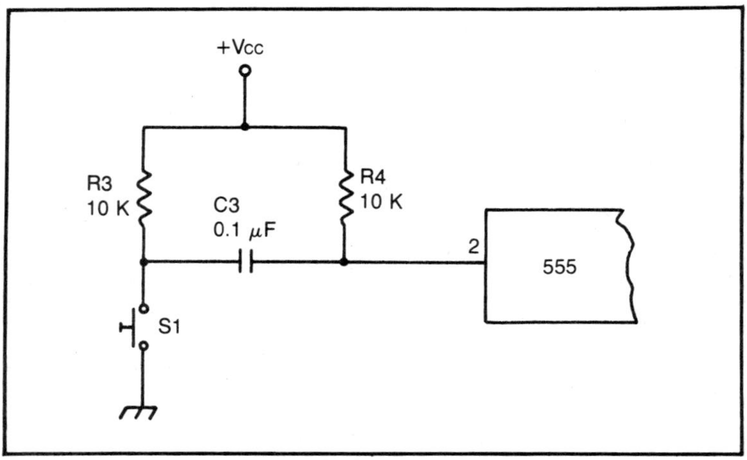

Fig. 8-12. Manual triggering of the 555.

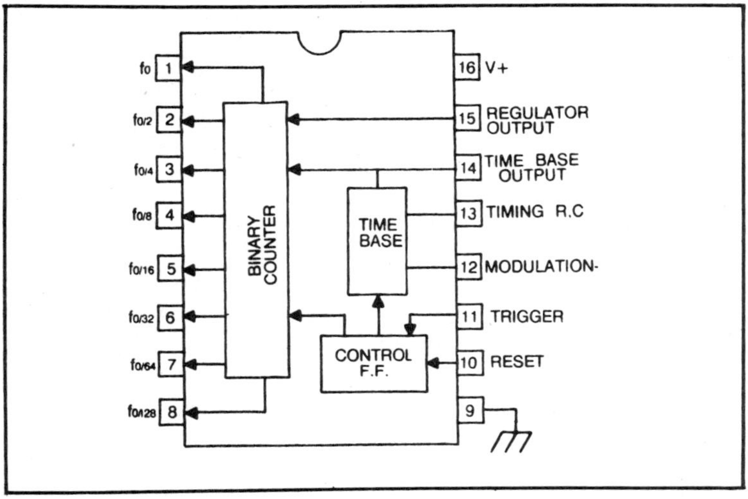

Fig. 8-13. Exar XR-2240 IC timer circuit.

555. Both 555 and 556 timers are so useful that it is an even bet you will someday use one or both if you do any circuit design work at all.

Another useful IC timer—in fact, a *family* of timers—is the Exar type XR-2240, XR-2250, and XR-2260. The internal circuitry of these is shown in Fig. 8-13. These devices are second sourced in the form of the Intersil 8240, 8250, and 8260, respectively. The 2240 is a binary counter with 8-bit outputs. The 2250 is a similar device, except that the outputs are in BCD code. The 2260 is also BCD, except that the most significant bit is 40, instead of 80. This allows the 2260 to be used as a timer with 60 count rather than 100. In the 2250 and 2260 devices, the regulator output terminal (pin no. 15) is used as an overflow that allows cascading of several stages. Table 8-1 shows the pinouts and their respective weighting.

These timers will operate over a supply voltage range of +4.5

Table 8-1. Pinouts of the Exar Timers.

Pin No.	2240/8240	2250/8250	2260/8260
1	1	1	1
2	2	2	2
3	4	4	4
4	8	8	8
5	16	10	10
6	32	20	20
7	64	40	40
8	128	80	(N.C.)

to +18 Vdc. The time base section is a clock circuit that is very similar to the 555 device. This similarity is shown in Fig. 8-14, which shows the internal circuitry of the XR-2240. One main difference between the 555 timer and the time base portion of the XR-2240 lies in the relative reference levels created by the respective internal voltage dividers (R1, R2, and R3 in Fig. 8-14). In the 555 timer all three resistors are of equal value, so comparator input terminals are at 0.33 V_{CC} and 0.66 V_{CC}. In the Exar chip, on the other hand, the reference levels are 0.27 V_{CC} and 0.73 V_{CC}, respectively. One result of this is simplification of the duty factor equation that gives the period of the waveform.

The binary counter section consists of a chain of J-K flip-flops connected in the standard manner, where each stage functions as a divide-by-2, or binary, counter. The binary counter chain is connected to the output of the time base section through an NPN open-collector transistor. The transistor collector is also connected to IC pin no. 14 (called *time-base output*) so that a 20 kohm pull-up resistor can be connected between the collector and the output of internal regulator pin no. 15.

Digital outputs from this counter are, in the usual fashion, given as voltage levels at a set of IC pins. Each output bit is delivered to a specific terminal of the IC package where it is connected to a pull-up resistor similar to that used for the time-base output terminal. The output terminals will generate a LOW condition when active. This may seem to be opposite to the usually accepted arrangement, but there is a method to this madness, and it does create a highly versatile, stable, long-duration counter. These are properties not usually associated with single-IC designs.

Figure 8-15 shows the basic operating circuit for the XR-2240 timer IC. This chip proves interesting because the sole difference between circuits for astable and monostable operation is the 51 kohm feedback resistor linking the reset terminal (pin no. 10) to the wired-OR outputs. The timer is set into operation by application of a positive-going trigger pulse to pin no. 11. This pulse is routed to the control logic and has several jobs to perform simultaneously: resetting the binary counter flip-flops, driving all outputs low, and enabling the time-base circuit. As was true in the 555 IC, this timer works by charging capacitor C1 through resistor R1 from a positive voltage source, V+ or V_{CC}. The period of the output waveform is given by:

$$t = R \times C$$

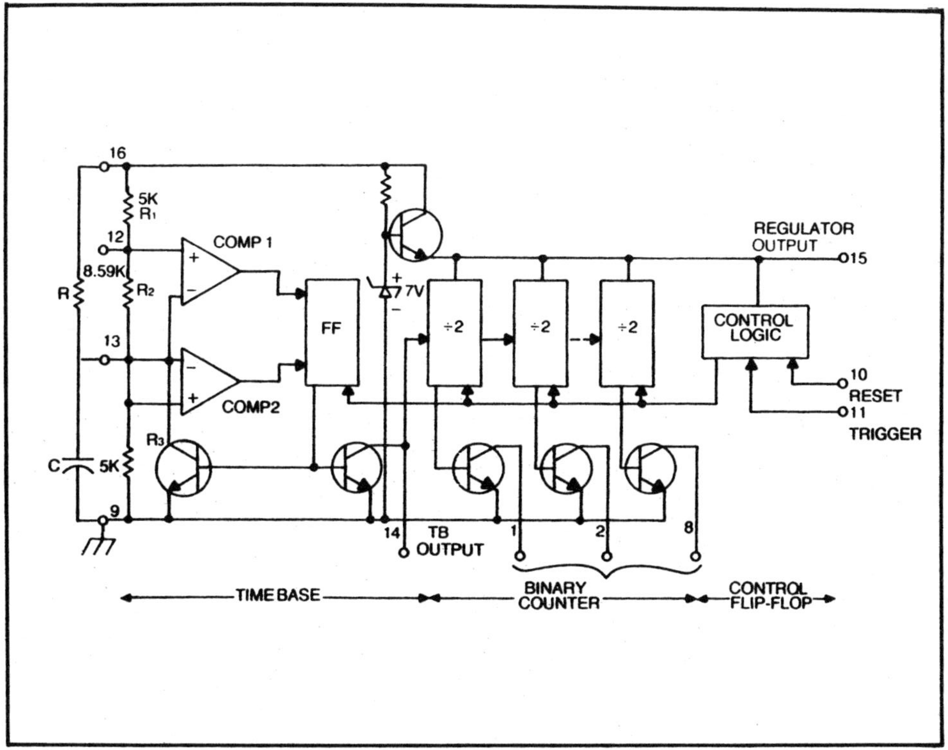

Fig. 8-14. Block diagram.

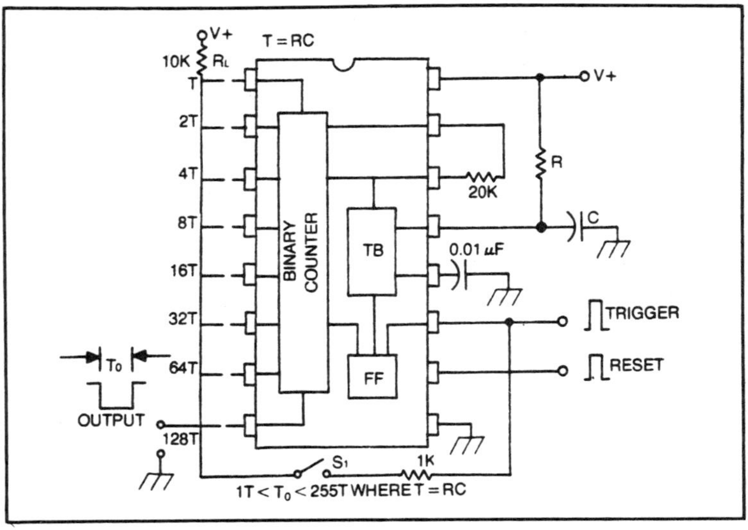

Fig. 8-15. Normal circuit for XR-2240.

Where:

t is the period in seconds.

R is the resistance of R1 in ohms.

C is the capacitance of C1 in farads.

The pulses generated in the time-base section are counted by the binary counter section, and the output stages change states to reflect the current count. This process will continue until a positive-going pulse is applied to the reset terminal.

Figure 8-16 shows the relationship between the trigger pulse, time-base pulses, and various output states. The reason for the open-collector output circuit is to allow the user to wire a perma-

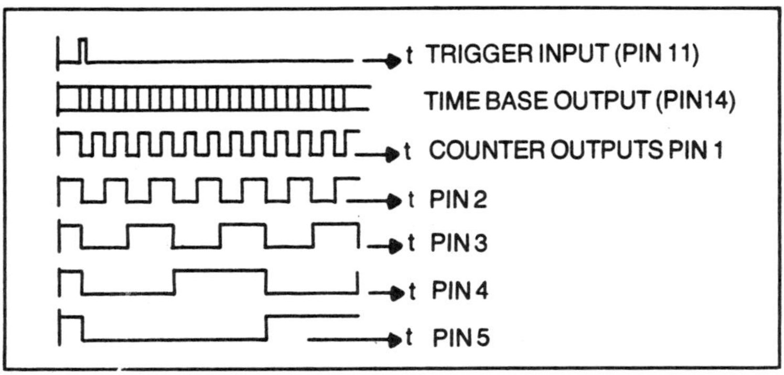

Fig. 8-16. Timing waveforms.

nent OR output so that the actual output duration can be programmed. Each binary output is wired in the usual power-of-two sequence: 1, 2, 4, 8, 16, 32, 64, and 128. If these are wired together, the output will remain LOW as long as *any one output* is LOW. This allows the output duration to be programmed from 1t to 255t, where t is defined as in the last equation, by connecting together those outputs whose sum equals the desired time period. For example, design a timer with a 57-second time delay. In the binary notation, decimal 57 is equal to:

$$32 + 16 + 8 + 1 = 57$$

Set t = R1C1 = 1 second wire and wire together the pins on the XR-2240 corresponding to these weights, then a 57-second time delay will be realized. The base diagram to this IC shows us that these pins are numbers 1, 4, 5, and 6. If those four pins are shorted together, the counter output will remain LOW for 57 seconds following each trigger pulse.

The time-base or the wired-OR terminals could be changed to vary the output periods. Of course, if the time-base frequency were doubled, the counter would reach the desired state in half the time. This feature allows programming of the XR-2240 to time durations that might prove difficult to achieve using conventional circuitry.

Each output must be wired to VCC through a pull-up resistor of 10 kohms, unless, of course, the wired-OR output configuration is used. In that case, a single 10 kohm resistor is used. Current through the output terminals must be kept at a level of 5 mA or less. This serves as a general guide to the selection of pull-up resistors.

The amplitude of the reset and trigger pulses must be at least two PN-junction voltage drops ($2 \times 0.7 = 1.4$ V). In most practical applications, it might be wise to use pulses greater than 4 V amplitude, or standard TTL levels, in order to guard against the possibility that any particular chip may be a little difficult to trigger near minimum values or that outside factors tend to conspire to actually reduce pulse amplitude at the critical moment.

Synchronization to an external time base or modulation of the pulse width is possible by manipulating pin no. 12. In normal practice, this pin, which is the noninverting input of comparator no. 1, is bypassed to ground through a 0.01 μF capacitor so that noise signals will not interfere with operation. A voltage applied to pin no. 12 will vary the pulse width of the signal generated by the time base. This voltage should be between +2 V and +5 V for a time base change multiplier of approximately 0.4 to 2.25, respectively.

If you wish to synchronize the internal time base to an external reference, connect a series RC network to IC pin no. 12. This forms an input network for sync pulses, and these should have an amplitude of at least 3 V at periods between 0.3t and 0.8t (see Fig. 8-17). Another way to link the count rate to an external reference is to use an external time base. This signal may be applied to the *time-base output* terminal at pin no. 14.

Each Exar XR-2240 has its own internal voltage regulator circuit to hold the dc potentials applied to the binary counters at a level compatible with TTL logic. This consists of a series-pass transistor, which has its base held to a constant voltage by a zener diode. If operation below 4.5 Vdc is anticipated, it becomes necessary to strap the regulator output terminal at pin no. 15 to Vcc (V+) at pin no. 16. The regulator terminal can be used to source up to 10 mA to external circuitry or an additional XR-2240.

LONG-DURATION TIMERS

Long-duration timers can be built using any of several approaches. You could, for example, connect a unijunction transistor (UJT) in a relaxation oscillator configuration, or use a 555. In almost all cases, though, there seems to be an almost inevitable error created by temperature coefficient and inherent tolerance limits of the high-value resistors and capacitors required. Also, a certain amount of voltage drop is across the capacitor due to its own internal leakage resistances and the impedance of the circuit in which it is

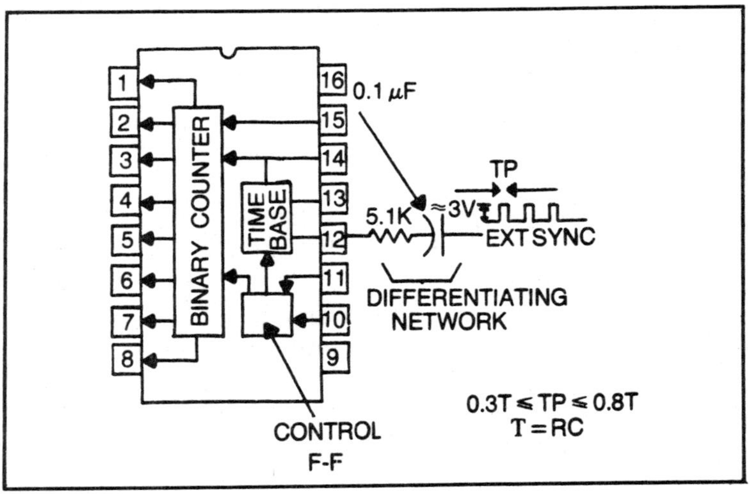

Fig. 8-17. External synchronization.

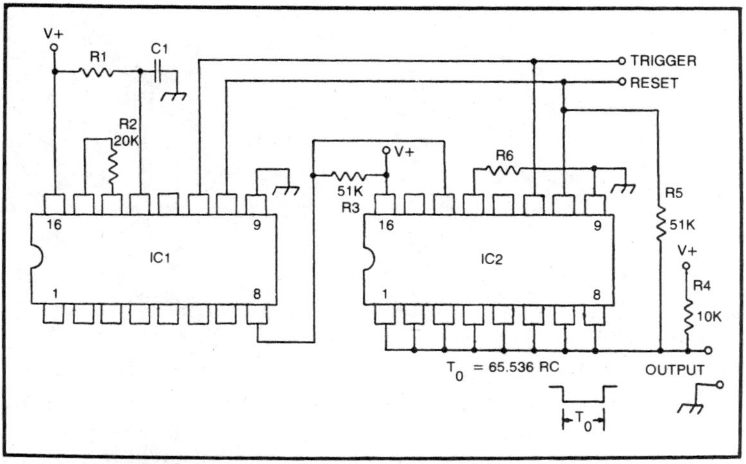

Fig. 8-18. Cascading to make a long-duration timer.

connected. The use of the XR-2240 IC timer will all but eliminate such problems because a higher clock frequency is allowed. Easier-to-tame component values will therefore satisfy the equation, $t = R \times C$. It is usually easier to specify and obtain high-quality, precision components in these lower values. This makes the period initially more accurate and results in less drift with changes in temperature.

An example of a long-duration timer is shown in Fig. 8-18; it consists of two XR-2240 IC timers cascaded to increase the time duration. In this circuit, the time-base output of IC2 (pin no. 14) is an input for an external time base. Timer IC1—also an XR-2240—is used as a time base for IC2. The most significant bit of IC1 (pin 8, weighted 128) is connected to pin no. 14 of IC2. This pin will remain LOW from the time IC1 is triggered until time $t_0 = 128R1C1$. It will then go HIGH and trigger IC2.

The binary counters in IC2 will increment once for every 128t. Timer IC1 is essentially operating in the astable mode because its reset pin is tied to the reset of IC2, and that point does not go HIGH until the programmed count for IC2 forces its output to go HIGH.

The total time duration for this circuit under the conditions shown (IC2 input from pin No. 8 of IC1, and with all IC2 outputs connected into the wired-OR configuration) is 256^2, or 65,536t. You can, however, manipulate three factors to custom program this circuit to your own needs: time-base period (R1C1), the output pin on IC1 used to trigger IC2, and the strapping configuration on IC2. In other words:

$$t_0 = R1C1 \times t_0' \times t_0''$$

Where:

t_0' is total period in seconds that the output is LOW.

t_0'' is the period that the selected output (s) of IC1 is LOW.

t_0 is the period that the selected output (s) of IC2 is LOW.

For example, assume that the product R1C1 is one second, as if R1 = 1 megohm and C1 = μF). If the output remains LOW for a period of 65536t = 65536 (R1C1) = t_0, it will remain LOW for 65536 seconds, or about 18 hours. This is on a one-second pulse! Of course, it is impossible to accurately generate a time delay this long using any of the other techniques. It is, for example, relatively common to find high-value electrolytic capacitors necessary in long-duration RC timers that are rated with a −20 percent to +100 percent tolerance in capacitance. This is incompatible with the goal of making a precision time base of long duration. In addition, most electrolytics—and that includes tantalum—will change value over time while in service. For most filtering and bypassing applications, this is acceptable; in timing, though, it is deadly. The use of cascaded XR-2240 devices allow us to select more easily managed values of resistance and capacitance.

PROGRAMMABLE TIMERS

By adding some external circuitry, the 2240, 2250, and 2260 timers can be made programmable; i.e., an operator can select the time duration at will. Figure 8-19 shows a circuit that allows timer control with a digital word (binary for the 2240 and BCD for the 2250 and 2260). In this circuit, the output terminals are connected to a pair of 7485 TTL magnitude comparators.

The 7485 device is composed of a set of TTL Exclusive-OR gates connected to compare two four-bit words and issue an output that indicates that word A is greater that B, A equals B, or that A is less that B. In this case, the output of the 2240/2250/2260 can be compared with the word applied by a microcomputer output port or an external binary or BCD thumbwheel switch. The timing diagram for Fig. 8-19 is shown in Fig. 8-20. Because word A programs the timer, the output durations are given in terms of the value of A, and the R1C1 time-base period.

When a trigger pulse is received, word A has previously been set to some value between 00000000_2 and 11111111_2. All this time, a word A is either greater than word B or equal to it if word A is also 00000000_2. In the former case, the A-greater-than-B output from

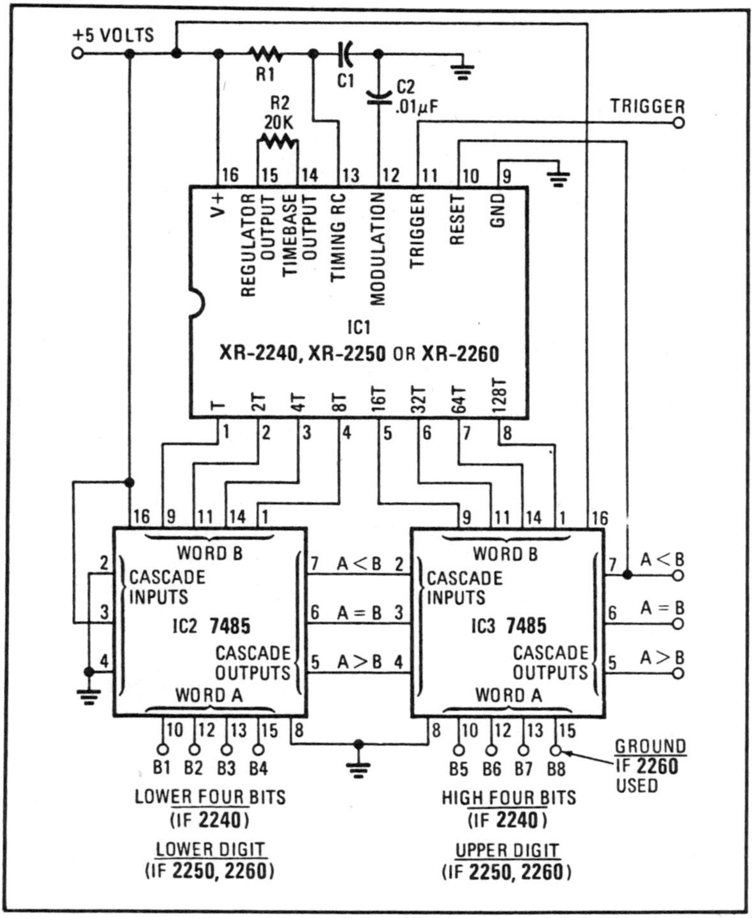

Fig. 8-19. Long-duration timer with preset count.

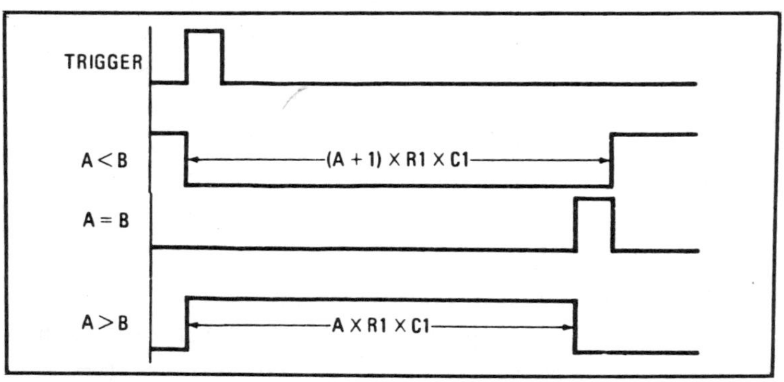

Fig. 8-20. Timing diagram of a long-duration timer.

IC3 is HIGH and the other two outputs are LOW. When the counter has incremented so that word A and word B are equal, then the A-equals-B output goes HIGH for a period of one clock pulse, after which word B is greater than word A, so the A-less-than-B output goes HIGH. The timing durations are (A + 1) R1C1 for A-less-than-B, and AR1C1 for A-greater-than-B. The A = B output produces a single pulse at time AR1C1.

Assume that the circuit shown in Fig. 8-19 is programmed so that the word A is 178_{10} (10110010_2y) and that the RC time constant is five seconds. What is the duration of the A-greater-than-B pulse? The duration is A × R1 × C1, or (178_{10}) × (5_{10} seconds), or 890 seconds (about 14.8 minutes).

Thumbwheel switches can be used with binary, BCD, or octal output encoding to program these IC timers. Figure 8-21 shows an XR-2250 timer connected to a BCD thumbwheel switch. The BCD output lines of the switch are connected to the timer output lines, and the switches are, in turn, connected to pull-up resistor R1. The switches provide a simple way to build circuits such as shown in Fig. 8-15 using convenient front panel switches to change the output duration as desired.

USING TRANSISTOR CRYSTAL OSCILLATORS IN TTL/CMOS CIRCUITS

Very few oscillators made from digital logic integrated circuits make accurate, precision-frequency clocks. Of course, RC timers are inherently less accurate and more subject to drift because of the nature of resistors and capacitors. Both elements, especially the resistors, tend to change value slightly as the temperature changes.

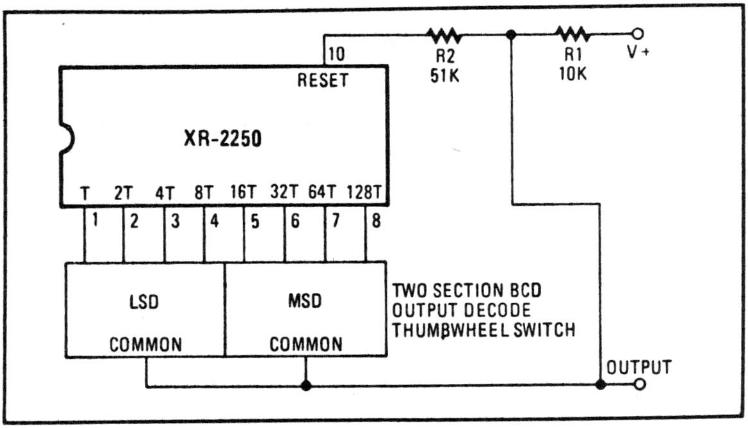

Fig. 8-21. Presettable one-shot.

Of course, a change in value will alter the frequency of the oscillator. This problem is of no importance in most cases because only an approximate frequency is important. If some application requires a precision clock circuit, however, it might behoove us to use a transistor oscillator circuit. Even when crystals are used with digital IC devices to make an oscillator, the result is often less than could be achieved with transistor elements. Not that transistor crystal oscillators are inherently more stable or more accurate than those made from digital ICs, mind you; it is that they can be *temperature compensated* more easily than most digital versions. Yet, while we need the precision of such an oscillator in some cases, we also need TTL or CMOS digital logic levels to drive logic elements.

Figure 8-22 shows two methods for using a transistor oscillator. Although there is a simple transistor crystal Colpitts oscillator shown in Fig. 8-22A, almost *any* oscillator could be used. In fact, if you are designing a circuit for an application where precision is needed, then it is likely that you will want to purchase a ready-built *temperature-compensated crystal oscillator* (TCXO) and interface it with digital logic elements.

Transistor Q1 in Fig. 8-22A is an NPN silicon unit that is selected so that it will oscillate at the frequency intended. For most cases, the unit should have a gain-bandwidth product (F_t) of 200 megahertz or more for oscillator frequencies up to 20 MHz. The frequency-determining element in Fig. 8-22A is a piezoelectric crystal cut for the frequency desired. The oscillating frequency of a crystal is partially dependent upon the circuit capacitance. By connecting a small series variable capacitor (C3) into the circuit we can provide a certain amount of control over the frequency. It is possible to change the frequency several kilohertz in some cases.

The circuit is identified as a Colpitts oscillator by the capacitor voltage divider network C1/C2, which is used to provide feedback. This network is semi-critical, but for most cases can use the values shown.

The interface between this oscillator and the digital devices to follow is accomplished by an LM-311 voltage comparator. The signal provided to the comparator is developed across emitter resistor R2 through capacitor C5.

A voltage comparator is basically an operational amplifier with no feedback resistor. This lack of feedback means that the gain seen by the inputs is the open-loop gain of the amplifier, which is at least 20,000 in junk op-amps and possibly over 1,000,000 in premium-grade devices. As a result, the output will be zero when the voltages applied to the inverting (−) and noninverting (+) inputs are equal. If the voltage applied to the inverting input is higher, however, then the amplifier thinks it is seeing a positive potential applied to the inverting input so the output will be negative. Because of the gain of the device, the output will saturate against the negative supply voltage rail when the input difference is more than a few millivolts. Similarly, when the voltage applied to the noninverting input is greater, then the output will be trying to go positive. Again, because of the immense gain of the amplifier, the output voltage will be hard against the positive supply rail when the input potentials are more than a few millivolts different.

The LM-311 contains an operational amplifier, but is specially configured for comparator service. When used in the monopolar mode (i.e., both pins 1 and 4 grounded) the output will be LOW when the two input potentials are the same, and HIGH when different so that the inverting input is higher than the noninverting. The output terminal of the LM-311 device is the so-called open-collector version. To make the LM-311 TTL-compatible we must connect a 2.2 k pull-up resistor between the output terminal and the +5 volt dc power supply. The rest of the LM-311 circuitry can operate from +5 volts also, but there is no reason why we cannot always make the oscillator work nicely at +5 volts. If we want, it is permissible to power the rest of the LM-311 from some positive potential higher than +5 volts, upscale R4 (in Fig. 8-22A) proportionally (i.e., approximately 5.6 kohms for +12 volts). If bipolar CMOS is used, then operate the LM-311 in its bipolar mode also.

The LM-311 device in Fig. 8-22A is connected so that the noninverting input is grounded, so it is at zero potential. The output will drop LOW every time the input voltage applied to the inverting input is either zero or negative, and will snap HIGH whenever the input voltage is positive. So, for half of each cycle produced by the oscillator the output of the LM-311 is HIGH, and for half of each cycle it is LOW. As a result, the output of the LM-311 is a square wave with the same frequency as the input sine wave. We have the stability and precision of the transistor oscillator or commercial TCXO, and the logic levels required by digital circuits.

A variation on the theme is shown in Fig. 8-22B. The interface device here is a *Schmitt trigger*. A Schmitt trigger is a circuit which will snap HIGH when a positive-going input signal crosses over one threshold point, and snaps LOW again when the signal then crosses over a lower threshold point. This operation is shown in Fig. 8-22C

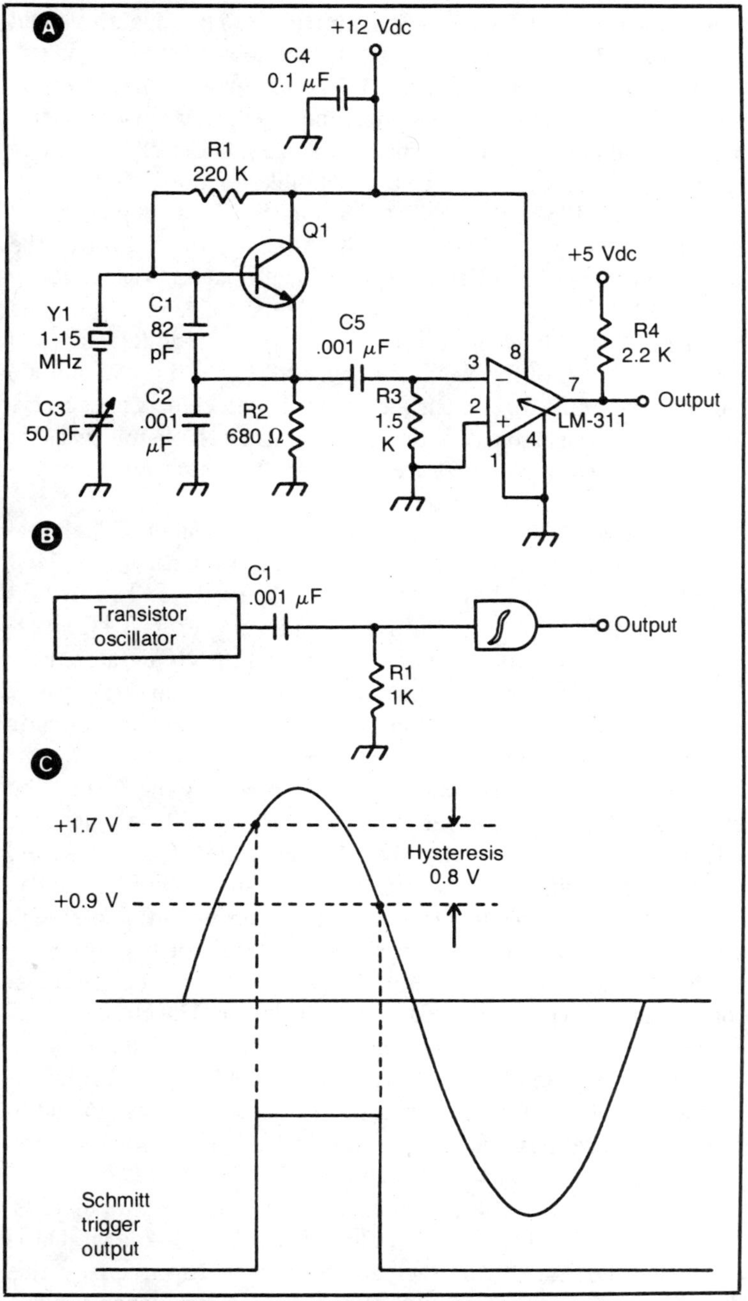

Fig. 8-22. (A) Transistor clock with TTL output, (B) Schmitt trigger output, (C) Schmitt trigger operation.

for the standard TTL 7414 hex Schmitt trigger. The upper threshold of the 7414 device is 1.7 volts, while the lower threshold is 0.9 volts; the hysteresis is therefore 1.7 – 0.9, or 0.8 volts. There is also a CMOS Schmitt trigger device, the 4093. That device has trip points that differ for different supply voltages, but are 2.9 volts and 2.3 volts for +5 volt (i.e., TTL-compatible levels) circuits.

The signal from the oscillator in Fig. 8-22B is coupled to the Schmitt trigger via an RC network. Although the values of the components may be different for certain frequency range, those shown in Fig. 8-22B are usable for most frequencies between 1 and 20 MHz. The voltage is developed across resistor R1, and will cause the output of the Schmitt trigger to toggle if the amplitude exceeds the limits for the particular device selected. If necessary, use another transistor amplifier state in order to ensure that this level is achieved. If both thresholds are not crossed, then the Schmitt trigger output will not toggle.

Figure 8-23 shows three basic clock circuits using TTL digital logic elements. Although two of these circuits appear to use 7400 NAND gate devices, they are connected as inverters. The circuits can, like the third of these circuits, also be built from TTL inverters. Keep in mind that both the NAND and the NOR gates (i.e., 7400, 7402, and CMOS equivalents) will operate as inverters *if both inputs are tied together*. Therefore, we will consider all three circuits of Fig. 8-23 to be constructed from inverter elements.

The circuit in Fig. 8-23A is an RC-timed ring oscillator. The frequency of oscillation is set by resistors R1 and R2, and capacitor C1. For most popular frequencies, i.e., those most frequently used as clocks in digital circuits, the capacitor will be 0.001 μF. The circuit shows a variable resistor as R1. The potentiometer will allow variation of the oscillation frequency, but is not strictly necessary when the frequency is fixed. We can, therefore, replace resistors R1 and R2 with a single fixed-value resistor. The circuit of Fig. 8-23A will oscillate (with the values shown) at frequencies between approximately 2 and 10 megahertz, assuming C1 = 0.001 μF. The frequency will *not* be stable because of the RC elements used to time the oscillations, but the circuit is still useful for some limited applications where such high stability/precision is wasted.

A pair of crystal-controlled oscillators are shown in Figs. 8-23B and 8-23C. The version in Fig. 8-23B is well-known, but is not always the best circuit to actually use in practical circuits. In this circuit, the active elements of the oscillator are two inverters made from 7400 NAND gate sections; a third NAND-gate inverter is used

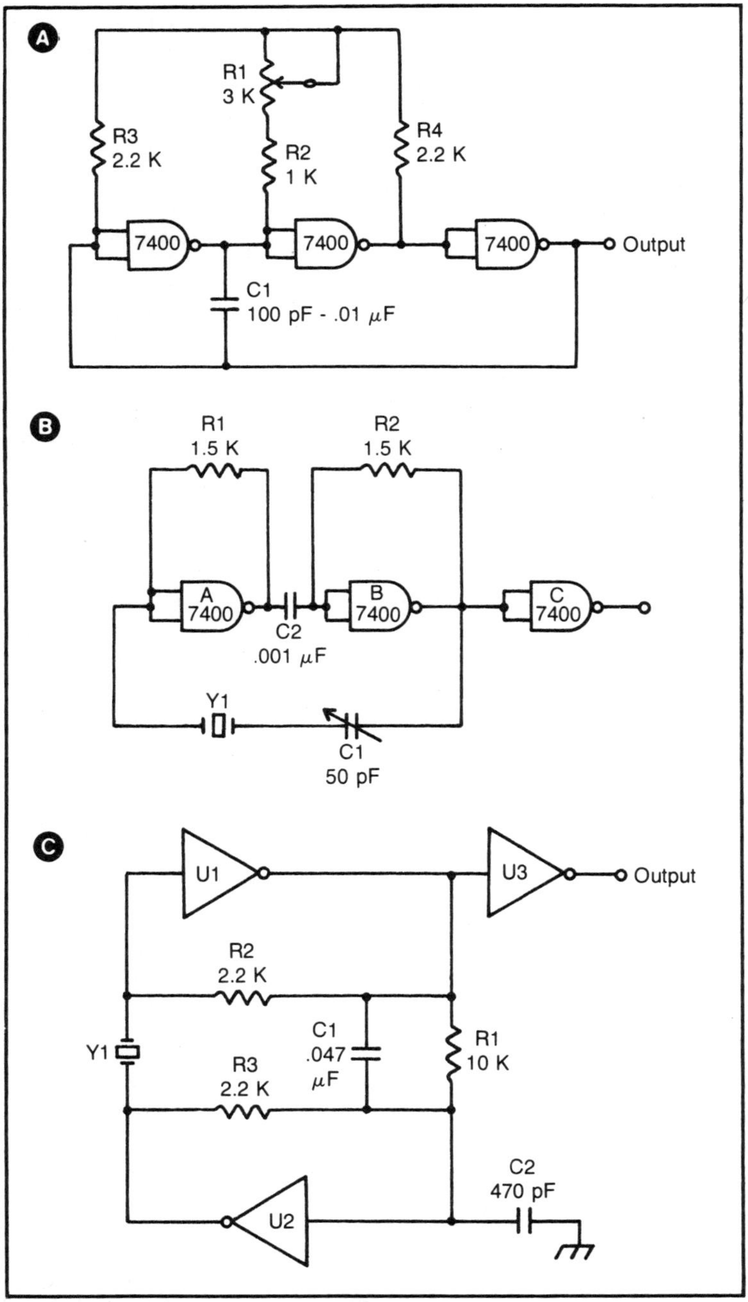

Fig. 8-23. (A) TTL RC clock, (B) TTL crystal clock, (C) TTL crystal clock with inverters.

as a buffer to isolate the oscillator from the outside world. The crystal element is in the feedback loop from the output of inverter "B" to the input of inverter "A." Since we have two inverters, each providing complementing (i.e., the binary equivalent of 180 degrees of phase shift), the feedback will be in-phase with the signal at the input of "A."

The frequency of oscillation is set by the interaction of crystal Y1 and variable capacitor C1. Since the frequency at which a crystal oscillates depends in part upon the circuit capacitance, varying C1 will also vary the frequency.

A second capacitor is used in the circuit to block the dc between the output of "A" and the input of "B." This isolation allows us to bias each inverter with fixed resistors, R1 and R2. The values of these resistors are not supposed to be critical, but experience shows that some amount of experimentation is, indeed, worthwhile.

The circuit in Fig. 8-23B is not trustworthy enough for practical consideration. It is popular, but tends to fail to start on occasion. I once had a kit-form microcomputer put together by a small outfit that went out of business a short while after its founding. The main system clock was one of these circuits, so on occasion the darn thing would fail to work. The outward appearance was a dead computer, but the root cause was that the main system clock was inoperative. A change in value for the feedback capacitor (C1 was fixed), and a small capacitance between the output of "B" and ground, all but cured the problem . . . but still occasionally the clock fails to start or turn on. A better circuit is shown in Fig. 8-23C.

Later I bought a more reputable microcomputer, this one a Rockwell AIM-65, which always starts when I turn it on. My computer is in my office, which is located in an alarmed backyard shed that is heated by electric space heaters. During the winters, that shed is cold for the first half-hour or so; in fact, it is freezing! Yet the little AIM-65 keeps on humming along nicely, and starts every time I throw power to 'er. The clock in that computer is shown in Fig. 8-23C. This circuit uses inverters, but could as easily be built from inverter connected NAND (7400) or NOR (7402) gates. Crystal *Y*1 is a 4 MHz unit in the computer, but the circuit should work well in the range 1 to 10 MHz.

MOTOROLA MC4024P

Perhaps the best solution for TTL-compatible signal sources is the Motorola MC4024P device. The IC is not one of the standard

"74xx" TTL chips, but operates from +5 volts dc and has TTL-controlled oscillator in one chip. There are two independent oscillators in one package, each with separate +5 volt and ground terminals. Note that there are also *package* power and ground terminals that also must be used whenever the MC4024P is operating. Pinouts are given below:

Function	OSC A	OSC B
Capacitor or crystal	10 & 11	3 & 4
Output	8	6
Control voltage	12	2
+5 volts	13	1
Ground	9	5
Package +5 volts: 14		
Package Ground: 7		

It is possible to shift the oscillating frequency of the MC4024P over a range of approximately 3:1 by changing the control voltage from +3 to +5 volts. If this control voltage is an ac or varying dc signal, then the output frequency will be frequency modulated by the signal. One application for this type of "FM" was once used in an analog modem that converted a ±1-volt human electrocardiograph signal into a varying tone (FM). An operational amplifier with a gain of −1 was connected to the control input, and an offset potentiometer connected to the summing junction so that the "normal" output potential with no input signal was +4 volts—right in the middle of the MC4024P control voltage range. The ECG signal would then vary ±1-volt for a total range of +3 to +5 volts, and a frequency-change of 3:1. It is important, by the way, that you not confuse the MC4024P device with a CMOS device with a similar number. The CMOS 4024 is a completely different IC and bears no relation to the Motorola MC4024P.

Figure 8-24A shows one popular circuit. Here we are only using one oscillator (OSC B), but the pin numbers can be changed should you want to use OSC A. The package +5 volt terminal, the OSC B +5-volt terminal and the control voltage terminal are all connected together at +5 volts; no frequency control is possible with this circuit. The package ground and OSC B ground are both connected to ground, so the device will operate. It is possible to use the OSC A or OSC B grounds to control the on/off operation of each oscillator, provided that the package ground is kept at zero volts

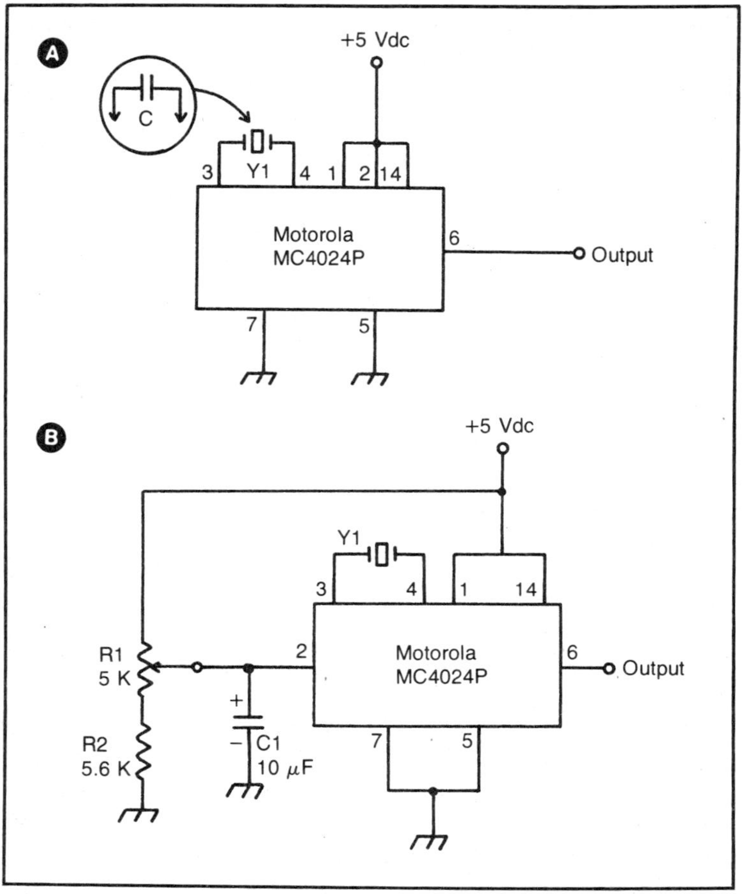

Fig. 8-24. (A) MC4024P TTL clock, (B) MC4024P TTL clock VCO control voltage.

potential. We can, therefore, lift pin no. 5 off ground in order to turn off this oscillator.

Frequency of oscillation is set by either a capacitor (see inset) or a piezoelectric crystal, Y1. If the crystal is used, it is necessary that the frequency be at least 2.5 MHz and less than 25 MHz in the fundamental mode. Crystals of lower frequency do not always oscillate in this circuit. If the capacitor is used, then the frequency is set very roughly according to the formula:

$$F_{Hz} = 300/C_{\mu F}$$

It will be necessary to experiment with the capacitor value in the circuit. Using a frequency counter and a handful of capacitors on my breadboard, I found that the formula given in another text on this same subject was in error by a factor of approximately ten. Even so, this formula will not yield any better than "ballpark" results, and some tweeking will be in order if the exact frequency is important to you.

The recommended circuit is shown in Fig. 8-24B. Here we have everything exactly the same as before, except that the control voltage applied to pin no. 2 is variable, rather than fixed. This circuit tends to work better with low frequency crystals, and is capable of providing a rather large degree of control. Note that pin no. 2 is bypassed to ground. Some people also like to use a 0.1 μF capacitor in parallel with the 10 μF electrolytic unit because the electrolytic capacitor is not able to work at high frequencies.

CMOS CLOCK CIRCUITS

In the previous sections of this chapter we have been discussing TTL compatible oscillator/clock circuits. Let's digress a bit and consider some of the CMOS versions. The circuit shown in Fig. 8-25A is a square-wave oscillator based on the 4093 Quad CMOS NAND-Gate with Schmitt trigger inputs. This IC will operate as a normal NAND gate in that a LOW on either input will force the output HIGH, and a HIGH on both inputs will force the output LOW. The difference between the 4093 NAND gates and ordinary NAND gates is that the inputs operate as Schmitt triggers. For +5 volt supplies, the positive-going trip-point is approximately +2.9 volts, while the negative-going trip-point is about +2.3 volts.

The oscillator can be turned on and off by manipulating one of the inputs. We have one input connected to V+ through a pull-up resistor (i.e., 10 kohms to 100 kohms). If we ground that input by closing switch S1, then the oscillator will stop; if, however, we leave S1 open thereby keeping the input HIGH, the oscillations continue. For applications where this control is not desired, simply strap the input HIGH permanently through the pull-up resistor.

The operating frequency of this circuit is set by the combination of R1 and C1. The operating frequency is given very approximately by the following expression:

$$F = 2.7/R1C1$$

Where:

F is the frequency in hertz.

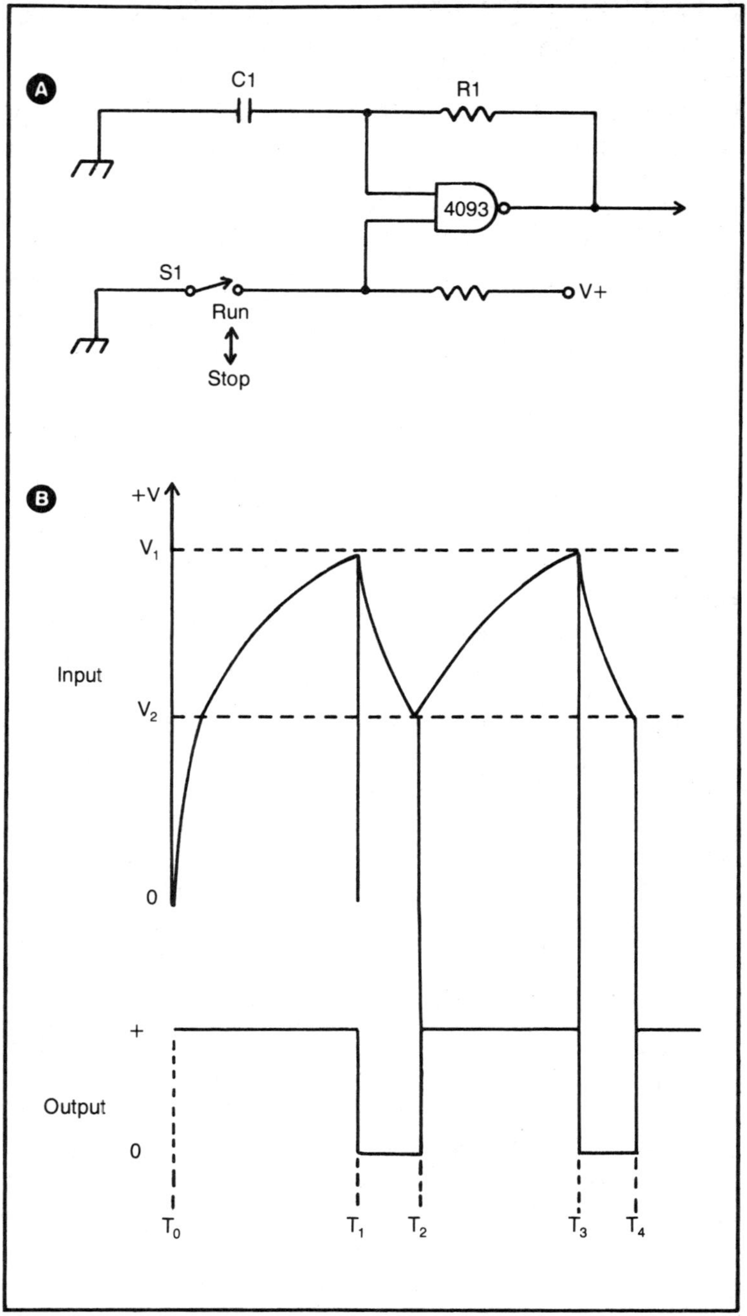

Fig. 8-25. (A) RC CMOS clock, (B) Timing waveform.

R1 is the resistance in ohms.

C1 is the capacitance in farads.

Constraints: V+ = 5-volts dc, V− = ground, C1 greater than or equal to 0.001 μF.

The timing waveform is shown in Fig. 8-25B. When the circuit is first turned on the voltage across capacitor C1, i.e., the voltage at the Schmitt trigger input, will be zero. This means the input is LOW, so the output will be HIGH. A HIGH on the Schmitt trigger output means that C1 will charge through resistor R1 at a rate determined by the Schmitt trigger output potential and the RC time constant. For practical reasons, the oscillation will not start instantly, but requires a delay of a few cycles. Eventually, though, the capacitor voltage climbs to the positive-going threshold, V1 (at time t_1). When this occurs, the output will snap LOW, allowing the capacitor to discharge through R1. When the capacitor voltage discharges down to the negative-going threshold, V2, which occurs at time t_2, then the output of the Schmitt trigger will snap HIGH again. This alternation occurs periodically from then on, and we have a square-wave oscillator with an unequal duty cycle.

Three other RC-timed CMOS oscillator/clock circuits are shown in Figs. 8-26A through 8-26C. The simplest version is shown in Fig. 8-26A, and consists of a pair of 4049A inverters and an RC timing network. Note that the 4049A device will drive TTL-compatible loads when the package voltage is +5 volts dc, even though the fan-out is limited to two.

The preferred CMOS RC oscillator is shown in Fig. 8-26B. This circuit uses a second resistor, and this permits the duty cycle to be nearer 50 percent. In most cases, we will want to make resistor R2, which is in series with one input of the 4001A device, about ten times the resistance of R1. If capacitor C1 is greater than or equal to 0.001 μF, then the frequency of oscillation is approximately:

$$F = 1/2.2\ R1C1$$

The 4001A NOR gates are connected in an inverter configuration. The NOR gate will act as an inverter under two circumstances: (a) when both inputs are tied together, and (b) when one input is permanently grounded. We can, therefore, make this circuit a gated oscillator by lifting one or both grounded 4001A inputs (G1 and G2). If the input is grounded, then the circuit will oscillate, but if it is held HIGH, then it will stop running.

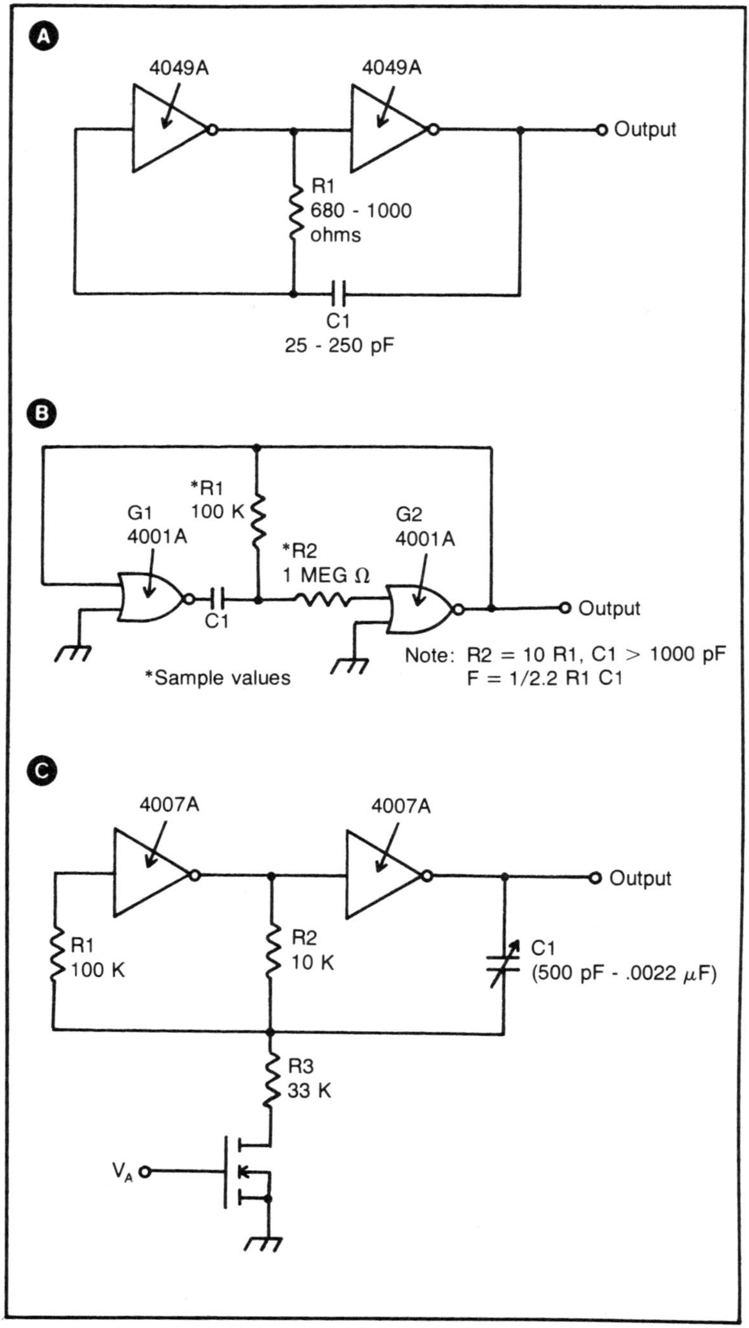

Fig. 8-26. (A) RC 4049 oscillator, (B) 4001 RC oscillator, (C) VCO using 4007.

A voltage-controlled pulse-width version of this same circuit is shown in Fig. 8-26C. Here we add two additional components, a resistor (R3) and a MOSFET transistor. The transistor can be one section of 4007 device. Voltage V_A applied to the gate of the transistor will vary its channel resistance, hence will affect the charge-discharge cycle of the RC network. We therefore gain pulse-width control as a function of voltage V_A.

The RC oscillators shown thus far will operate to frequencies of several megahertz, if you are lucky, but "under 1 MHz" is more like realistic, especially with the Schmitt trigger version. Selected devices will allow operation at frequencies greater than 1 MHz, but don't count on it for run of the mill CMOS devices.

Our last circuit is the crystal controlled oscillator of Fig. 8-27. This circuit will operate to frequencies of 2 MHz, or so, with good precision and reliable operation. It uses a pair of 4001A NOR gates connected as inverters. As in the previous circuits, these devices can be replaced with either inverter-connected NAND gates or actual inverters. A variable capacitor allows some control over operating frequency. At some settings of C2, however, starting might tend to be a little flakey, so you will have to set the capacitance to a value that will ensure good starting, 100 percent of the time.

OTHER CLOCKS

We have discussed a number of approaches to providing TTL

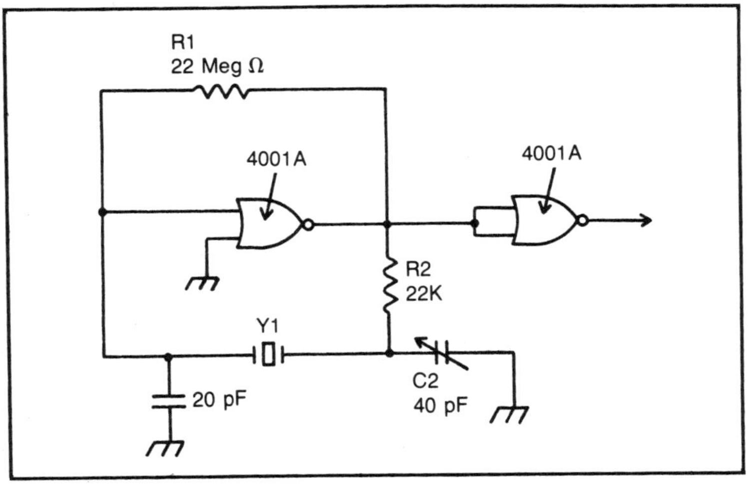

Fig. 8-27. 4001 crystal oscillator.

and CMOS clock signals with an assortment of oscillators and astable multivibrators. There are other circuits and devices as well, but time and space simply do not permit covering all of them. I recommend, therefore, that you examine the most recent data books for devices most suited for your own needs. If you are designing a microcomputer or anything around one of the popular microprocessor chips, be aware that most semiconductor makers in that business offer special purpose clock chips that are especially suited for that particular application.

Chapter 9

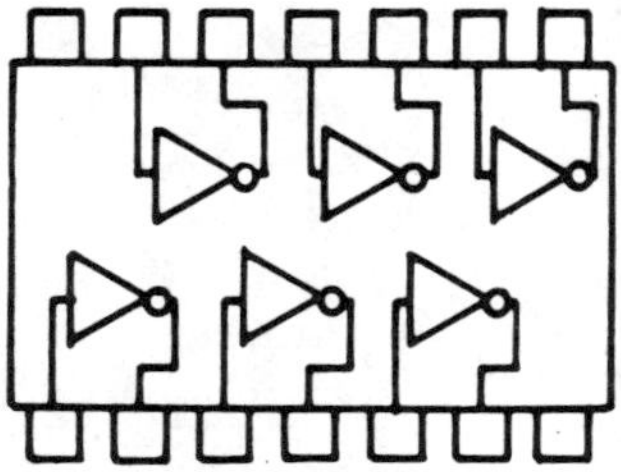

Digital Displays and Display Drivers

A digital display is actually a decade or hexadecimal display in most cases, and is used to indicate a value to the "outside world." The binary arithmetic/logic of digital circuits is incomprehensible to most users (although there are those who actually think in base-2, and I know one chap who keeps his checkbook in octal!) who prefer base-10 displays. Computerniks, on the other hand, prefer hexadecimal displays (base-16) because they typically have to express binary values or address locations in a "shorthand" form. Of course, it is a lot easier to write "8A" in hexadecimal than "10001010" in binary!

There have been a lot of different displays over the years. Assorted technologies have also been used: incandescent lamps, neon lamps, gas plasma, flourescent, light-emitting diodes and liquid crystals. There have also been several different display formats.

Figure 9-1 shows a crude display method that was current as recent as the late 1960s. In fact, the August, 1969 edition of *QST* magazine (an amateur radio journal) featured a frequency counter that used this type of display (I know, because I have the printing plate used for the cover of that issue, and the counter was on the front cover!). An indication of how electronics has changed since those days is seen in the price of the main counter chip used in each stage of that project. The SN7490 counter chip cost $14.00 each in those days, and several were needed. Today, that chip can be bought for around $0.75 each and is obsolete. In fact, if one is

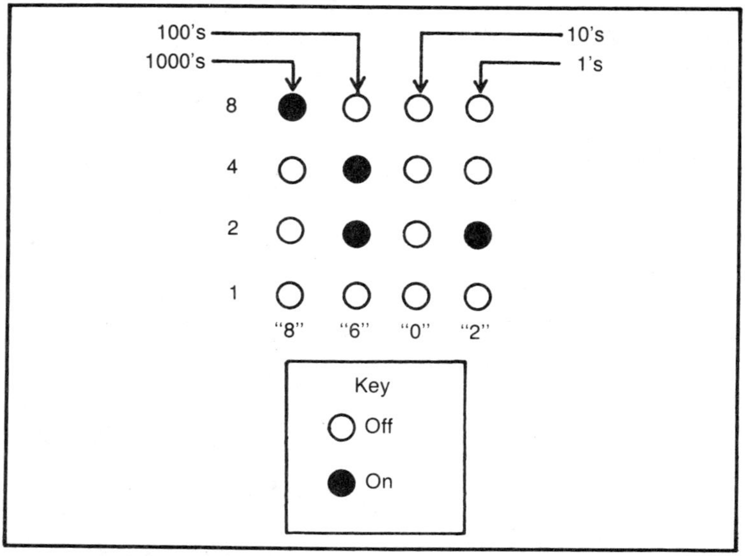

Fig. 9-1. Obsolete weighted display.

seriously considering building a frequency and/or period counter it might be wiser to use a special LSI frequency/period counter chip such as those marketed by *Intersil.*

The display in Fig. 9-1 uses a column of incandescent lamps, one column of four lamps for each digit. In this example, we have four columns so up to four digits can be displayed: 1000's, 100's, 10's and 1's.

The display shown here depends upon the fact that each binary digit ("bit") of the four-bit word that forms the *binary coded decimal* (BCD) number system is weighted according to position. The following protocol is used:

$$a_3 2^3 + a_2 2^2 + a_1 2^1 + a_0 2^0$$
$$a_3 8 + a_2 {}^4 + a_1 2 + a_0 1$$

Where:

a_0 through a_4 are either 1 or 0. The coefficients of each "a" term, then, give the position weights appropriate to a base-2 number. The coding scheme for BCD presentation of decimal digits, therefore, is:

D C B A
8 4 2 1

In Fig. 9-1, we have the "8" weighted lamp at the top of the column, and the "1" weighted lamp at the bottom. If a specific lamp is turned "on," then the "a" term is 1, and if off the "a" term is 0. Hence, we have to work the little arithmetic expression given above in order to find the correct value of the decimal digit being displayed. In the example shown in Fig. 9-1 we are displaying the base-10 number "8602." This means that the "8" lamp in the *most significant digit* (MSD) column must be lit, the "4" and "2" lamps in the MSD-1 column (4 + 2 = 6, you see), none in the MSD-2 column and the "2" lamp is lit in the least significant digit column.

The type of display shown in Fig. 9-1 was once popular, but only because it was the only way to do the job cheaply. It seems that most flip-flop circuits used in counters in those days were difficult to decode, and the decoders that were possible in some circuits were terribly expensive. I can recall a decoder used in a pioneer Hewlett-Packard counter that used photoresistive strips and neon lamps in a complicated decoding scheme. Except for the per digit cost of decoded BCD counters, all would have used a different display method than Fig. 9-1—the columns and work needed to understand what they meant was a bit cumbersome.

An advance came when some counters appeared on the market with decimally decoded display lamps. Figure 9-2 shows the display format that became popular by the mid-sixties and continued in use even after certain display devices became available. In this display, each column represents a decimal digit, but there are then separate lamps each of which uniquely represents one of the decimal digits 0, 1, 2, 3, 4, 5, 6, 7, 8, and 9. In the example shown in Fig. 9-2, we are displaying the decimal number "64," so the "6" lamp in the MSD column is lit while the "4" lamp in the LSD column is lit.

Even though the decimal display was an improvement over earlier four-bit methods (e.g., Fig. 9-1), it was still clumsy because the digits were not in-line when read unless all were the same (i.e., 555555555). It still took a moment or two to read the value and record it. What followed, however, was a series of different display devices which were a lot more human-oriented.

Perhaps the earliest "human engineered" decimal display device was the Burroughs *Nixie*® tube (*Nixie* is a registered trademark of Burroughs). This tube used ionized neon gas to make the light seen by the user, and operated in much the same manner as neon lamps (NE-2, etc.). Inside the glass envelope of the *Nixie*® tube were ten numerals each formed separately of wire filaments. The wire-filament numerals were fashioned such that they face the same

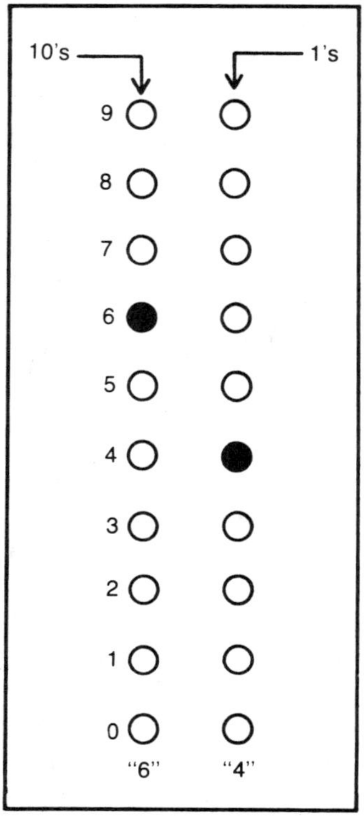

Fig. 9-2. Obsolete decade display.

direction. A wire from each wire numeral to an outside pin was used to turn on that numeral. The other pole of the neon "lamp" was connected to a positive voltage, often as much as +170 volts dc. Since dc is used, only one pole of the "neon lamp" formed by the anode and the numeral cathode would light up—the numeral would glow with the dull orange of ionized neon.

The voltage levels used in *Nixie®* tubes were appropriate to some of the early decoders (such as that neon lamp/photoresistor device), but was incredibly high as far as TTL levels were concerned. As a result, it was rare that TTL devices would directly drive *Nixie®* tubes, despite the fact that one or two decoders would tolerate up to 60 volts. The usual procedure, shown in Fig. 9-3, was to use a transistor with a high collector voltage rating to switch the *Nixie®* tube on and off, and a TTL decoder output to drive the base of the transistor. The usual decoder has active-LOW 1-of-10 outputs, so PNP lamp driver transistors were often used. When the output ("5" shown here) would go LOW, the base of Q1 will be at a lower potential than the emitter, so Q1 will turn on hard. This action places the "5" cathode at ground potential, thereby turning on the "5" digit.

One problem associated with the *Nixie*® tube is the tremendously high voltages needed to operate the display. This requirement means that a separate power supply is needed, and also increases the probability of damage to TTL devices in the event of faults in the circuit or service-connected foul-ups. The solution to this problem was to find some display method that used a lower voltage element. Hence, the seven-segment and 5×7 dot matrix displays were born.

Perhaps the most familiar digital display is the seven-segment readout shown in Fig. 9-4A. This display has been immortalized in millions of electronic hand-held calculators, as well as many millions of electronic instruments of other sorts. Several different technologies have been used to make seven-segment readout/display devices. The first were made with incandescent filaments in RCA *Numitron* displays. Two sizes were originally available, one

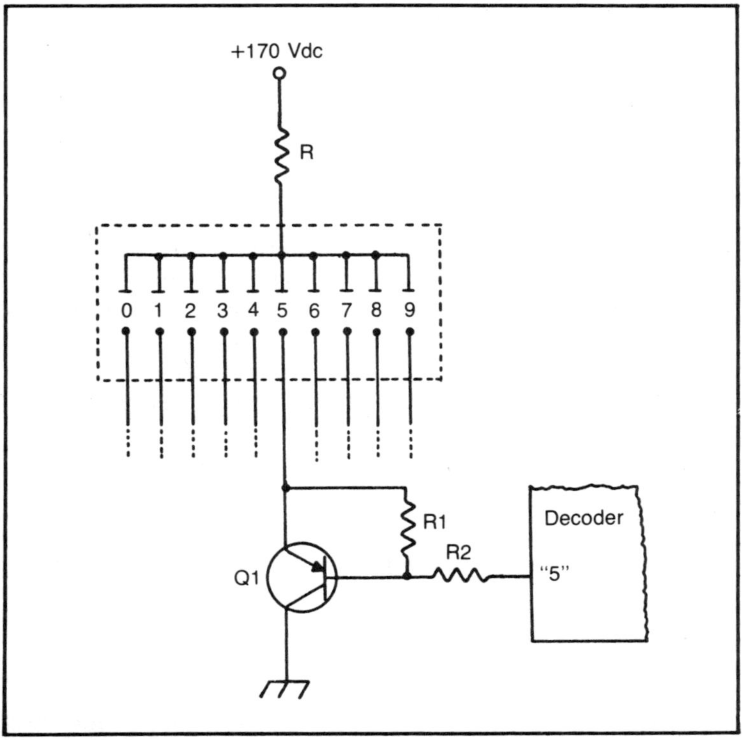

Fig. 9-3. *Nixie*® tube display.

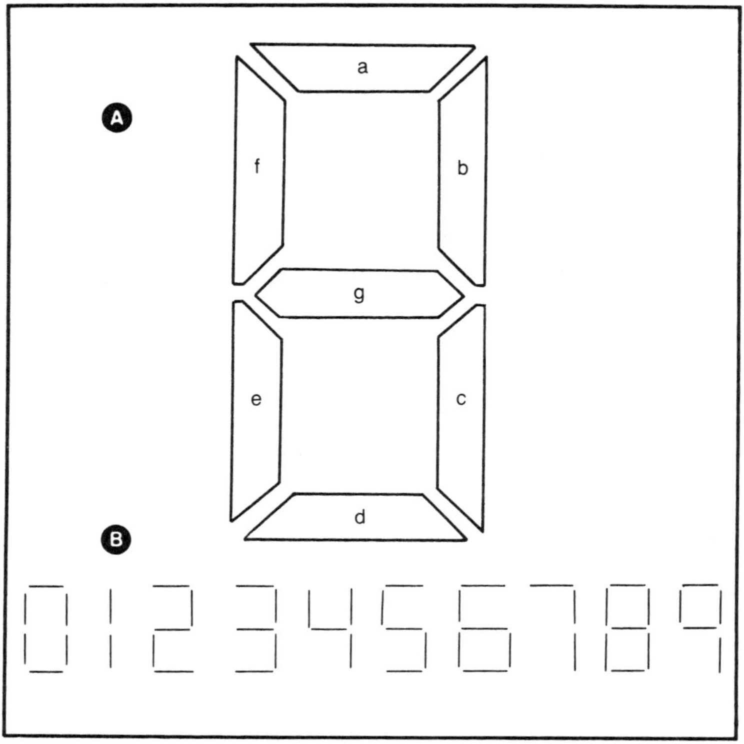

Fig. 9-4. (A) Seven-segment display, (B) seven-segment display for digits 0-9.

that was housed in the seven-pin vacuum-tube envelope while the larger was housed in the nine-pin vacuum-tube envelope. Later versions, and perhaps the most popular until recently, used light-emitting diodes (LED) for each bar in the seven-segment readout. When the LED was turned on, the bar segment was lit. Overtaking the LED readout today is the liquid-crystal display (an example of which is found on LCD wristwatches). Both the LED and liquid-crystal devices, however, suffer from an inability to be seen under certain lighting conditions. The LCD, for example, fades out fast as the light intensity goes down, while the LED fades out in bright light (try and read a calculator output display in sunlight). In certain applications where seeing the digit was critical, some other means had to be provided. A partial solution was blue-green fluorescent displays, while another was orange-red gas plasma displays such as the *Panaplex.*

The seven-segment display device shown in Fig. 9-4A gives the standard identification labels (a through g) normally used in schematic diagrams. The segments that are lit for the ten digits of the decimal, base-10, numbers systems are shown in Fig. 9-4B. Note that the seven-segment readout can only display these ten digits in a comfortable way, but are often used to also display the digits of the hexadecimal, base-16, numbers system. These digits are 0, 1, 2, 3, 4, 5, 6, 7, 8, 9, A, B, C, D, E, and F. While "A," "E," and "F" can be displayed with some accuracy, "B" and "D" are always flaky!

Among LED versions of the seven-segment readout are two different types: *common-anode* and *common-cathode*. In each case, the name tells us whether the anodes or cathodes of the LEDs forming the segments (a-g) are connected together. In the case of a common-anode display, the anode will be connected to V+, requiring each segment to be grounded through a current-sinking circuit in order to turn on. The common-cathode is exactly the opposite: the decoder will supply current (i.e., will source current) to the LED.

Each segment of the LED seven-segment readout requires from 10 to 25 milliamperes when turned on. If the digit displayed is "8," then the total current required will be 80 to 200 milliamperes. Adding more digits will only exacerbate the problem. For example, a digital frequency counter with 12 digits will require from 960 to 2400 milliamperes of current. Obviously, the display current is significant when LED readouts are used. Of course, these requirements are less than the 40 milliamperes required for incandescent filament sections.

Figure 9-5 shows the 5×7 dot matrix type of digital display. Each dot in the matrix is a lamp which can be turned on or off according to the needs of the decoded digit. This type of display is the old-fashioned athletic scoreboard display. In the scoreboard, of course, the lamps are 100-watt incandescent light bulbs not small, low-power LEDs.

There are several forms of 5×7 dot matrix displays that are not discrete LED displays. For example, the printhead used in a lot of computer peripherals use either thermal or impact printing methods based on a 5×7 dot matrix. Another example is in TV lettering used on computer CRT displays. Each light point is a spot where the raster-scan is unblanked for a brief instant. Several manufacturers offer 5×7 dot matrix displays with as many as twenty digits in a row.

DISPLAY DECODERS

The digital display is useless unless there is some means of

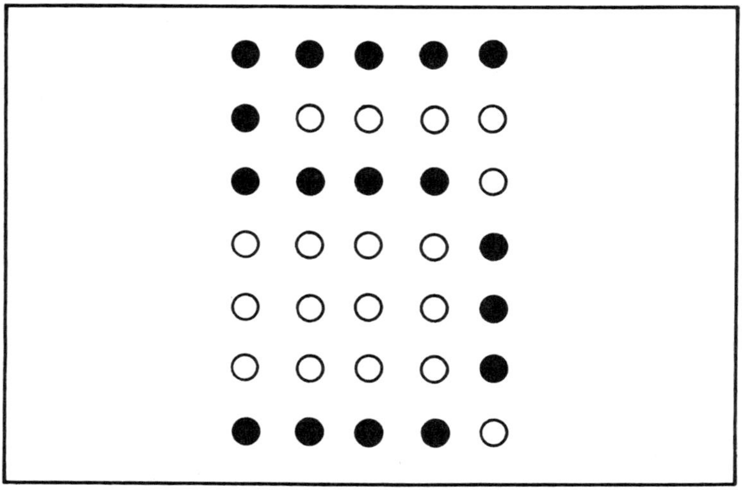

Fig. 9-5. 5×7 dot matrix numerical display.

decoding the four-bit binary coded decimal word into a code compatible with the seven-segment readout. The 7447 is one such display decoder. A model equivalent circuit for a seven-segment decoder is shown in Fig. 9-6. Here we show the display device in its equivalent form with each light-emitting diode shown individually. The letters *a* through *g* correspond to the *a* through *g* segments of the LED display.

A resistor must be connected in series with each segment in order to limit the current flowing in the LED. Otherwise, the display may burn out. The value of the resistor is selected to keep the current within the specifications for the display—15 milliamperes is a common figure. In the case of a +5 volt dc power supply, the 330 ohm value shown in Fig. 9-6 is sufficient. Scale the resistor value upward proportionally for higher voltages. The idea is to maintain the 15 milliampere current flow. In other words: R = (V+)/0.015.

The decoder can be modeled as a bank of switches that are turned on and off in accordance with BCD data to produce the seven-segment code. In the case shown as an example, the decimal digit "4" is being displayed, so we will want the following conditions on segments *a* through *g*:

a OFF
b ON
c ON

d OFF
e OFF
f ON
g ON

Accordingly, the following outputs of the decoder are LOW (i.e., the switch closed): *b, c, f,* and *g.* This condition places the cathodes of the respective segments at ground potential, so the segment turns on.

Perhaps the most popular seven-segment decoder over the past two decades has been the 7447 device, shown with an appro-

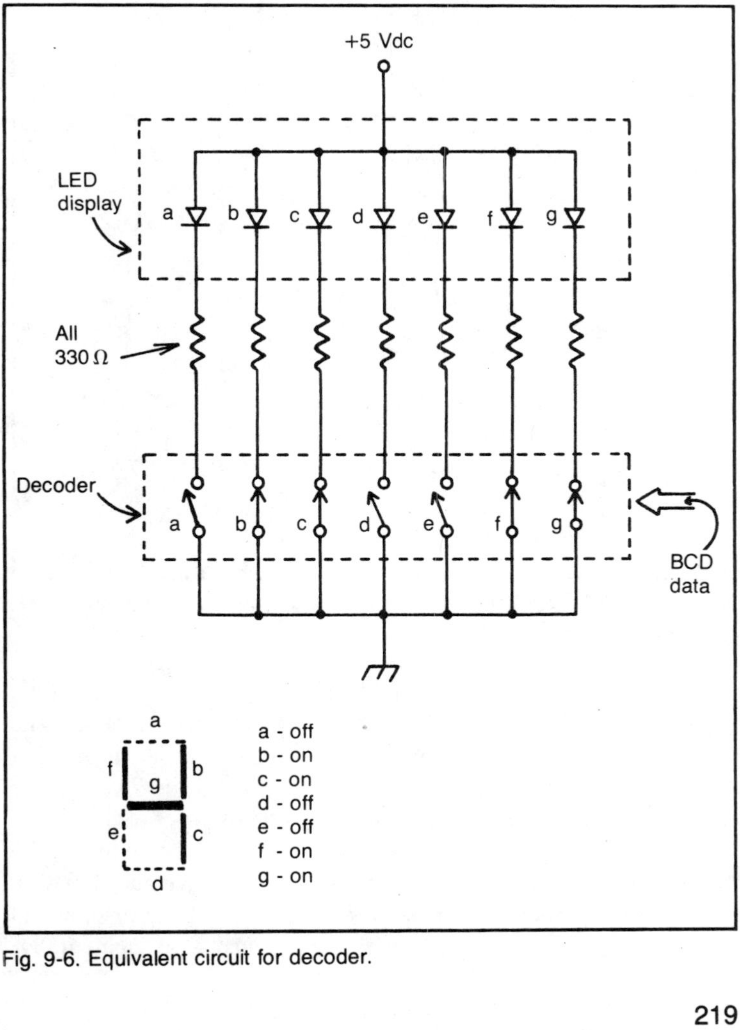

Fig. 9-6. Equivalent circuit for decoder.

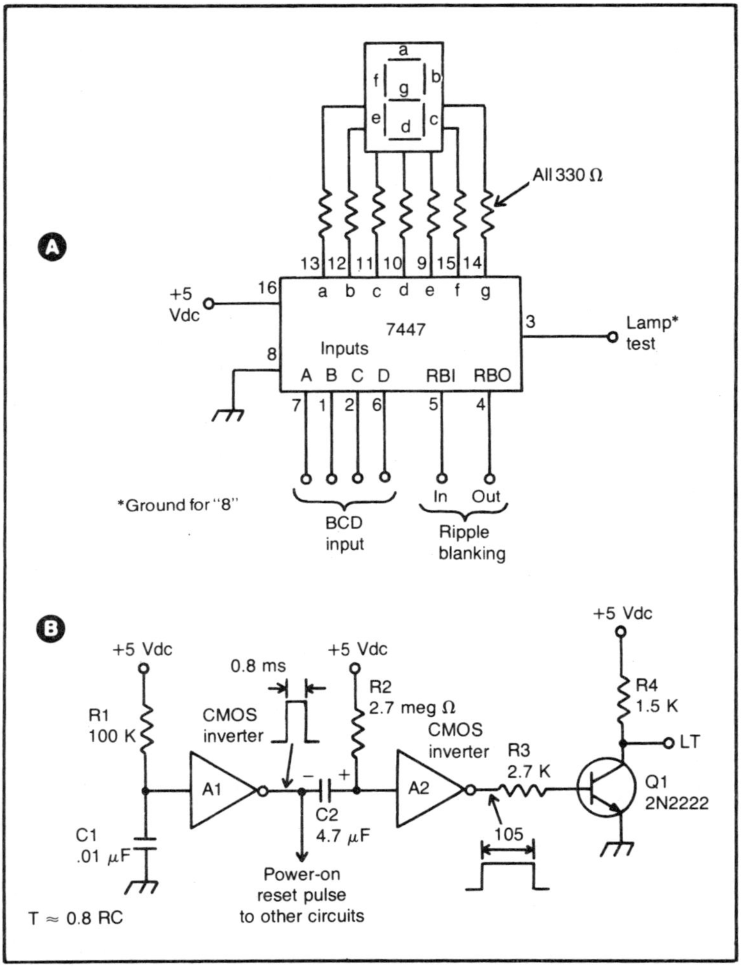

Fig. 9-7. (A) 7447 seven-segment decoder with LED display, (B) power-on reset lamp test.

priate circuit in Fig. 9-7. The official name for this device is *BCD-to-seven-segment-decoder*, which implies that it will examine the four-bit BCD input and issue a seven-segment code output that corresponds to the selected digit. The outputs of the 7447 device are active-LOW, so are compatible with common-anode LED displays as shown in Fig. 9-7. The line from each segment to the corresponding 7447 output, by the way, must have the 330-ohm current-limiting resistor, or damage to both the LED display and 7447 may result. There are some display decoders, incidentally, which have a constant current source (15 mA) that eliminates the need for the resistors. These are especially nice when a large number of digits are used. After all, with seven resistors per digit a twelve digit display would require 7 × 12, or 84 330-ohm, ¼-watt resistors!

The outputs are labeled as shown in the figure and below:

Pin	No.
a	13
b	12
c	11
d	10
e	9
f	15
g	14

The ABCD inputs (pin nos. 7,1,2, and 6, respectively) are weighted in the usual A = 1, B = 2, C = 4 and D = 8 system. The decimal equivalent of the code applied to these inputs will be converted to active-LOW seven-segment code at the outputs. In actual fact, I suppose, this means the 7447 is not a decoder but a recorder!

The *lamp test* terminal (pin no. 3) is used to turn on all segments of the LED seven-segment display. One problem with those displays is the fact that a burned out segment is difficult to detect unless the correct digit is known to be present—but isn't, or until some ridiculous display is seen. If the "g" segment is burned out, then an "8" will appear to be "0." With the lamp test function, we can design into a digital instrument a means for testing each LED of the seven-segment display. This feature can be activated voluntarily, with a switch, or automatically. Many digital instruments display all "8's" for a few seconds every time the power is applied. How would you like your digital blood pressure machine to read "110" instead of "190" because segments *a, f* and *g* are burned out; don't laugh and say that's unlikely because I have seen it on a piece of hospital equipment I was servicing!

In normal non-test operation, the lamp test terminal on the decoder is held HIGH (i.e., +5 volts), but is brought LOW when a test is being performed. Making pin no. 3 LOW causes all 7447 outputs to also go LOW, which is what turns on numeral "8" and allows checking the digit.

Figure 9-7B shows a method for making an automatic test circuit that will turn on the lamp test ("LT") terminal for ten seconds. The drive capacity of this circuit is sufficient to allow up to fifteen LT terminals (i.e., on a fifteen digit counter). The circuit of Fig. 9-7B uses a pair of half-monostable multivibrators made from CMOS inverters such as 4010 or 4049. Inverter A1 is connected as a *power-on reset* pulse generator (as used in most microcomputers). At power turn-on, the voltage across capacitor C1 is zero, so the input of inverter A1 is at ground potential (i.e., LOW). Since this is an inverter, the output of A1 will be HIGH at this time. As the capacitor charges, however, the voltage at the input of A1 will also rise. When this potential reaches a level of +2.5 volts (i.e., one-half V+), then the output drops LOW. The purpose of this circuit is to generate a pulse of approximately 1 millisecond duration at power turn-on. This pulse is used to reset any flip-flops, counters, or other synchronous devices in the circuit. As mentioned before, this method is often used in microcomputers to force the computer to reset (a *reset* pulse in a computer is essentially a hardware "JUMP to location 0000_{16}" instruction, so will initialize the operation of the microcomputer). If there are no other circuits which require a reset pulse at power-on, then eliminate stage A1 and use only A2. In that case, the negative end of capacitor C2 is grounded.

The stage that actually drives the lamp test is A2. This stage is a 10-second half-monostable multivibrator, so its output line will go HIGH for approximately 10 seconds. During this period, the base of transistor Q1 will be forward biased, so transistor Q1 will be saturated. In that condition, its collector—which delivers the drive signal to the lamp test line—is LOW so the 7447 devices will all display eight. When the output of A2 is LOW, on the other hand, the base-emitter junction of Q1 is unbiased so the transistor is turned off. The collector of Q1 is HIGH (+5 volts dc), so the lamp test line is also HIGH.

The *ripple blanking input* (RBI) and *ripple blanking output* (RBO) terminals on the 7447 device are used to blank, i.e., turn off, leading zeros. If we did not turn off unneeded leading zeros, we would not only waste current (important in portable instruments), but also create a situation that is difficult to read. For example, if we had an eight-digit display that was displaying the number "347," then the non-blanked display would read "00000347." If we turn off those unneeded leading zeros, then the display would read "-----3 4 7," where " " indicates a dark digit. The *ripple blanking input* (RBI) terminal controls the operation of the 7447 in this respect. If

RBI is left HIGH, then the display will recognize zeros. But, if the RBI is made LOW, then the 7447 will extinquish the display in the case of a zero being displayed (in other words, if the ABCD input is 0000, then all seven-segment output lines are HIGH).

The *ripple blanking output* (RBO) tells the next stage whether or not to extinquish the leading zero. The RBO terminal of a less significant digit is connected to the RBI of a more significant digit, and the RBI terminal of the most significant digit is permanently grounded. We can also use the RBO terminal to turn off the display, or, in display multiplexing applications. If the RBO terminal is grounded, then the display will turn off regardless of the BCD word applied to the inputs.

If the display decoder is connected directly to the output of the BCD counter, then the display will "roll." This means that the display will follow the counter output as it counts up to the final digit. For example, if the count is to be "7," then the display will roll up 0-1-2-3-4-5-6-7 before coming to rest at "7." We can correct this by connecting a four-bit memory device between the counter output and the ABCD inputs of the 7447. Figure 9-8 shows the use of a 7475 quad-latch chip as a four-bit display memory. The 7475 is a quad-latch arranged in a two-by-two array, each of which is controlled by an external strobe signal. A "latch" is nothing more than a type-D flip-flop in which the *clock* line is redesignated "strobe." When the strobe is made HIGH, then the Q output will follow the data applied to the D-input. But, when the strobe clock drops LOW, then the Q-output will remain at the last valid state existing before the strobe went LOW. As a result, we can hold the formerly valid data as the counter updates, and then transfer the data to the display at the critical time.

The circuit in Fig. 9-8 interposes a 7475 latch between the BCD outputs of the counter and the BCD inputs of the decoder. The two strobe lines are connected together so that they can be operated simultaneously. As mentioned above, this strobe line is active-HIGH, so will transfer the counter output to the decoder input when brought momentarily HIGH.

Figure 9-9 shows the use of a dual quad-latch device, the TTL 74100, to latch data to two 7447/display combinations. The 74100 device contains two independent quad-latches that are capable of storing four-bits each. Sometimes, the strobe lines (S1 and S2) are connected together so that both digits are updated simultaneously—the usual situation in frequency counters.

The lamp test circuit in Fig. 9-9, incidentally, is the way that a

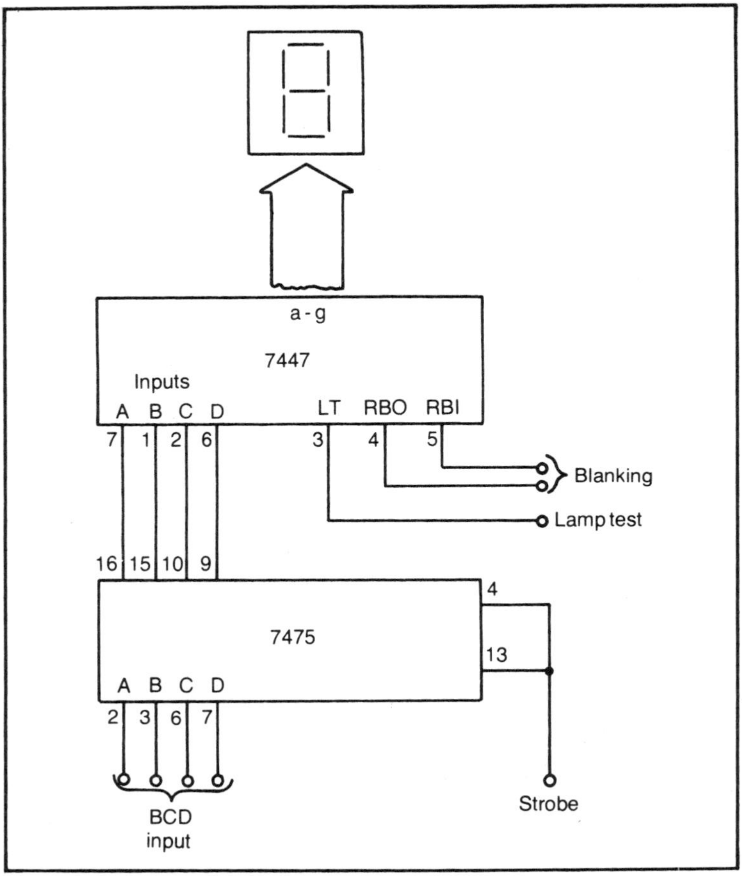

Fig. 9-8. Latched display.

manual test function is provided. A 1 to 3 kohm resistor is connected from the LT terminals (pin no. 3) to +5 volts. This resistor is a "pull-up" resistor that keeps the terminal HIGH unless the switch is closed. As long as switch S1 is open, then the LT line is HIGH, but when S1 is closed then LT line is LOW and the outputs of the 7447 are all LOW—displaying an "8."

There are a number of new TTL and CMOS devices that contain display decoders. There is also an LED display (made as a component by Hewlett-Packard that contains its own BCD-to-seven-segment-decoder built-in. One of the newer devices is the 74143 device shown in Fig. 9-10. This device contains a counter, latch and decoder all in one 24-pin DIP package. There are two sets of outputs on the 74143 device, *seven-segment* and *latched data outputs*. The seven-segment outputs are the active-LOW outputs that turn on the seven-segment LED display device, while the latched data outputs are the BCD (four-bit) data lines.

There are actually two similar devices in the TTL line. The 74143 device is identical to the 74144, except that the 74143 uses a constant-current (15 mA) output scheme that permits us to connect the LED display directly without the need for current-limiting resistors. The 74144 device uses seven-segment outputs more like the 7447, so 330-ohm, ¼-watt resistors are needed in series with each output line. The 74143/74144 devices can be connected in cascade to make higher order decade counter circuits.

LED seven-segment displays require a lot of current, and this can be costly in portable instruments. Even in fixed instruments, a large decade display with lots of digits will cause a need for a larger dc power supply, hence is more expensive. The larger current requirements are also responsible for greater heating of the circuit—a distinct reliability problem.

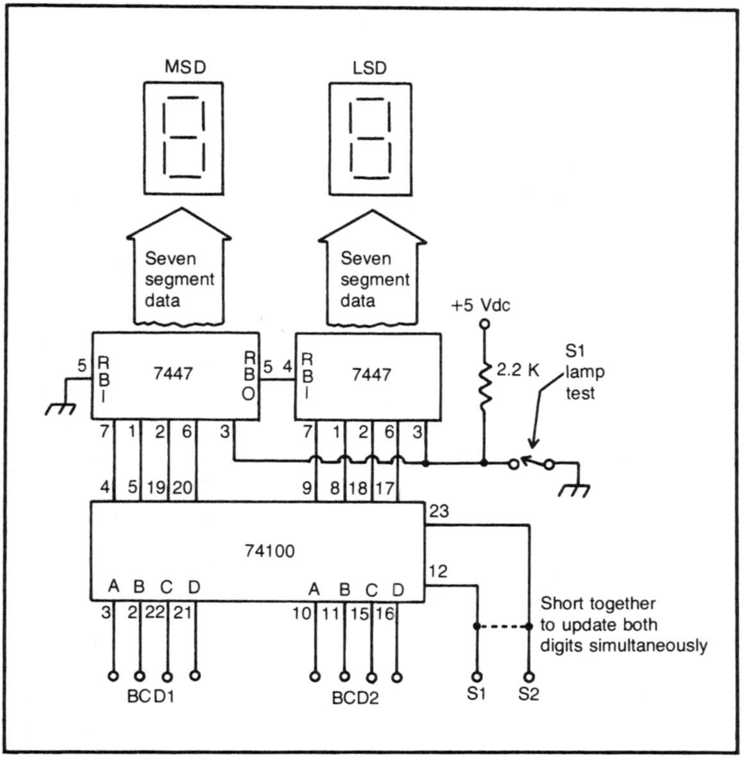

Fig. 9-9. 74100 latch circuit.

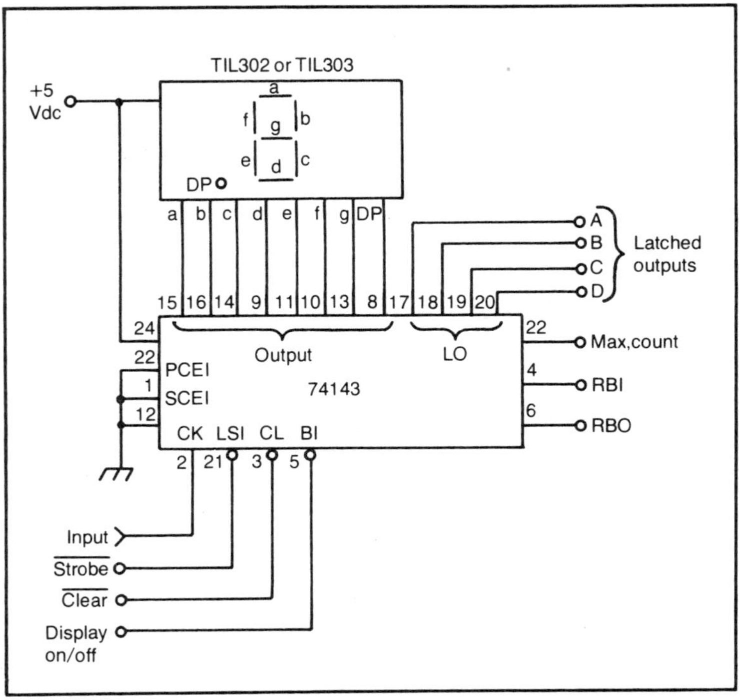

Fig. 9-10. 74143 display.

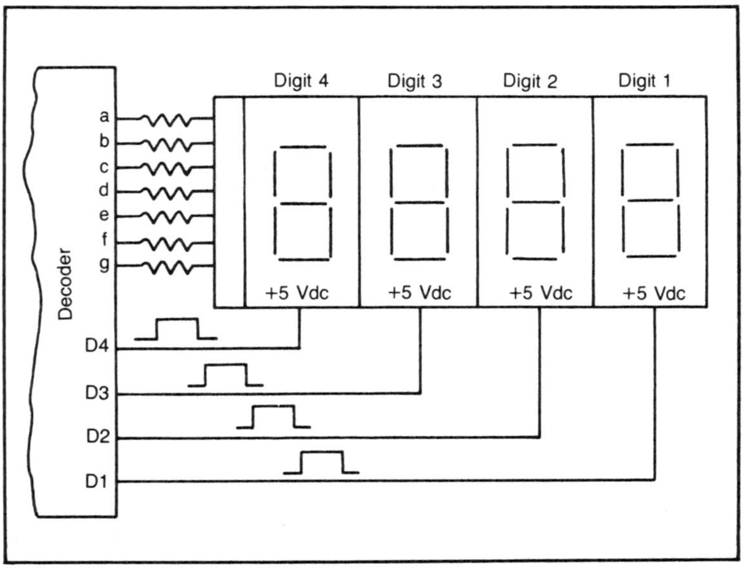

Fig. 9-11. Multiplexed display.

A solution to the problem is to time multiplex the display. This means that we will sequentially turn on and off each digit at a high rate of speed. In a large array, then, the total time the display is on is quite low. Let's consider a 10-digit display as an example. If we turn on only one digit at a time, then the display will be on only 1/10 of the time. By stepping through all displays fast enough, the human eye is decieved into thinking that the display is constant.

A typical display multiplex is shown in Fig. 9-11. Here we have the displays connected to something like a counter or clock chip in which a common multiplexed seven-segment code bus drives all digits simultaneously, but digit on lines (D1-D4) keep only one turned on at a time. The following truth table applies:

Digit On	D1	D2	D3	D4
1	1	0	0	0
2	0	1	0	0
3	0	0	1	0
4	0	0	0	1

If we scan this array at a frequency of 50 kHz, then each digit will turn on every 50,000/N seconds (where N is the number of digits). In this case, therefore, a 50 kHz scan rate will result in each digit turning on 12,500 times per second. The total on-time is ¼-second, but in 1/12,500 second bursts (i.e., every 80 microseconds.

Chapter 10

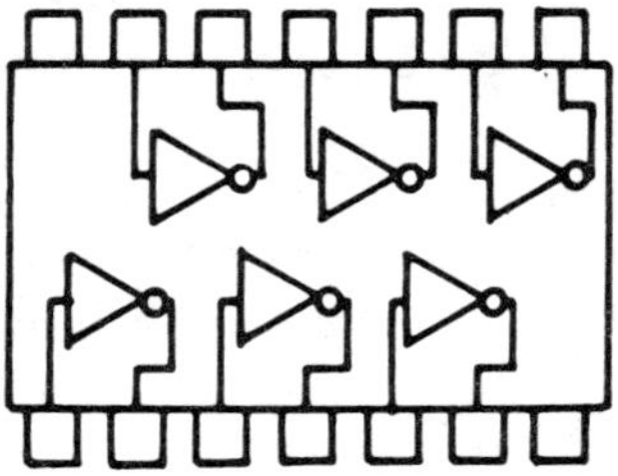

Counters

The most basic counter is the simple J-K flip-flop connected so that J and K are tied HIGH. This makes the output produce *one* output pulse for every *two* input pulses. It is therefore a *binary*, or *divide-by-two*, counter. We can generalize, saying that a digital counter is a circuit that produces a single output pulse for x number of input pulses.

You are probably familiar with digital frequency counters, which are test instruments normally used to measure frequencies from radio transmitters. These instruments contain *decade*, or *divide-by-ten*, counters. The generic term "counter," however, applies to a wide range of circuits that fit the basic definition.

There are two basic classes of digital counter circuits, *serial* and *parallel*. The serial types are all examples of *ripple* counters, which means an input change must ripple through all stages of the counter to its proper point. Parallel counters are also called *synchronous* counters. Ripple counters are operated serially, which means that the output of one stage becomes the input for the following stage.

The basic element used in counters is the J-K flip-flop, which is shown in Fig. 10-1. Note in the figure that the J and K inputs are tied HIGH and therefore remain active.

A timing diagram (Fig. 10-2) shows the action of this circuit. When a J-K flip-flop is connected as in Fig. 10-1, its output changes state on the negative-going transition of the clock pulse. In Fig. 10-2, the first negative-going clock pulse causes the Q output to go

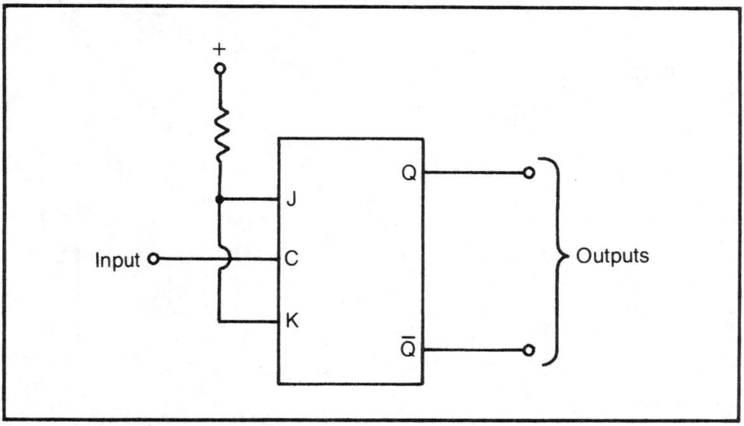

Fig. 10-1. J-K flip-flop.

HIGH. It will remain HIGH until the input sees another negative-going clock pulse, at which time the output will drop LOW again.

The actions required to complete the output pulse take two input clock pulses. This J-K flip-flop, therefore, divides the clock frequency by two, making it a *binary* counter.

A binary ripple-carry counter can be made by cascading two or more J-K flip-flops, as shown in Fig. 10-3. This particular circuit uses four J-K flip-flops in cascade. Any number, however, could be used. A problem with this simple type of counter is that only division ratios that are powers of two (2^n) can be accommodated. In all cases, the division ratio will be 2^n, where n is the number of flip-flops in cascade. In this circuit, therefore, there is a maximum

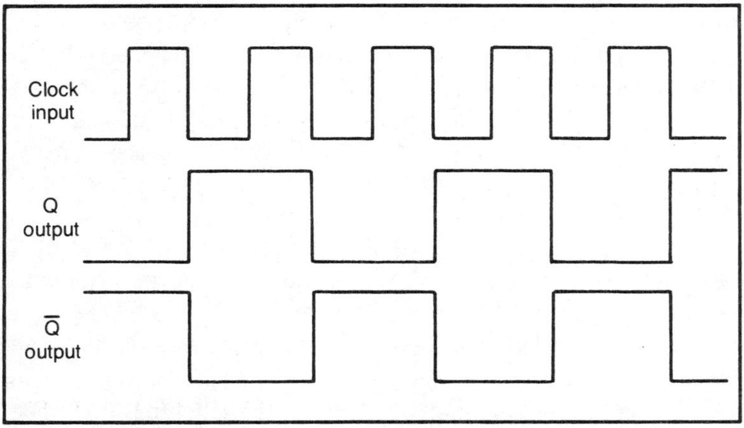

Fig. 10-2. Timing diagram for binary division.

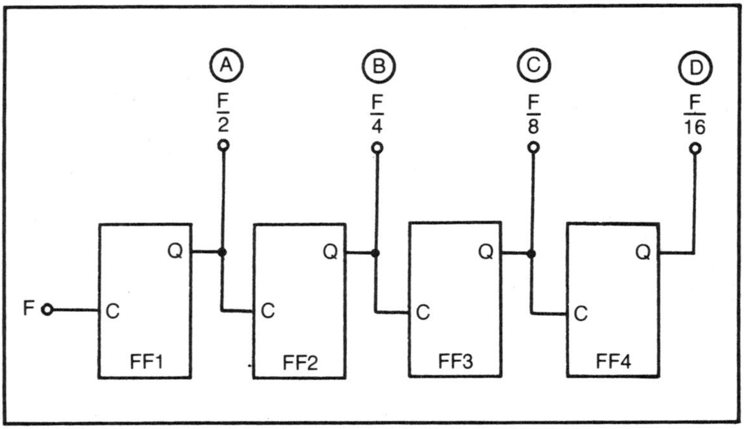

Fig. 10-3. Four-bit counter.

division ratio of 2^4, or 16. Of course, if we can get to the Q outputs of all four J-K flip-flops, we can create frequency division ratios of 2, 4, 8, or 16.

Frequency division is one used for such a counter. You may, for example, want to build a *prescaler* or some other type of divider. You can use a circuit such as shown in Fig. 10-3 for any division ratio that is a power of two (2, 4, 8, 16, 32 . . .).

But prescaling and most other simple frequency division jobs are but one example of counter applications. We can also use the circuit to store the total number of input pulses; the job most people mean when they think of "counters." Consider again the circuit of Fig. 10-3 and the timing diagram of Fig. 10-4. Outputs A, B, C, and D are coded in *binary*, with A being the least significant bit and D being the most significant bit. These outputs are weighted in the following manner: $A = 2^0$, $B = 2^1$, $C = 2^2$, $D = 2^3$. This is called the 1-2-4-8 system because these figures evaluate to A = 1, B = 2, C = 4, and D = 8. Recall from our discussion in Chapters 2 and 3 that these are the weights assigned to the binary number system. By arranging the digits in the form DCBA, a binary number is created that denotes the number of pulses that have the input.

Consider the timing diagram of Fig. 10-4. Note that all Q output changes occur following the arrival of each pulse. After pulse no. 1 has passed, the Q_A line is HIGH and all others are LOW. This means that the binary word on the output lines is 0001_2 (1_{10}); *one* pulse is passed.

Following pulse no. 2, we would expect 0010_2 (2_{10}) because *two* pulses have passed. Note that Q_B is now HIGH and that all others

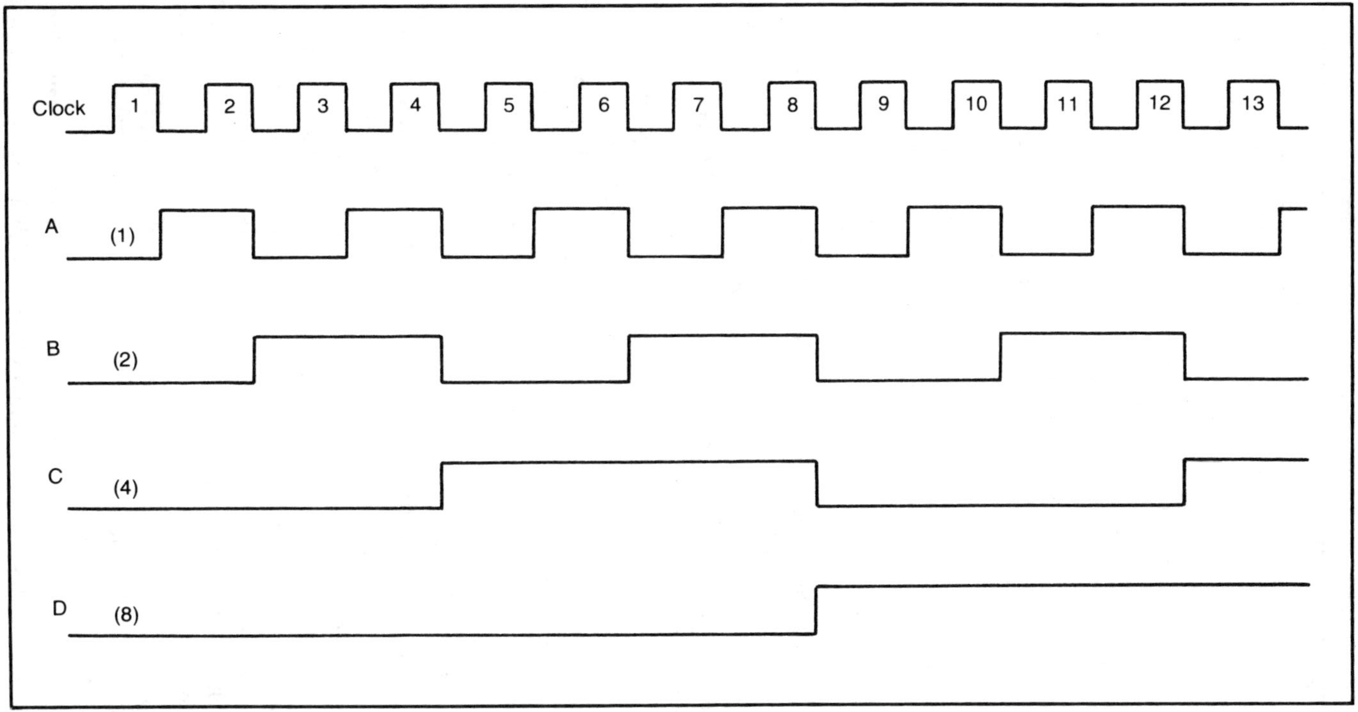

Fig. 10-4. Timing waveform.

Table 10-1. Decimal-Binary-Hexadecimal Codes for 0-15.

Decimal	Binary	Hexadecimal
0	0000	0
1	0001	1
2	0010	2
3	0011	3
4	0100	4
5	0101	5
6	0110	6
7	0111	7
8	1000	8
9	1001	9
10	1010	A
11	1011	B
12	1100	C
13	1101	D
14	1110	E
15	1111	F

are LOW. The digital word is, indeed, 0010_2. If you follow each pulse, you will find the binary code to be as shown in Table 10-1.

The counter shown in Fig. 10-3 could be called a *modulo-16* counter, a *base-16* counter, or a *hexadecimal* counter. All of these terms mean substantially the same thing.

The output of a hexadecimal counter can be decoded to drive a display device that indicates the digits 0 through F—the hexadecimal digits. In most applications where humans read the output, however, a *decimal* counter is needed. We human beings use the decimal number system because of our ten fingers.

DECIMAL COUNTERS

A decimal counter operates in the *base-10*, or *decimal*, number system. The most significant bit of a decimal counter produces one output pulse for every 10 input pulses. Decimal counters are also sometimes called *decade* counters. The decimal counter forms the basis for digital-event, period and frequency-counter instruments.

The hexadecimal counter shown in Fig. 10-3 is not suitable to decimal counting unless modified by adding a single 7400 NAND gate. Recall that a TTL J-K flip-flop uses inverted inputs for the *set* and *clear* functions. As long as the *clear* input remains HIGH, the flip-flop will function normally. When the clear input is momentarily brought LOW, the Q output of the flip-flop goes LOW.

The decade counter shown in Fig. 10-5 is connected so that all four *clear* inputs are tied together to form a common clear line. This line is connected to the output of a TTL NAND gate, which is one section of a 7400 IC device. Recall the rules of operation for the TTL NAND gate: If either input is LOW, then the output goes HIGH, but if both inputs are HIGH, then the output goes LOW.

The idea behind the circuit of Fig. 10-5 is to clear the counter to 0000 following the tenth input pulse. Let's examine the timing diagram of Fig. 10-6 to see if the circuit performs the correct action. Up until the tenth pulse, this diagram is the same as for the base-16, or hexadecimal, counter discussed previously.

The output of the NAND gate will keep the *clear* line HIGH for all counts through 10. The inputs of this gate are connected to the B and D lines. The D line stays LOW, which forces *clear* to stay HIGH, up until the eighth input pulse has passed. At that time (t_0 in Fig. 10-6), D will go HIGH, but B drops LOW. We still have at least one input of the NAND gate (line B) LOW, so the *clear* line remains HIGH.

The *clear* line remains HIGH until the end of the tenth input pulse. At that point (t_2), both B and D are HIGH, so the NAND gate output goes LOW, clearing all four flip-flops. This forces them to go

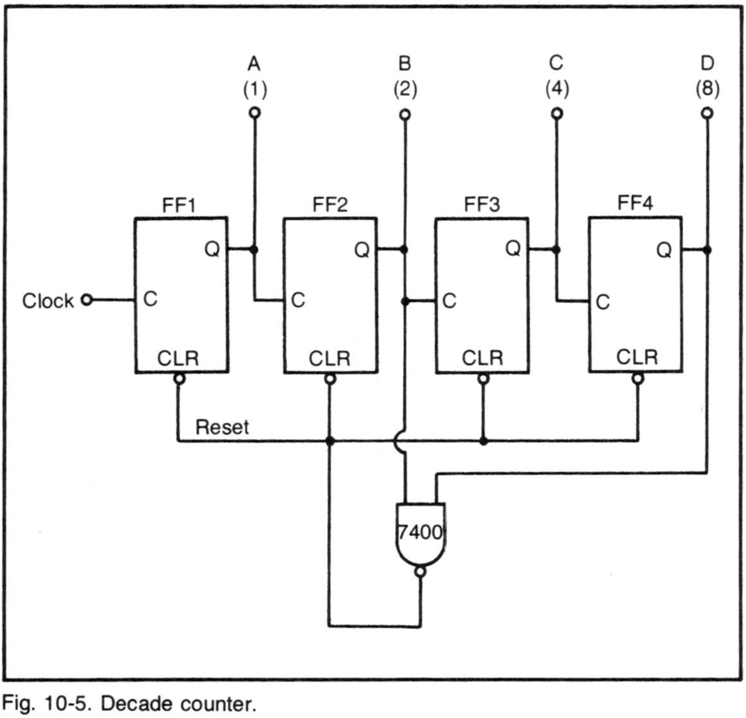

Fig. 10-5. Decade counter.

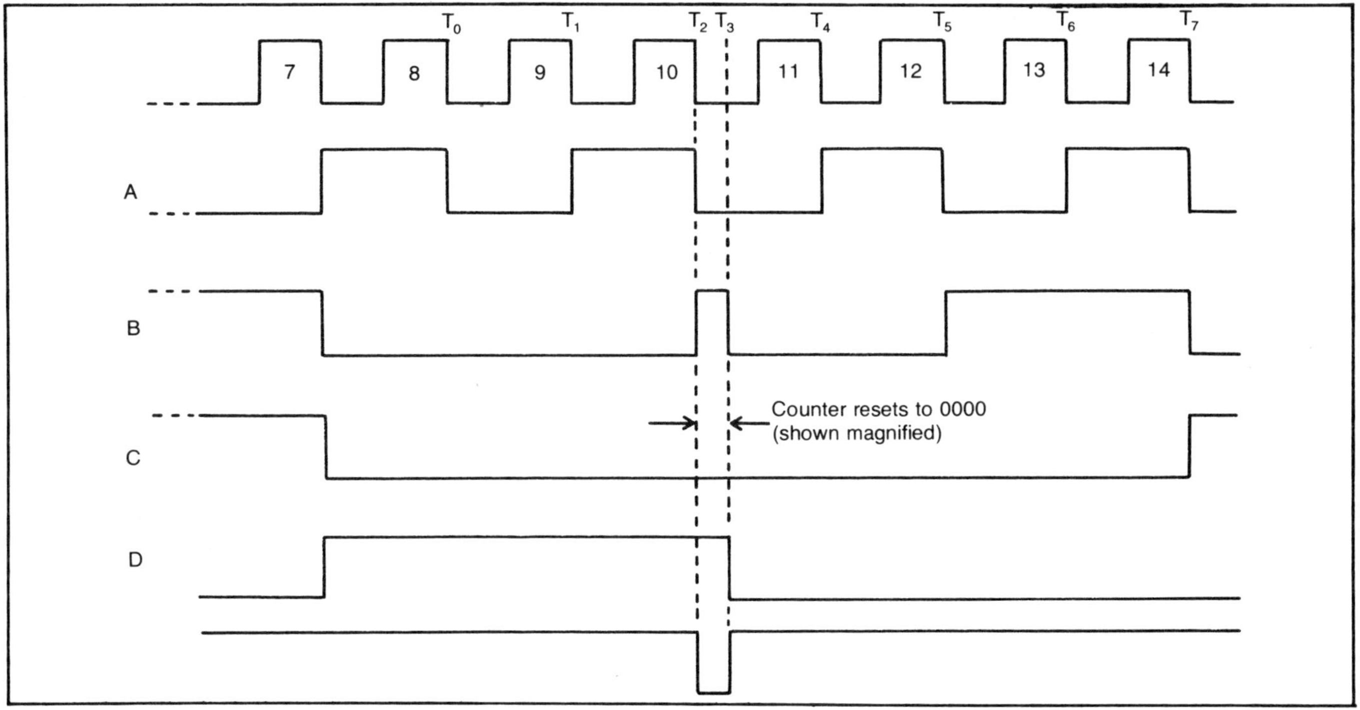

Fig. 10-6. Timing waveform.

to the state where all Q outputs are LOW. The counter is now reset to 0000.

The reset counter produces a 0000 output, so both B and D are now LOW, and this forces the *clear* line HIGH once again. The entire reset cycle occurs in the period (t_3-t_2). This period has been expanded greatly in Fig. 10-6 for simplification. Actually, the reset cycle takes place in nanoseconds, or at the most, microseconds—whatever time is required for the slowest flip-flop to respond to the clear command.

The eleventh pulse will increment the counter one time, so the output will indicate 1 (0001). This counter, then, will count in the sequence 0-1-2-3-4-5-6-7-8-9-0-1 . . . The output code is a 10-digit version of four-bit binary, or hexadecimal. It is called *binary coded decimal* (BCD).

SYNCHRONOUS COUNTERS

Ripple counters suffer from one major problem: speed. The counter elements are wired in cascade, so an input pulse must ripple through the entire chain before it affects the output. A synchronous counter feeds the clock inputs of all flip-flops in parallel. This results in a much faster circuit.

Figure 10-7 shows the partial schematic for a synchronous binary counter. Synchronous operation is accomplished by using four flip-flops with their clock inputs tied together, and a pair of AND gates.

One AND gate is connected so that both Q1 and Q2 are HIGH

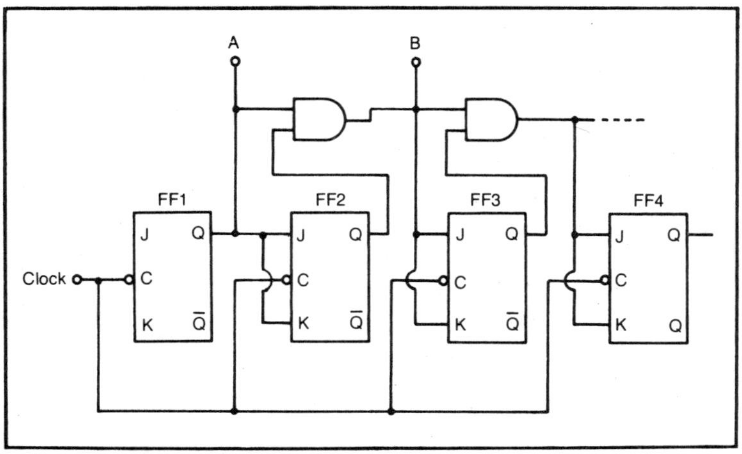

Fig. 10-7. Preset counter.

before FF3 is active. Similarly, Q2 and Q3 must be HIGH before FF4 is made active. On a clock pulse, any of the four flip-flops scheduled to change will do so *simultaneously*. Synchronous counters attain faster speeds, although ripple counters seem to predominate the common counter applications.

PRESET COUNTERS

A preset counter increments from a preset point other than 0000. For example, suppose we wanted to count from 5_{10} (0101_2). We could preset the counter to 0101_2 and increment from there. The counter output pattern will be:

5	0101	9	1001	3	0011
6	0110	0	0000	4	0100
7	0111	1	0001	5	0101
8	1000	2	0010		

Figure 10-8 shows a common method for achieving preset conditions: the *jam input*. Only two stages are shown here for the sake of simplicity, but adding two additional stages will make it a four-bit counter. Of course, any number of stages may be cascaded to form an N-bit preset counter.

In Fig. 10-8, the preset count is applied to *A* and *B*, and both bits will be entered simultaneously when clock line CP2 is brought HIGH. Line CP2 is sometimes called the *enter* or *jam* input. Once the preset bit pattern is entered, the counter will increment from these with transitions of clock line CP1.

DOWN AND UP-DOWN COUNTERS

A down counter decrements, instead of increments, the count for each excursion of the input pulse. If the reset condition is 0000, then the next count will be 0000-1, or (1111). It would have been 0001 in an ordinary up-counter. The count sequence for a four-bit down counter is:

0	0000
15	1111
14	1110
13	1101
12	1100
11	1011
10	1010

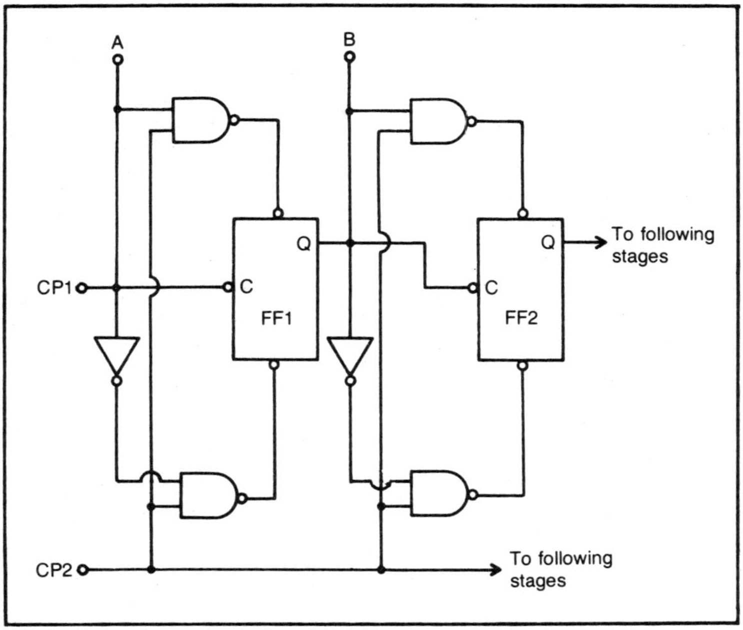

Fig. 10-8. JAM input counter.

9 1001
8 1000
7 0111
6 0110
5 0101
4 0100
3 0011
2 0010
1 0001
0 0000

Basically the same circuit is used as before, but they toggle each flip-flop from the not-Q of the preceding flip-flop. An example of a four-bit binary down counter is shown in Fig. 10-9. Note that the outputs are taken from the Q outputs of the flip-flops, but toggling is from the not-Q.

The preset inputs of the flip-flops are connected together to provide a means to preset the counter to its initial (1111) state. This counter is also called a *subtraction* counter, because each input pulse causes the output to decrement by one bit.

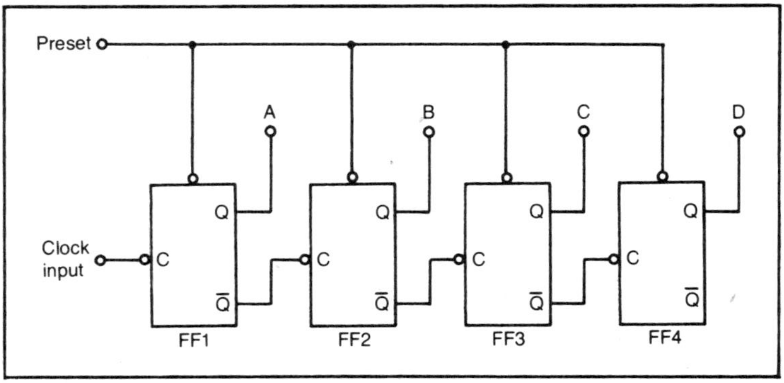

Fig. 10-9. Preset counter.

A *decade* version of this circuit is shown in Fig. 10-10. As in the case of the regular decade counter (the up counter), a NAND gate is added to the circuit to reset the counter following the tenth count. The states are detected when *C* and *D* are HIGH, and the two middle flip-flops are cleared. This action forces the output to 1001_2 (9_{10}). The counter then decrements from 1001 in the sequence:

9	1001	5	0101	1	0001
8	1000	4	0100	0	0000
7	0111	3	0011	9	1001
6	0110	2	0010		

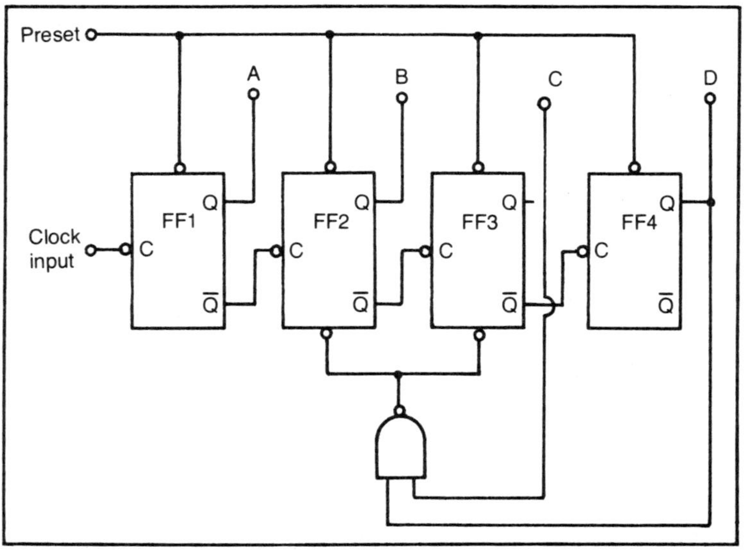

Fig. 10-10. Down counter.

UP/DOWN COUNTERS

Some counters will operate in both up and down modes, depending upon the logic level applied to a *mode* input. Figure 10-11 shows a representative circuit, in which the first two stages of a cascade counter are modified by the addition of several gates. If the mode input is HIGH, then the circuit is an UP counter. If the mode input is LOW, however, then the circuit is a DOWN counter.

TTL/CMOS EXAMPLES

Very few digital circuit designers construct counters from individual flip-flops. Too many ready-built IC counters are available in all of the major IC logic families.

There are three basic counters commonly available in the transistor-transistor logic (TTL) line: 7490, 7492, and 7493. The 7490 is a decade counter, the 7492 is a divide-by-12 (also called modulo-12) counter, and the 7493 is a base-16 (hexadecimal) binary counter. All three of these counters are of similar construction, and their respective pinouts are shown in Fig. 10-12.

The 7490 is a biquinary type of decade counter. This means that it contains a single, independent, divide-by-two stage. This is followed by an independent divide-by-five stage. Decade division is accomplished by cascading the two stages.

Both 7492 and 7493 follow similar layout schemes. In both, the first stage is a single divide-by-two flip-flop, followed by divide-by-six (7492) or divide-by-eight (7493) stages. These form divide-by-12 and divide-by-16 counters, respectively.

The 74142 is a special function TTL IC that contains a divide-by-10 counter (BCD), a four-bit latch circuit, and a display decoder suitable for driving a *Nixie*® (registered trademark of the Burroughs

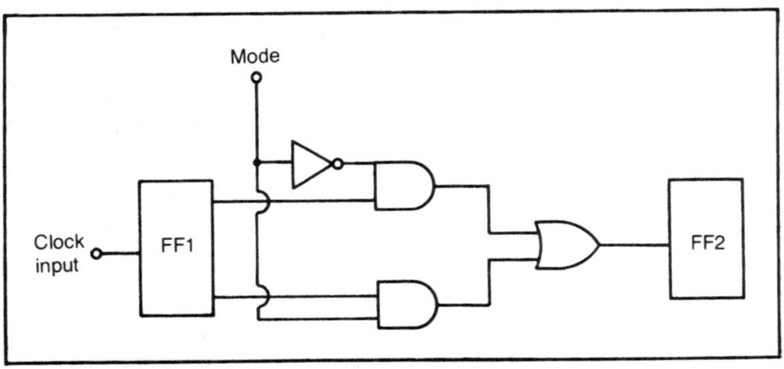

Fig. 10-11. Up/down counter.

DECADE COUNTERS

90 DIVIDE-BY-TWO AND DIVIDE-BY FIVE

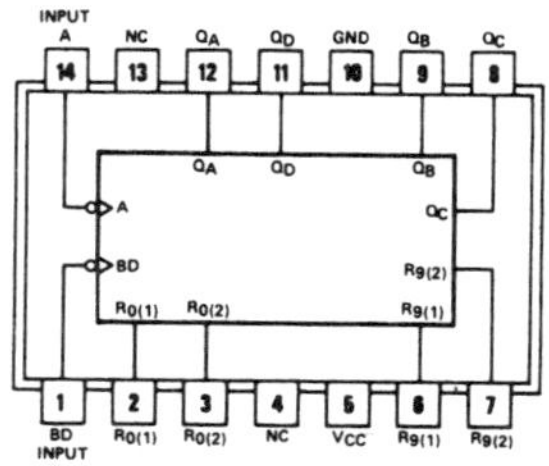

SN5490A (J, W) SN7490A (J, N)
SN54L90 (J, T) SN74L90 (J, N)
SN54LS90 (J, W) SN74LS90 (J, N)

NC — No internal connection

DIVIDE-BY-TWELVE COUNTERS

92 DIVIDE-BY-TWO AND DIVIDE-BY-SIX

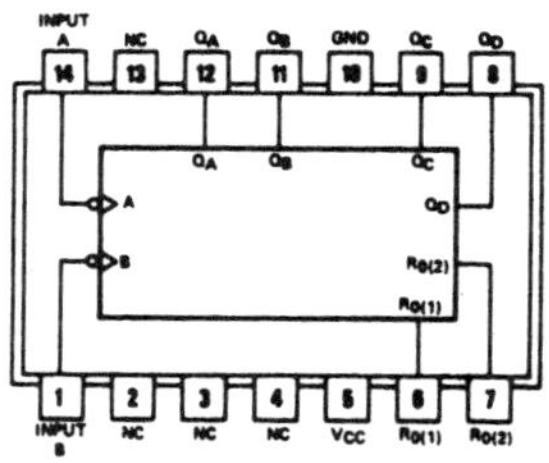

SN5492A (J, W) SN7492A (J, N)
SN54LS92 (J, W) SN74LS92 (J, N)

NC—No internal connection

4-BIT BINARY COUNTERS

93 DIVIDE-BY-TWO AND DIVIDE-BY-EIGHT

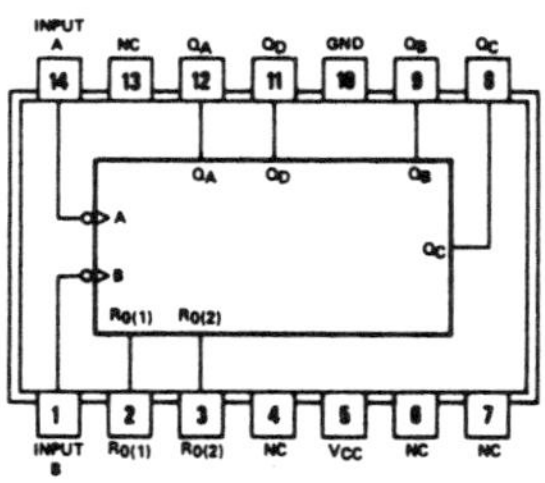

SN5493A (J, W) SN7493A (J, N)
SN54LS93 (J, W) SN74LS93 (J, N)

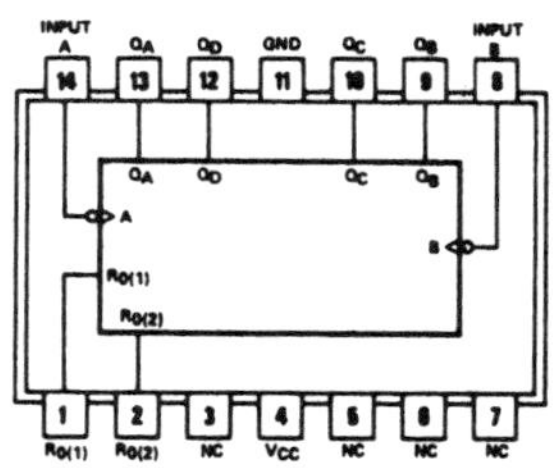

SN54L93 (J, T) SN74L93 (J, N)

NC—No internal connection

Fig. 10-12. 7490, 7492, and 7493 pinouts.

Corporation). The 74142 is housed in a standard 16-pin DIP package (see Fig. 10-13).

Figure 10-14 shows the pinouts for the TTL type 74160 through 74163 devices. These are BCD and binary four-bit synchronous counters:

- ☐ 74160—Decade (BCD) synchronous, direct-clear.
- ☐ 74161—Binary, synchronous, direct-clear.
- ☐ 74162—Decade (BCD), fully synchronous.
- ☐ 74163—Binary, fully synchronous.

These counters typically operate to 32 MHz and dissipate approximately 325 mW. All are housed in standard 16-pin DIP packages. These counters are discussed in detail elsewhere.

These four counters are different from those that have been discussed previously, because they are divide-by-N counters, where N is an integer. The value N is applied to the *data inputs* and loaded into the counter when the *load* terminal is momentarily brought LOW.

Examples of basic CMOS counters are shown in Fig. 10-15. Again, these examples are not exhaustive, but merely representative of those commonly used in electronic circuits. Not one of them is a multidigit decimal counter such as the eight-digit, 10-MHz device made by Intersil, Inc.

☐ 4017—This device is a fully synchronous decade counter, but the outputs are decoded 1-of-10. The active output is HIGH while the inactive outputs are LOW. The 4017 is positive-edge triggered. The *reset* and *enable* inputs are normally held LOW. If the *reset* is momentarily brought HIGH, the counter goes immediately to the zero state. The *enable* input is used to inhibit the count

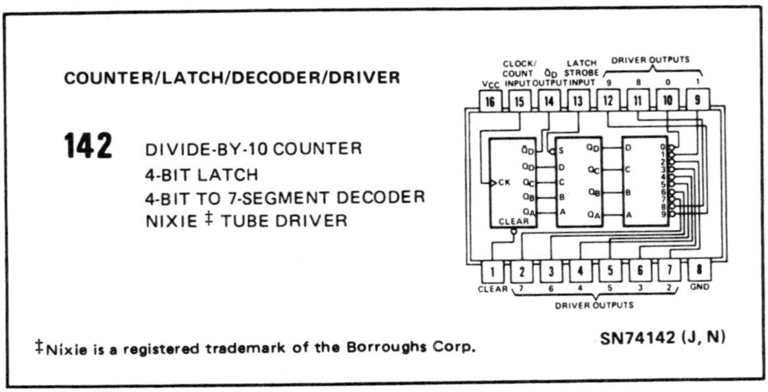

Fig. 10-13. 74142 pinouts.

SYNCHRONOUS 4-BIT COUNTERS

160 DECADE, DIRECT CLEAR

161 BINARY, DIRECT CLEAR

162 DECADE, SYNCHRONOUS CLEAR

163 BINARY, SYNCHRONOUS CLEAR

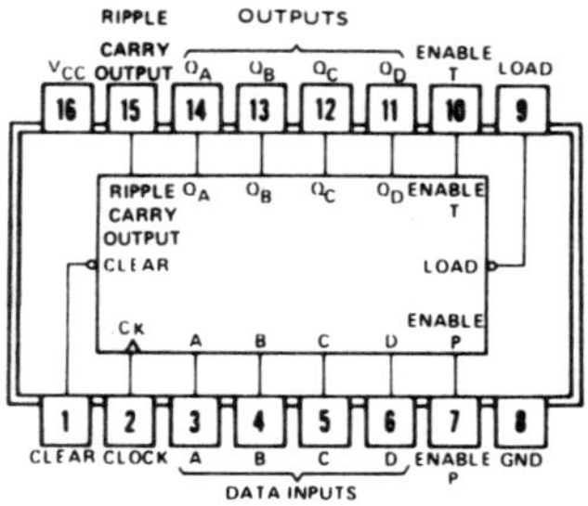

SN54160 (J, W)	SN74160 (J, N)
SN54LS160A (J, W)	SN74LS160A (J, N)
SN54161 (J, W)	SN74161 (J, N)
SN54LS161A (J, W)	SN74LS161A (J, N)
SN54162 (J, W)	SN74162 (J, N)
SN54LS162A (J, W)	SN74LS162A (J, N)
SN54S162 (J, W)	SN74S162 (J, N)
SN54163 (J, W)	SN74163 (J, N)
SN54LS163A (J, W)	SN74LS163A (J, N)
SN54S163 (J, W)	SN74S163 (J, N)

Fig. 10-14. Pinouts for 74160-163.

without resetting the device; that is, if the *enable* input is made HIGH, the count ceases and the output remains in its present state. The output terminal produces a pulse train of F/10, which is HIGH for counts 0, 1, 2, 3, and 4, and LOW for 5, 6, 7, 8, and 9.

□ 4018—This device is a synchronous divide-by-N counter, where N is an integer of the set 2, 3, 4, 5, 6, 7, 8, 9, and 10. It is difficult to decode the outputs of this counter, so its principal use is in frequency division. For normal running, the *reset* and *load* inputs must be held LOW. The 4018 is positive-edge triggered. The N-code for determining the division ratio is set by connecting the input terminal to an appropriate *output* or in certain cases, an external AND gate. For *even* division ratios, no external gate is needed; merely connect the *input* terminal as shown in Table 10-2.

Table 10-2. Even Division Table.

Even division ratio	Connect input pin 1 to:
2	Q1 (Pin 5)
4	Q2 (Pin 4)
6	Q3 (Pin 6)
8	Q4 (Pin 11)
10	Q5 (Pin 13)

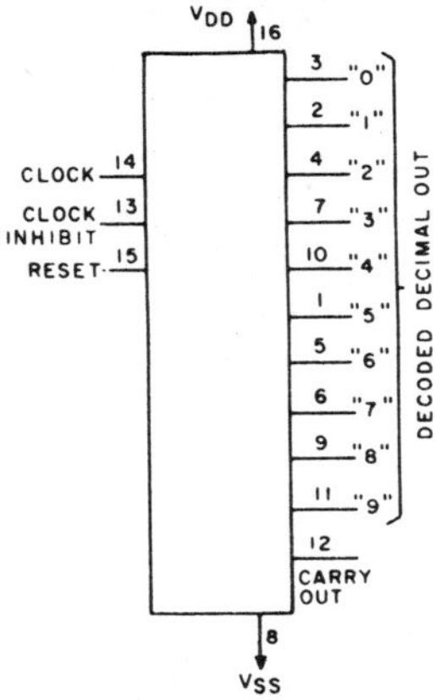

Decade Counter/Divider with 10 Decoded Decimal Outputs

CD4017A
CD4017B

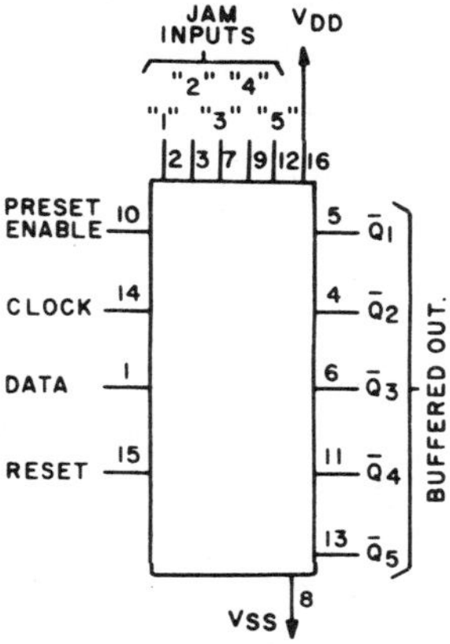

Presettable Divide-by-"N" Counter Fixed or Programmable

CD4018A
CD4018B

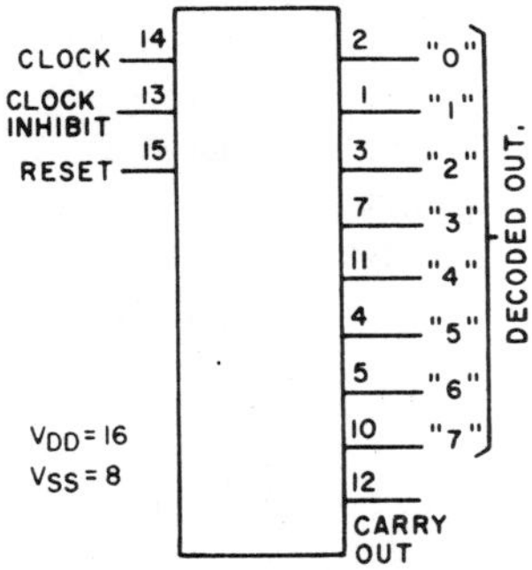

Divide-by-8 Counter/Divider with 8 Decoded Decimal Outputs

CD4022A
CD4022B

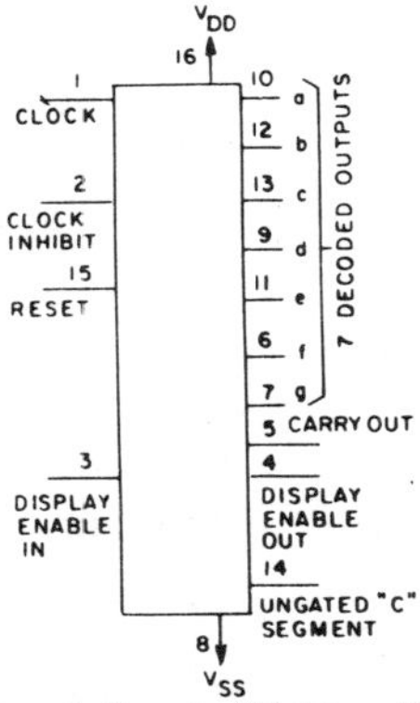

Decade Counter/Divider with 7-Segment Display Outputs and Display Enable

CD4026A
CD4026B

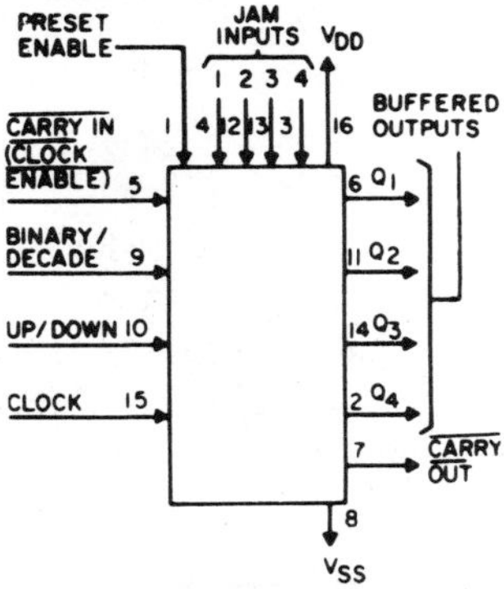

Presettable Up/Down Counter, Binary or BCD-Decade

CD4029A
CD4029B

Fig. 10-15. Pinouts for CMOS devices.

Table 10-3. Odd Division Table.

Odd division ratio	Connect input Pin 1 to:
3	Q1 (Pin 5) & Q2 (Pin 4)
5	Q2 (Pin 4) & Q3 (Pin 6)
7	Q3 (Pin 6) & Q4 (Pin 11)
9	Q4 (Pin 11) & Q5 (Pin 13)

(external gate needed)

The odd division ratios, such as 3, 5, 7, and 9, require an external, two-input AND gate. The 4018 outputs are connected to the AND gate inputs, and the AND gate output is connected to the 4018 input. See Table 10-3.

The feedback line just described is also the main output from the counter. If, for example, "input" pin no. 1 will be (by the table above) 1/6 of the clock frequency.

The 4018 can also be parallel loaded using the jam terminals P1 through P5. These terminals will program the 4018. A LOW on the jam input forces the related Q output HIGH, and vice versa. For example, if a LOW is applied to P2, it will force Q2 HIGH.

☐ 4022—This device is an octal, or divide-by-eight, counter that provides 1-of-8 decoded outputs. The 4022 is very nearly the same as the 4017, which is a decade version (see the discussion for the 4017).

☐ 4026—The 4026 is a decade counter that produces uniquely decoded outputs for seven-segment displays. The 4026 is a positive-edge triggered device that is fully synchronous. This chip is similar to the 4017, in that it provides an F/10 output in addition to the seven-segment decoded outputs. The decoded outputs are HIGH for *active*.

There are two *enable* inputs. One is a *clock enable* input, which will cause the count to cease when brought HIGH. The counter outputs, however, remain in their present state when the clock is inhibited. The other enable input is a display enable terminal. A HIGH on this input will turn the display on, and a LOW will turn the display off. A possible use of this terminal is to turn off the high-current display when it is not needed.

☐ 4029—The 4029 is an up-down counter that will divide by either 10 or 16, depending upon whether pin no. 9 is HIGH or LOW.

A HIGH on pin no. 9 causes the 4029 to be a base-16 binary counter. A LOW causes it to be a base-10 decade counter.

The count direction (up/down) is determined by the level applied to pin no. 10. If pin no. 10 is HIGH, then the 4029 operates as an up counter, but if pin no. 10 is LOW, then it operates as a down counter.

Chapter 11

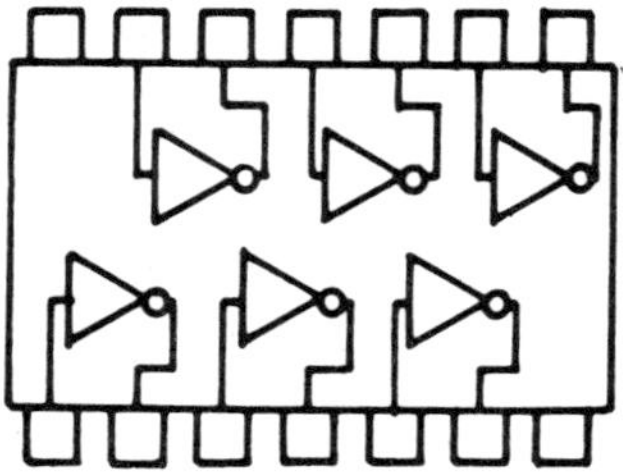

Shift Registers

A flip-flop is able to store a single bit of digital data. When two or more flip-flops are organized into some configuration to store multiple bits of data, they constitute a *register*. Most registers are merely specialized arrays of ordinary flip-flops.

There are several different circuit configurations that one would call registers. We classify these according to the manner in which data is input to and output from the register. We have, for example, *serial-in-serial-out* (SISO), *serial-in-parallel-out* (SIPO), *parallel-in-parallel-out* (PIPO), and *parallel-in-serial-out* (PISO) registers.

SISO AND SIPO REGISTERS

Figures 11-1 and 11-2 represent both SISO and SIPO shift registers. The only really significant difference is the parallel output lines used on the SIPO device; these lines would be absent on the SISO register. Note that some registers are both SISO and SIPO, depending upon how you use them. The SIPO shift register consists of a cascade chain of type-D flip-flops, with the clock lines tied together; i.e., they share a common clock line.

Recall one of the rules for type-D flip-flops: Data can be transferred from the D-input to the Q-output only when the clock input is HIGH. This rule can be applied to the situation shown in Fig. 11-2, where the transmission of a single data bit from left to right through the SISO shift register is shown.

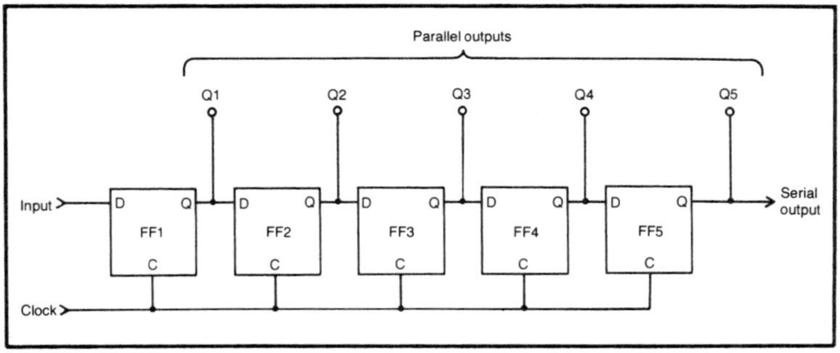

Fig. 11-1. Five-stage shift register.

At the occurrence of the first clock pulse, the input is HIGH. This point is the D-input of FF1, so a HIGH is transferred to the Q1 output. This HIGH, which is also applied to the D input of the second flip-flop (FF2), remains after the clock pulse vanishes.

When the second clock pulse arrives, FF2 sees a HIGH on its D-input, and FF1 sees a LOW on its D-input. This situation causes a LOW at Q1 and a HIGH at Q2.

The third clock pulse sees a LOW on the D-inputs of both FF1 and FF2, and a HIGH at the D-input of FF3. The third clock pulse, then, causes Q1 and Q2 to be LOW and Q3 to be HIGH.

Note that the SISO input remains LOW after the initial HIGH during clock pulse no. 1. This means that the single HIGH will be propagated through the entire SISO shift register, one stage at a time. The HIGH bit will shift one flip-flop to the right each time a clock pulse arrives. If the data at the input had changed, then the bit pattern at the input will be propagated through the shift register.

The shift register shown in Fig. 11-1 is a five-bit, or five-stage, shift register, although any bit-length could be selected. On the sixth clock pulse, therefore, the HIGH is propagated out of the register, so all flip-flops are LOW; i.e., Q1 = Q2 = Q3 = Q4 = Q5 = 0.

The SISO shift register can be made into a SIPO device by adding parallel output lines, one each for Q1, Q2, Q3, Q4, and Q5. One use for the SIPO shift register is serial-to-parallel code conversion. For economic reasons, digital data is usually transmitted from device to device as a serial stream of bits, especially if long distances are involved. In other words, the bits of the digital word are sent over a communications channel one at a time. But most computers and other digital instruments use data in the more efficient parallel form.

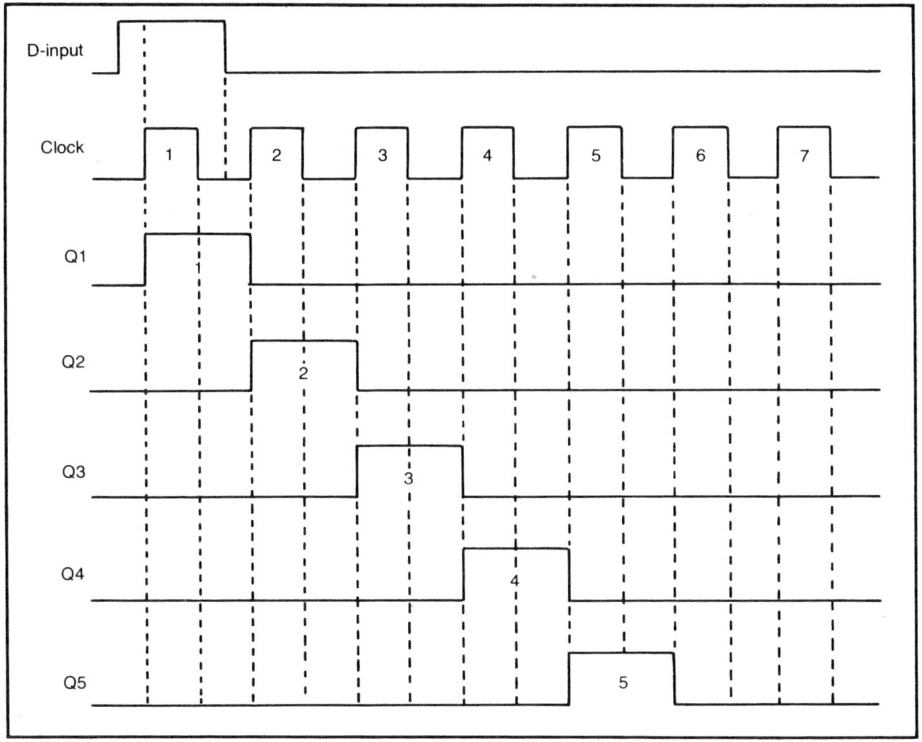

Fig. 11-2. Timing diagram.

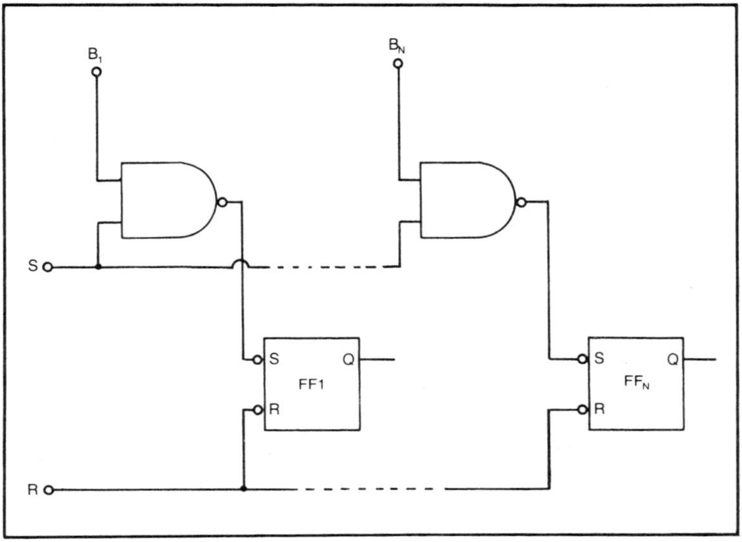

Fig. 11-3. Parallel loading.

This format is both faster and more expensive; hence the use of serial-in transmission. As an example, consider an eight-bit binary data system. We would need an eight-stage SIPO to convert the eight-bit serial binary code to eight-bit parallel form. The code is entered into the SIPO register one bit at a time, so that after eight clock pulses, the first bit will appear at Q8, and the last bit at Q1.

PARALLEL

Parallel entry shift registers are faster to load than the serial input types because a given bit in the shift register can be changed directly, without applying it through all of the other stages. The new bit need not ripple through all of the preceding stages to reach its destination. There are two basic types of parallel entry: *parallel-direct* and *jam*.

In parallel-direct entry, or simply parallel entry, shown in the partial schematic of Fig. 11-3, the register must first be cleared by setting all bits to LOW and first bringing the *reset* line momentarily LOW. The data that is applied to inputs B1 through BN can then be loaded into the register by momentarily bringing the *set* line HIGH.

The jam entry circuit of Fig. 11-4 is able to load data from the B1 through BN inputs in a single operation. This job is accomplished by loading the complements of the B1 through BN data into the other inputs. The actual IC shift registers using this technique, inverter

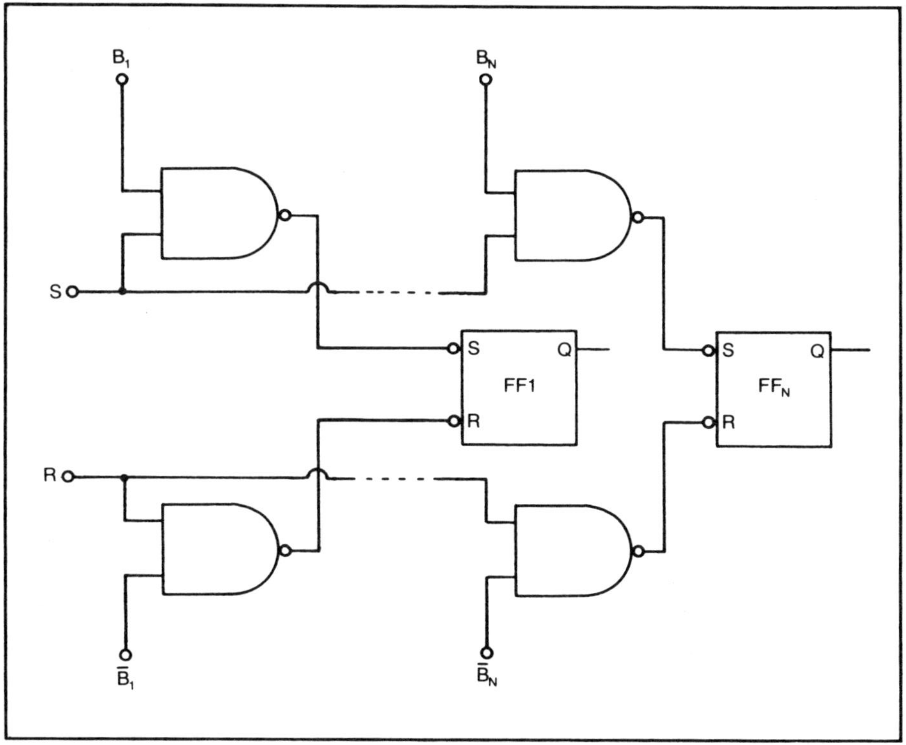

Fig. 11-4. Loading.

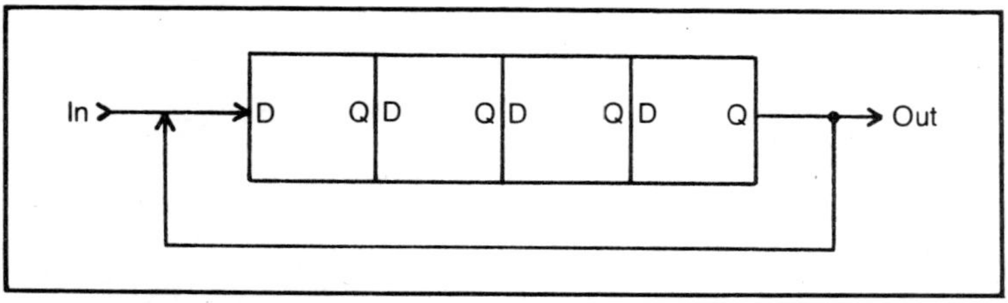

Fig. 11-5. Recirculating shift register.

stages are connected internally to the inputs to automatically complement the input lines. The outside user never sees the complementing process.

A recirculating shift register circuit is shown in Fig. 11-5. These registers are able to read out the data in serial format and then re-enter it so that it is retained. This is done by connecting the output of the SISO register, it is destroyed and lost forever; however, we sometimes wish to retain the data and be able to read it out, too. The solution here is to read it right back in as it is read out. Although only three bits are shown here for simplicity, the bit length could be almost anything.

IC EXAMPLES

Few designers would bother to use a series of cascade flip-flops in making a shift register. There are simply too many different types of IC shift registers available on the market. Both TTL and CMOS examples abound.

Chapter 12

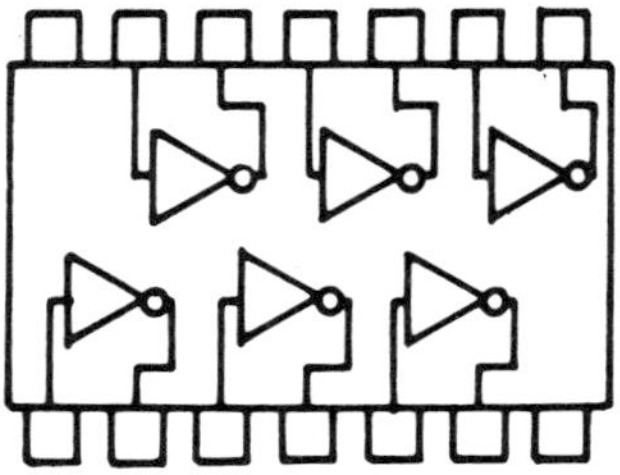

Data Multiplexers and Selectors

Multiplexing is the art of interleaving data so that it can be transmitted together over a single line or through a single data communications channel. Two basic forms of multiplexing exist: frequency domain and time domain. The most common examples of frequency domain are color television, FM stereo broadcasting, and subcarrier telephony. All of these use one or more subcarriers to transmit information. Digital systems, however, use time domain multiplexing.

Figure 12-1 shows a simplified model of a time domain multiplexer circuit. Here two rotary switches select from any of six different data channels. These switches are rotated synchronously so that both are always in the same position; i.e., S1 is connected to channel no. 3 at the same time that S2 is connected to channel no. 3. This situation allows us to transmit data from six channels serially along the communications path by using techniques that will be discussed more fully in the next section.

Figure 12-2 shows a system in which four eight-bit devices are connected to a single eight-bit input port. If they were connected in parallel, 4 × 8, or 32, wires would be necessary. Because the system is being multiplexed, however, only one I/O port (instead of four I/O ports) and 10 wires are needed. Eight of the wires are used to form the eight-bit data bus over which all of the data is to pass. The remaining two wires are used for control of the multiplexer. In the circuit shown, the multiplexer can be viewed as a programmable switch. These two wires do the programming, according to the

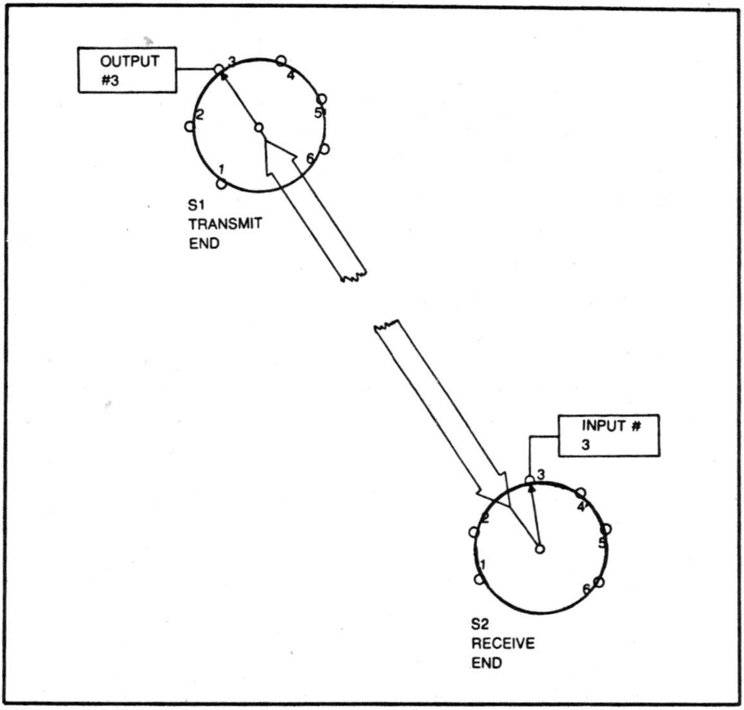

Fig. 12-1. Electrical circuit model of time domain multiplexers.

rules shown in Table 12-1. If the code is 00, then channel 0 is selected and connected to the data bus. If the code is 01, then channel 1 is selected, and so on. A simple two-bit binary counter, or a counter program if a computer is used, will create the codes. It will sequence 00-01-10-11-00 . . . until turned off or disabled.

The circuit shown in Fig. 12-2 assumes that the data at the inputs will always be ready, because it operates in a quasiasynchronous mode. Synchronous operation would require at least one more wire. Of course, it must also be assumed that the device receiving the data will not change the data selector code until it has successfully completed the current data transfer. It will not, for

Channel	Code
0	00
1	01
2	10
3	11

Table 12-1. Channel-select code.

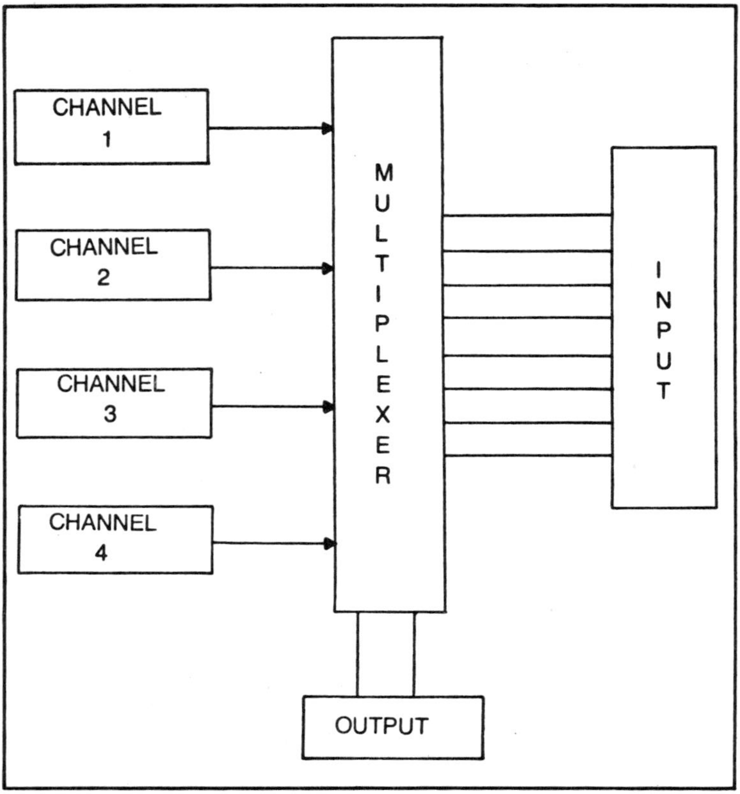

Fig. 12-2. Multiplexed system of four channels.

example, switch to channel no. 2 until it has successfully input the data on channel no. 1.

Figure 12-3 shows another form of digital multiplexing, used extensively in calculators and other portable digital instruments. LED and incandescent seven-segment display readouts (see Chapter 10) draw a considerable amount of current from the power supply. Some devices draw up to 40 mA per segment. In the four-digit display shown in Fig. 12-3, when all segments are turned on (the displayed value is 8888), the current drawn is:

$$\frac{0.04 \text{ amperes}}{\text{segment}} \times \frac{8 \text{ segments}}{\text{digit}} \times 4 \text{ digits} = I$$

$$0.04 \times 8 \times 4 \text{ amperes} = 1.28 \text{ amperes}$$

If this were a 12-digit portable calculator, even more current capability (almost 4 amperes) would be needed. Once you begin to require so much current on a continuous basis, battery life drops rapidly. If a battery large enough to allow lengthy operation is used, then portability becomes questionable, at the least.

The answer to the battery life dilemma is to multiplex the digits. Only one digit at a time will be turned on, but if the switching rate is fast enough, the persistence of the human eye will make the display appear constant.

Figure 12-3 shows a multiplexed digital display, while Fig. 12-4 shows the timing diagram. Recall that a seven-segment readout requires one wire for each segment; these are usually labeled a, b, c, d, e, f, and g. In a typical device, power is applied to the anode of the digit, and the decoder will ground the terminals to those segments that are to be lighted. To display a 3, for example, ground terminals a, b, c, d, and g (see the inset of Fig. 12-3). The instru-

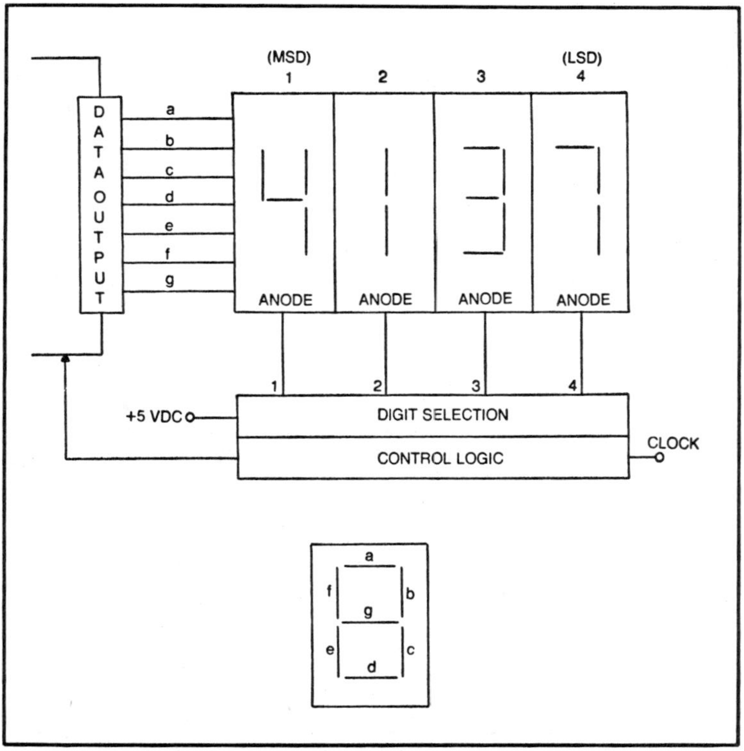

Fig. 12-3. Multiplexed seven-segment display.

ment will place the proper seven-segment code for the digit to be displayed on the seven-bit data bus at the appropriate time. Timing is critical to the proper operation of this system, and it is handled by the control logic section.

The digit selector is a circuit that enables each digit in sequence. In the crude example shown here, a digit is selected by applying +5 Vdc to the common anode. This circuit has four output lines, one for each digit. When a selector line is HIGH, then +5 V is applied to its digit.

In the situation shown in Fig. 12-3, the four-digit number 4137 is displayed. The sequential operation is as follows:

☐ During time t1, the seven-segment code that represents "4" is placed on the data bus. Simultaneously, the digit selector applies +5 Vdc to digit no. 1. This causes digit no. 1 to display a "4."

☐ During time t2, the seven-segment code representing "1" is placed on the data bus. Simultaneously, the digit selector applies +5 Vdc to digit no. 2 after having disconnected all other digits.

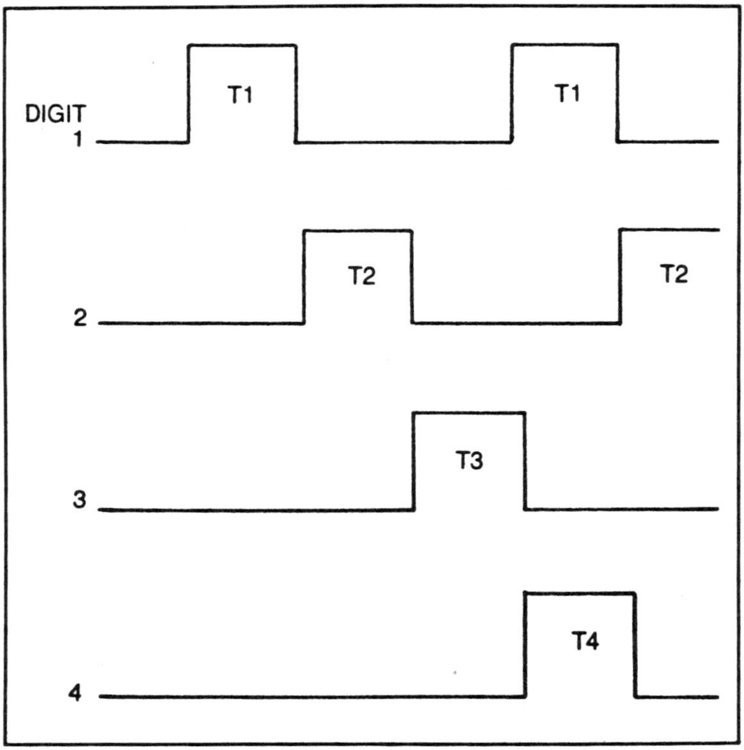

Fig. 12-4. Timing diagram of the multiplexed seven-segment display.

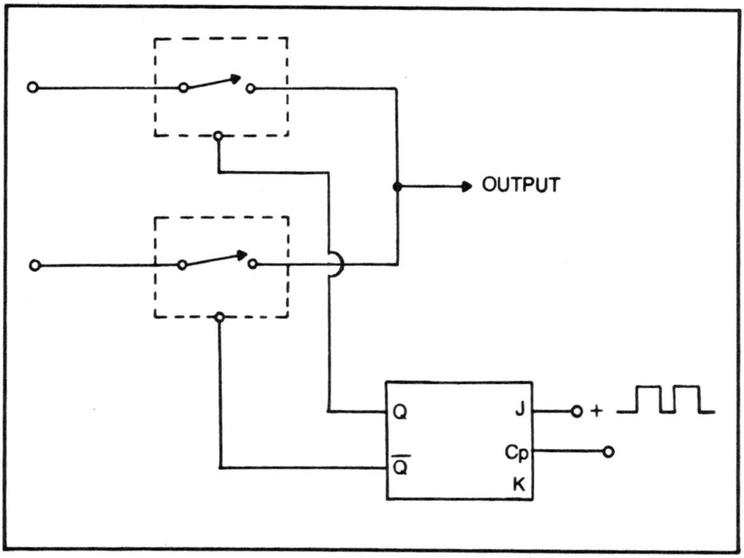

Fig. 12-5. Two-channel multiplexer based on 4016/4066 device.

☐ During time t3, the seven-segment code representing "3" is placed on the data bus. Simultaneously, the digit selector applies +5 Vdc to digit no. 3, which turns off the other three digits.

☐ During time t4, the seven-segment code representing "7" is placed on the data bus, and the digit selector turns on digit no. 4. This sequence happens continuously. Only one digit is turned on at any one time, but they all turn on sequentially many times per second.

A simple example of a two-channel multiplexer is shown in Fig. 12-5. This circuit uses a common CMOS digital integrated circuit, the 4016, or its newer and preferred cousin, the 4066. Both 4016 and 4066 devices are easily available through mail-order electronics houses specializing in IC sales. Note that the CMOS electronic switches used in these devices are bilateral, meaning that current will pass in either direction. This means that they can be used as either analog, where bipolar signals might be encountered, or digital multiplexers. Each of the switches inside the 4016/4066 devices is completely independent of the others. They share only common power-supply terminals.

The control pins are 5, 6, 12, and 13. When the voltage applied to these pins is equal to the voltage on pin no. 7, the switch selected by that particular pin is turned off. In digital systems, pin no. 7 will most likely be grounded, so the switches are turned off when their control pins are grounded (zero volts). The voltage applied to pin no. 7 can be anything in the range of 0 V to −5 V (−7 V in some models), but the negative voltages are only seen occasionally in digital circuits.

Applying a voltage equal to the voltage on pin no. 14 to a control pin will turn on the switch. In most digital circuits, this will be +5 V, although CMOS devices can operate up to +18 Vdc. The rationale is to make the whole circuit TTL compatible.

The terms "on" and "off" refer to the series impedance of the switch. In the off condition, the series impedance is very high, usually well over a megohm. In the on condition, however, the series impedance of the switches drops very low. It will be in all cases less than 2000 ohms, and in some devices less than 100 ohms.

In the "mixer" of Fig. 12-5, only two of the switches are used. This circuit, incidentally, is also useful for making a single trace oscilloscope into a dual-trace scope. Note that one terminal of both S1 and S2 are connected together to form a common output. The remaining terminal on both switches goes to the respective channels. Pin no. 1 is connected to channel no. 1, and pin no. 4 is connected to the channel no. 2.

The switching action is performed by a J-K flip-flop connected in the clocked binary-divider method. The Q and not-Q outputs on a J-K flip-flop are complementary. This means that one output will be HIGH when the other is LOW. When the state of the FF changes, the HIGH output becomes LOW and the LOW output becomes HIGH. The timing diagram shown in Fig. 12-5 shows this relationship more clearly.

During interval T1, the Q output is HIGH and not-Q is LOW. This turns on switch S1 and turns off S2 (pin no. 5 is HIGH and pin no. 13 is LOW). According to the rules for operation of the 4016/4066, channel no. 1 is then connected to the output.

Following the next clock pulse (J-K flip-flops operate on the negative-going edge of the clock pulse), the condition of the outputs becomes reversed. S1 is now turned off and S2 is turned on. This will connect channel no. 2 to the output. The switches are turned on and off alternately. This means that the output will be connected to the respective inputs alternatively.

In the case of an oscilloscope switch, only this one circuit is needed to make things happen. In data communications systems, however, the signals are now interleaved, and that makes things a hopeless nightmare at the other end of the transmission path unless some form of decoding is provided. A demultiplexer must then be provided at the receive end of the path.

Happily, the 4016/4066 devices are bilateral. Basically the same circuit is useful for "demuxing." The serial data stream coming into the receiver is fed to the common terminal between the two switches (used as an output in the mux end and an input in the demux end).

Three additional standard CMOS devices are useful as multiplexers and demultiplexers: 4051, 4052, and 4053. All three of these CMOS devices have chip enable pins that allow the chip to be turned on and off at will.

The 4051 shown in Fig. 12-6 is a 1-of-8 selector and can be viewed as analogous to a single-pole, eight-position rotary switch.

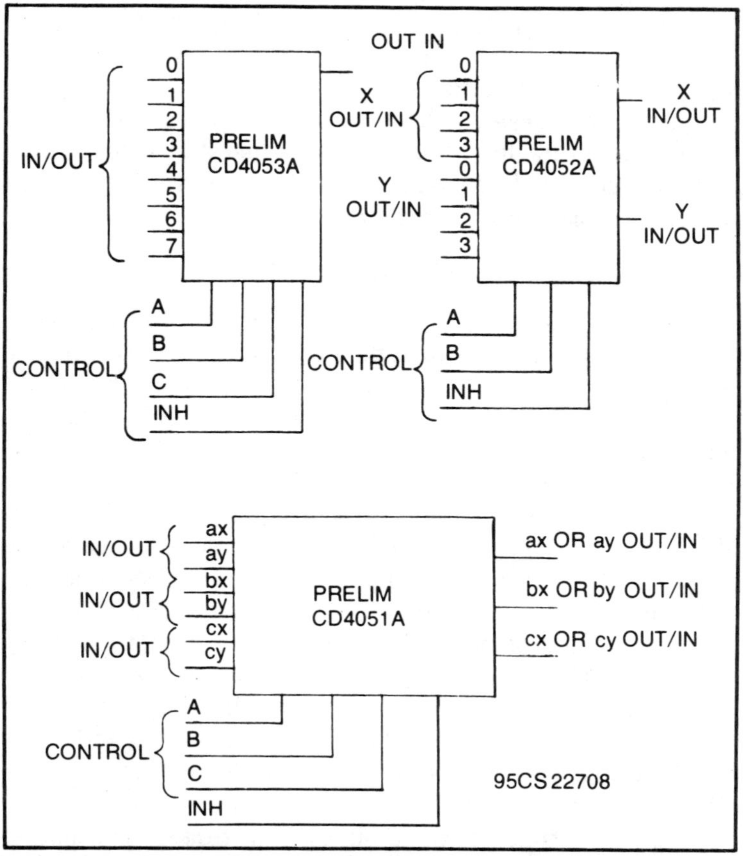

Fig. 12-6. 4051, 4052, and 4053 analog/digital muxers.

Again, like the 4016/4066 devices, the switches are bilateral. Three control terminals (A/B/C) select which one of the switches is turned on at any given time. These are programmed according to the ordinary binary sequence for the digits 0 through 7, such as 000 on A/B/C causes channel 0 to be on, 001 causes channel 1 to turn on, and so on. A binary counter, driven by a clock, will sequence through all eight possible conditions.

The 4052 device shown in Fig. 12-6 is a little different. It contains two independent 1-of-4 switches. The independence is not total, however. There is only one set of control terminals: A/B. These respond to the codes for 0 through 3. When A/B sees 000, both channel 0s are turned on. When the code is 01, both channel 1s are turned on. This process continues. We can view the 4052 as a pair of single-pole, four-position rotary switches.

The 4053 device shown in Fig. 12-6 is a triple 1-of-2 switch. It is, therefore, analogous to a three-pole, double-throw switch, or, perhaps, three SPDT switches ganged together.

All three of these CMOS devices can be used as multiplexers and demultiplexers. In most cases, though, the 4051 would be used for a single channel. If you were to multiplex together more than one data source, the other devices could be used.

IC MULTIPLEXERS/DEMULTIPLEXERS

The previous examples have been regular 4000-series CMOS devices pressed into service as mux/demux devices. The trend in recent years had been to use specialized integrated circuits that employ FET technology for the muxer/demuxer. In this section, certain commercial examples that have become popular are covered.

Figure 12-7 shows the Precision Monolithics, Inc. (PMI) MUX-88 multiplexer/demultiplexer device. It is a monolithic eight-channel switch using junction field-effect transistors, instead of the MOSFETs used in CMOS devices. This substantially reduces the susceptibility of the device to damage by static electricity discharges. The MUX-88 is versatile enough to operate from either TTL or CMOS logic levels and can therefore be used in almost any existing application. The switching action is make-before-break.

As in the simple example of Fig. 12-5, the output of the MUX-88 is connected to one side of all eight switches. The remaining terminal of each switch will go to the respective channels. Like the 4051, this device responds to a three-bit (octal) binary code that determines which switch is to be turned on at any given time.

FUNCTIONAL DIAGRAM AND TRUTH TABLE

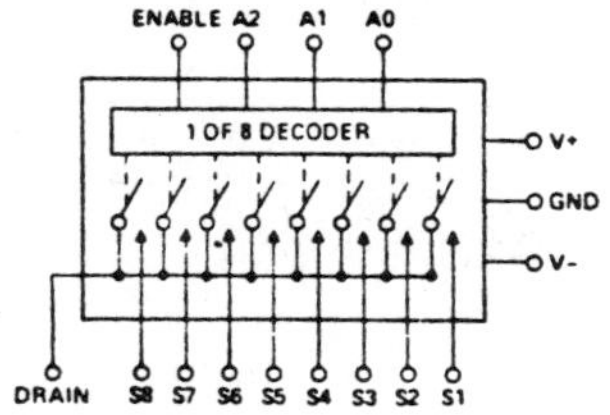

A_2	A_1	A_0	E_N	"ON" CHANNEL
X	X	X	L	NONE
L	L	L	H	1
L	L	H	H	2
L	H	L	H	3
L	H	H	H	4
H	L	L	H	5
H	L	H	H	6
H	H	L	H	7
H	H	H	H	8

PIN CONNECTIONS AND ORDERING INFORMATION

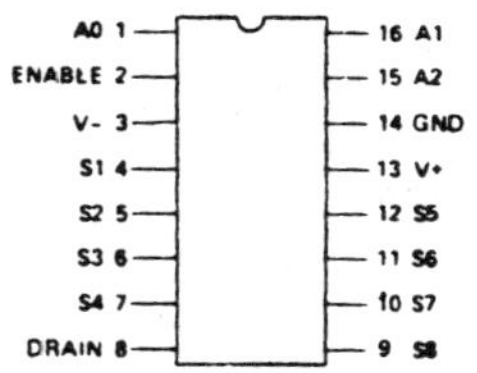

TOP VIEW

16 PIN HERMETIC DUAL IN LINE (Q Suffix)

MODEL	TEMP RANGE
MUX-88AQ	-55°C TO +125°C
MUX 88BQ	-55°C TO +125°C
MUX 88EQ	-25°C TO +85°C
MUX 88FQ	-25°C TO +85°C

Fig. 12-7. Precision monolithics MUX-88.

There is also a chip enable terminal that turns on and off the MUX-88.

The Datel MX-series of multiplexer/demultiplexers is shown in Fig. 12-8. This family of IC devices is useful in a variety of applications because it is compatible with TTL, CMOS, and even the old DTL logic system. Like the MUX-88, these are monolithic

Table 12-2. MX 1606 Channel Addressing.

8	4	2	1	Inhib.	On Channel
X	X	X	X	0	NONE
0	0	0	0	1	1
0	0	0	1	1	2
0	0	1	0	1	3
0	0	1	1	1	4
0	1	0	0	1	5
0	1	0	1	1	6
0	1	1	0	1	7
0	1	1	1	1	8
1	0	0	0	1	9
1	0	0	1	1	10
1	0	1	0	1	11
1	0	1	1	1	12
1	1	0	0	1	13
1	1	0	1	1	14
1	1	1	0	1	15
1	1	1	1	1	16

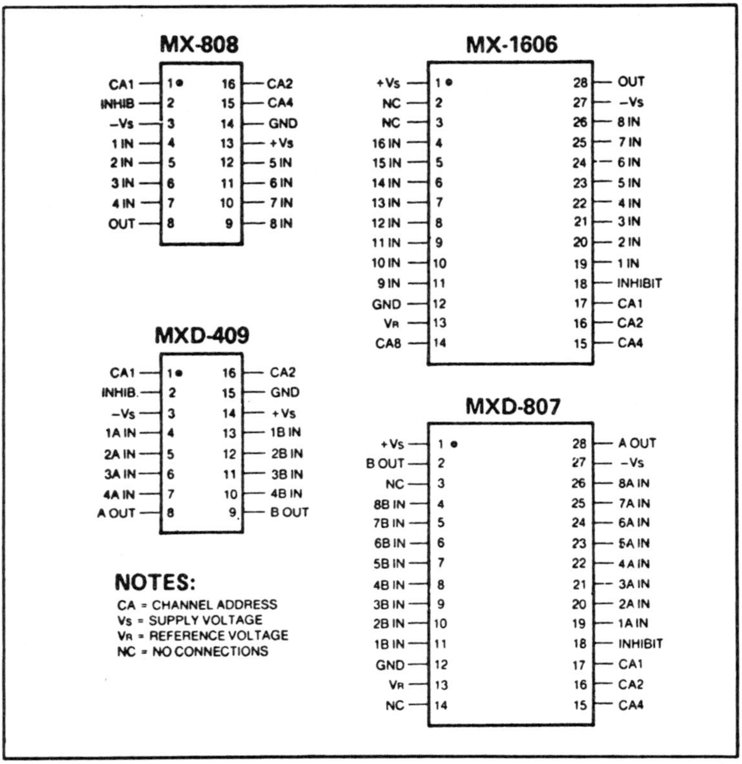

Fig. 12-8. Pinouts for MX-series muxers.

integrated circuits, rather than discreet or hybrid technology devices. There are different sizes of multiplexers in this family, but all use either 2-, 3-, or 4-bit binary codes as the switch select signals. Tables 12-2 through 12-4 show channel addressing.

One advantage the specialized CMOS devices has over standard devices is that transfer accuracy is generally guaranteed more

4	2	1	Inhib.	On Channel
X	X	X	0	NONE
0	0	0	1	1
0	0	1	1	2
0	1	0	1	3
0	1	1	1	4
1	0	0	1	5
1	0	1	1	6
1	1	0	1	7
1	1	1	1	8

Table 12-3. MX 1606 Channel Addressing.

2	1	Inhib.	On Channel
X	X	0	NONE
0	0	1	1
0	1	1	2
1	0	1	3
1	1	1	4

Table 12-4. MX 1606 Channel Addressing.

closely. In the Datel MX-series, for example, 0.01 percent transfer accuracy is obtainable at sample rates up to 200 kHz. The devices will tolerate bipolar signal excursions over the range ±10 V, so they can be used for either analog or digital muxers.

This series requires a power supply in the range of ±5 to ±20 Vdc. Interestingly, the power consumption is only 7.5 mW (standby condition). Even at a high sampling rate, in the 100-kHz range, the power consumption is only 15 mW.

Be careful about the total power consumption and the total package dissipation. The former refers to the amount of power required from the dc supply in order to power the internal circuits of the device. The total dissipation refers to the amount of power that can be dissipated by the device and includes the power loss in the switches. Because each switch has a series resistance—even though small—there will be a signal loss, and this translates as a power dissipation. In the Datel MX-series, the total allowable dissipation is on the order of 725 mW for the MX-808 and 1200 mW (that's 1.2 watts!) for the MX-1606 and MXD-807 devices.

The MX-1606 is a 1-of-16 channel signal-ended multiplexer. You might view it as an electronic SPDT switch. It uses a four-bit address select code; four bits are required for 16 channels because $2^4 = 16$. Similarly, the four-channel device (MXD-409) and the eight-channel device (MXD-807) use two-bit and three bit select codes, respectively.

The transfer accuracy of any electronic switching multiplexer depends upon the source and load resistances. The output voltage is given by the standard voltage divider equation:

$$E_{out} = \frac{E_m R_L}{R_s + R_{on} + R_L}$$

Where

E_{out} is the switch output voltage, measured across load resistor R_L.

E_{in} is the input voltage, or the open-circuit voltage of the input signal source.

R_L is the load resistance.

R_s is the output impedance of the signal source.

R_{on} is the resistance of the switch when on.

It doesn't take a mathematical genius to figure out that it is wise to keep the R_s and R_{on} terms of the equation as low as possible, and to keep the R_L term as high as possible. We can meet the low-R_{on} criterion by using an operational amplifier with a low output impedance at the input of the electronic switch. The load resistance requirement can also be met by using an operational amplifier. It should have an input impedance that is greater than 10^7 ohms. This is not too hard to obtain because some low-cost BiFET or BiMOS input op amps boast input impedances in the 10^{12} ohms range.

Figure 12-9 shows a scheme for using four 16-channel muxes, such as the Datel MX-series (MX-1606), to make a 64-channel multiplexer. A 6-bit select code is required to determine the chan-

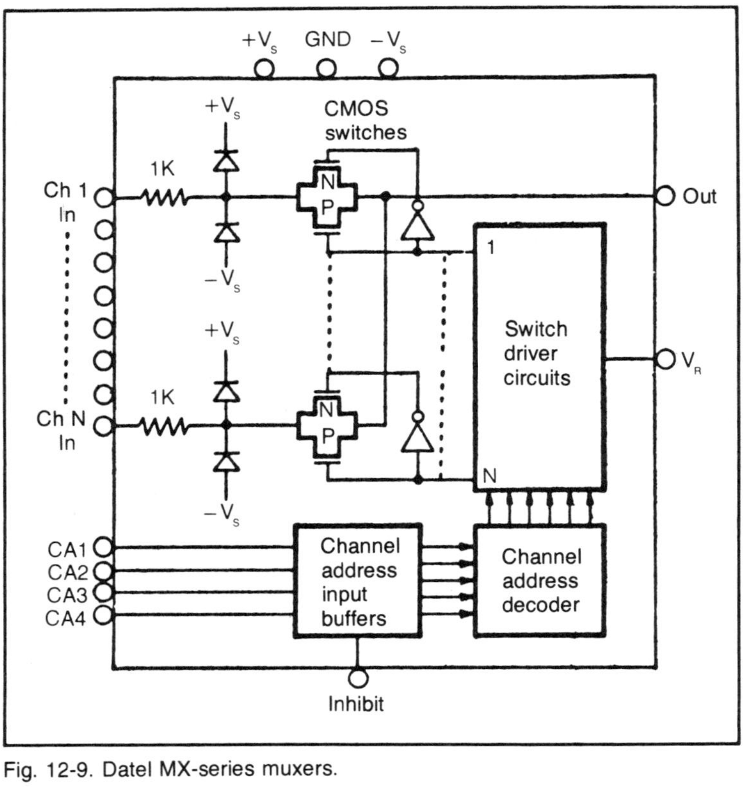

Fig. 12-9. Datel MX-series muxers.

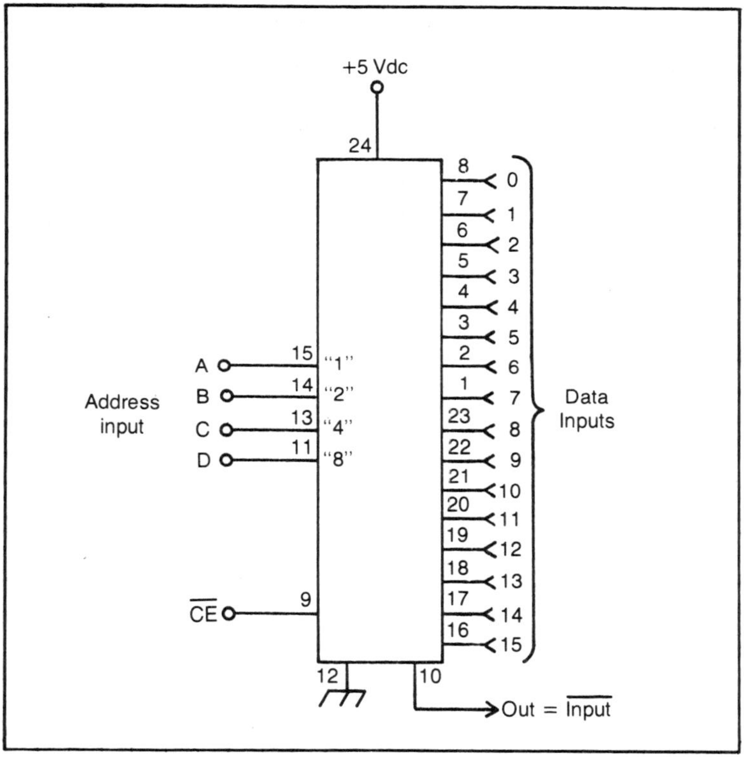

Fig. 12-10. 74150 IC.

nel that is turned on at any one time. To minimize loading, the common outputs are connected together at the noninverting input of a unity gain operational amplifier.

There are several different devices in the TTL line of digital logic ICs that are specifically designed as data selectors or data distributors, and these chips are intended for use in multiplexing and other applications. In this section we will consider only two of them, the 74150 Data Selector and the 74154 Data Distributor, with the understanding that you can look at Chapters 3 and 4 for other examples.

A circuit showing the application for the 74150 Data Selector is shown in Fig. 12-10. The 74150 device has sixteen separate input lines, and one output line. A four-bit binary word (0000-1111) applied to the ABCD inputs (weighting is the standard 1-2-4-8) will select one and only one input. The data (HIGH or LOW) applied to the selected input will appear at the output (pin no. 10) in complemented form. In other words, if a LOW is applied to the selected input, then a HIGH will appear on the output; similarly, when a HIGH is applied to the selected input a LOW appears on the output terminal.

The input selection is according to the following truth table:

A B C D	Input Selected	74150 pin no.
0 0 0 0	0	8
0 0 0 1	1	7
0 0 1 0	2	6
0 0 1 1	3	5
0 1 0 0	4	4
0 1 0 1	5	3
0 1 1 0	6	2
0 1 1 1	7	1
1 0 0 0	8	23
1 0 0 1	9	22
1 0 1 0	10	21
1 0 1 1	11	20
1 1 0 0	12	19
1 1 0 1	13	18
1 1 1 0	14	17
1 1 1 1	15	16

The 74150 chip has a *chip enable* terminal that will force the output terminal HIGH when inactive, and allows it to operate in the manner described above when active. Since $\overline{CE}$ is an active-LOW input, the device will be operative when pin no. 9 is grounded (LOW), and inoperative when pin no. 9 is at a potential greater than +2.4 volts (HIGH).

The opposite number for the Data Selector is the 74154 Data Distributor shown in Fig. 12-11. This integrated circuit will examine a data input (i.e., either HIGH or LOW level), and then distribute it to one of sixteen lines. Which of the sixteen lines is selected is determined by the four-bit address word applied to the A-B-C-D inputs. This action occurs in accordance with the same code given above for the 74150 device. When an output is selected, it will follow the data level applied to *Data In*.

One application for this IC is selection of I/O ports, memory banks or other devices. We can, therefore, use the 74154 to generate up to sixteen unique device select pulses. Whether it is an active-LOW or active-HIGH device select pulse will depend upon whether *Data In* is made HIGH or LOW. In some circuits, we will

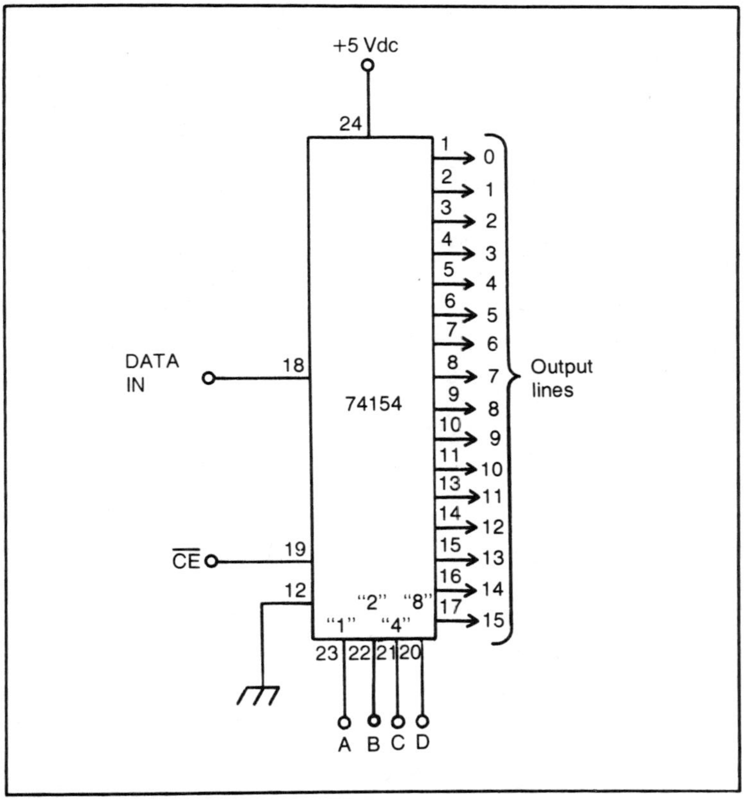

Fig. 12-11. 74154 IC.

normally keep *Data In* either HIGH or LOW. Should we want to select active-HIGH or active-LOW on command, we can add a switch at *Data In* that either grounds the terminal or connects it to +5-volts dc (i.e., HIGH). In some cases, we will want a programmable control over whether it is active-HIGH or active-LOW. In that instance we place a type-D flip-flop (or some similar circuit) at *Data In,* and then either *set* or *reset* depending upon how we want it programmed.

Chapter 13

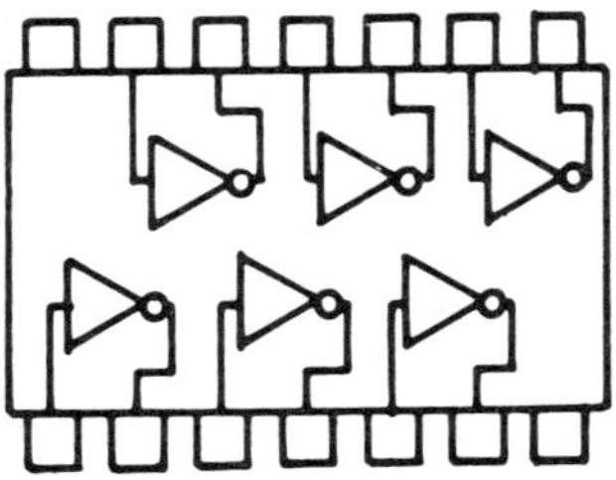

Rate Multipliers and Monostables

Books are supposed to be designed in a logical, free-flowing and rational manner so that the reader is lead by the hand through the subject matter with greatest ease. After writing thirty-three books and more than 200 magazine articles and engineering papers I can state categorically that it isn't always easy to meet that ideal goal! That is the reason for this chapter—it is an effort to cover (with as little pain as possible) some topics that were missed because they don't fit anyplace very neatly. The rate multipliers discussed below are a case in point. There are sufficiently few times when the main use of the rate multiplier is used, i.e., arithmetic circuits, that it didn't seem worthwhile to cover that topic in the limited space available. Hence, I decided to cram the topic of rate multipliers in where it might be a little uncomfortable.

The main topic of this chapter is the monostable multivibrator circuit. This circuit is also sometimes called the *one-shot* or *pulse stretcher* (a misnomer, even though very graphically illustrates what seems to happen!). We will find many and varied applications for the ubiquitous one-shot, so you should find this chapter highly useful. We will cover the making of one-shots from CMOS and TTL digital logic ICs, special-purpose ICs that were designed as one-shots (e.g., 74123 and 4528), and will even discuss how to make a TTL or CMOS compatible one-shot from an operational amplifier and a few miscellaneous components.

RATE MULTIPLIERS

The main function of the rate multiplier is to output a fixed number of pulses for every tenth input pulses. If we apply a count of ten pulses to the input, then we will output from 0 to 9 output pulses, depending upon programming. Figure 13-1 shows a TTL rate multiplier, the 74167 device. Programming is accomplished through a Binary Coded Decimal (BCD) four-bit word applied to the ABCD inputs. These inputs are weighted in the normal 1-2-4-8 manner.

There are two outputs on the 74167 device, pins 5 and 6. The normal output is pin no. 5, while the complemented output ($\overline{OUT}$) is pin no. 6. The *clear* line is normally held LOW for proper operation. If *clear* is brought momentarily HIGH, then the decade counter inside of the 74167 device is reset to zero, so the output drops LOW and $\overline{OUT}$ goes HIGH. The rate multiplier follows the rule:

$$F_{out} = M\,F_{in}/10$$

Where:

$M = D2^3 + C2^2 + B2^1 + A2^0$ for all decimal digits between 0 and 9.

Applications for the rate multiplier include both fixed and variable-rate frequency division, BCD arithmetic, D/A conversion, A/D conversion and certain other conversions of codes. The truth table for this device is complex, and I recommend that you see the Texas Instruments spec sheet if you want to use it (see *The TTL Data Book for Design Engineers–2nd Edition*, published by TI).

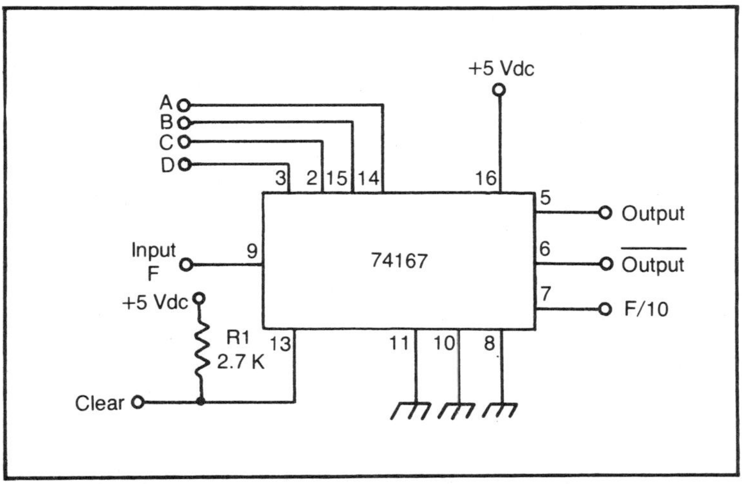

Fig. 13-1. 74167 circuit.

MONOSTABLE MULTIVIBRATORS

The monostable multivibrator, also sometimes called a *one-shot* circuit or (erroneously) a *pulse-stretcher* circuit, is designed to have but *one stable output state*. Normally, the output terminal will be in the dormant state, which (depending upon the design) will be either HIGH or LOW. When a trigger input pulse is received, then the output will go to the unstable state (the opposite of the dormant state) for a predetermined period of time after which it reverts to the dormant stable state. The trigger pulse can be either positive-going or negative-going, depending upon design, and in a few circuits both positive and negative pulses can cause a trigger action.

There are two basic forms of monostable multivibrator: *non-retriggerable* and *retriggerable*. The difference between these two categories lies in whether or not the circuit will respond to additional trigger pulses before the output period from the first pulse expires. Figure 13-2A shows the operation of the normal non-retriggerable monostable multivibrator. Initially, the output (Q) is

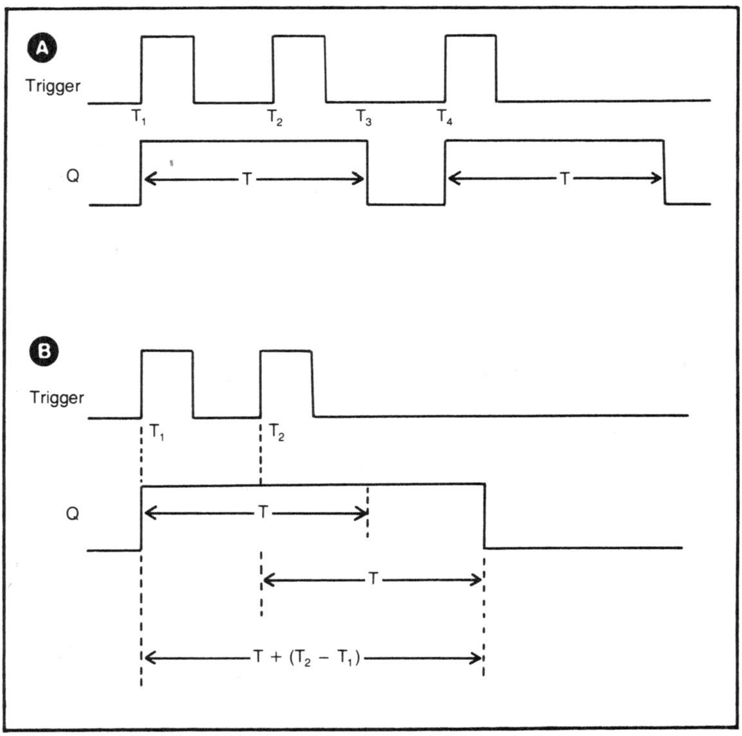

Fig. 13-2. Timing waveforms.

dormant, which in this case is the LOW condition. At time t1, however, a positive-going trigger pulse is received so the output snaps HIGH. The HIGH state is unstable, so will only last for a predetermined period of time, *t*. At time t2 another trigger pulse is received. Since the output is already in the unstable state the t2 pulse has no effect on the output timing; it is essentially "wasted." At time t3, the output time expires so the output level drops LOW again. A third trigger pulse is received at time t4, so the output level snaps HIGH again for another period *t*.

The principal action noted in the above description of the nonretriggerable monostable multivibrator is that additional trigger pulses will not cause additional action on the output timing. There are several advantages to this action. Switches, for example, do not make a firm, absolute closure, but tend instead to "bounce." Contact bounce causes a string of short duration pulses to occur on closure. This phenomenon is not terribly important in most circuits, but can be disastrous in digital circuits because each contact bounce pulse will be seen as a valid input pulse by the digital circuit. As a result, some designers will use a nonretriggerable one-shot as a switch contact debouncer. In that application, the time period of the one-shot is set to be long enough to allow contact bounce to settle down, yet short enough that human operators will not notice the delay in operation once the switch is depressed. To the human, the operation seems instantaneous, but to the digital circuitry it is actually delayed a few milliseconds.

The operation of the retriggerable monostable multivibrator is shown in Fig. 13-2B. In this case, the output is triggered HIGH by t1 and would normally remain HIGH for period *t*. But at time t2 a second trigger pulse is received which causes the period of the one-shot to be held for an additional period *t*. The result is the one-shot output remains HIGH for a period equal to *t* plus the expired portion of the original period (i.e., $t + (t_2 - t_1)$).

An application of the retriggerable monostable multivibrator is in alarm circuits. For example, consider a heart-rate meter in a medical ECG monitor. The alarm is to sound if the patient's heart stops beating so that the staff can perform whatever medical miracles are appropriate. The alarm would be held *off* as long as the one-shot is in the unstable state (e.g., HIGH in Fig. 13-2). Each time a trigger pulse is received the one-shot retriggers for another period *t*. If the trigger pulses are generated by the ECG signal (an easy trick, actually), then the one-shot will not time out as long as the patient's heart is beating. But, if the patient tries to die, then the

ECG pulses stop and this allows the one-shot to time out. At the end of the last *t* period the output drops LOW and thereby sounds the alarm.

QUASI-MONOSTABLES

There are a number of circuits that will give us either true or quasi-monostable operation. In this section we will consider the so-called quasi-monostable circuits. These are useful for many applications, but suffer from the constraint that the ***trigger pulse must have a duration longer than the output pulse***. If this criterion is met, then the circuit will operate in quasi-monostable fashion.

Figure 13-3 shows two quasi-monostable circuits in one. Either inverting (A1) or noninverting (A2) CMOS devices can be used, and will produce opposite polarity output states. In this case, the CMOS 4049B device is used for the inverting type of circuit, while the CMOS 4050B is used for the noninverting. Note that these chips are "hex" devices, meaning that there are six independent stages inside of each IC package. We will, therefore, find six noninverting buffers inside of the 4050B device, and six inverters inside of the 4049B; only one is needed so the remaining five stages can be used for other purposes or left unused.

The timing of the output pulse duration is set by resistor-capacitor network R1C1. The duration is given approximately by:

$$t = 0.8\ R1C1 \qquad \textbf{(13-1)}$$

Where:

t is the time in *seconds* (s).
*R*1 is expressed in *ohms*.
*C*1 is expressed in *farads*.

Example

Find the duration of a monostable multivibrator that uses a 100 kohm resistor for R1, and a 0.1 μF capacitor for C1. Note that 0.1 μF is 0.0000001 F (i.e., 1×10^{-7} F).

Solution

$t_{sec} = 0.8\ R1C1.$
$t_{sec} = (0.8)\ (10^5 \text{ ohms})\ (1 \times 10^{-7} \text{ F}).$
$t_{sec} = (0.8)\ (10^{-2})$ seconds.
$t_{sec} = 0.008$ seconds $= 8$ milliseconds.

Of course, in the usual fashion of textbooks we have presented you with the wrong form of Eq. 13-1. In most cases, we will know the approximate duration desired from other aspects of the circuit design and will want to calculate the component values that will result in that duration. This determination requires a little algebraic manipulation of Eq. 13-1 as follows:

If the capacitor value is known:

$$R1 = t/(0.8\ C1) \qquad \textbf{(13-2)}$$

or, if the resistor is known:

$$C1 = t/(0.8\ R1) \qquad \textbf{(13-3)}$$

In most practical cases, we will select the capacitor value and then calculate the resistor value. This procedure is done because there are fewer standard capacitor values than resistor values, so we are constrained a little tighter. The usual method is to select some capacitor value and then use Eq. (13-2) to find the resistor value that will yield the correct duration. This procedure may have to be reiterated several times using different values of C1 until a combination is produced in which both resistor and capacitor are standard values. We have a little more latitude in the resistor value because there are so many available, especially in precision resistor lines. Even in nonprecision resistors, however, we can oftentimes select a value from a standard value close to the normal. A population of ±10 percent resistors is likely to yield a precise value close to the desired value if enough are tested. It is also possible that we can tolerate a little slop in the actual period *t*. Sometimes, the value of t need only be a certain minimum (as in the case of contact debouncing), or, can be anything within a large range between minimum and maximum values.

Figure 13-3B shows the timing waveform for this circuit. Keep in mind that the CMOS inverter will change output states when the input trigger potential is midway between V+ and V−; or, in the case where V− is ground (zero volts), then the transition point for the output will be when the input is ½(V+). When the trigger pulse reaches this potential, therefore, the output will snap to the unstable condition, i.e., HIGH for V_{02} and LOW for V_{01}.

When the trigger pulse is received, it causes capacitor C1 to begin charging. When the capacitor voltage reaches a point where it

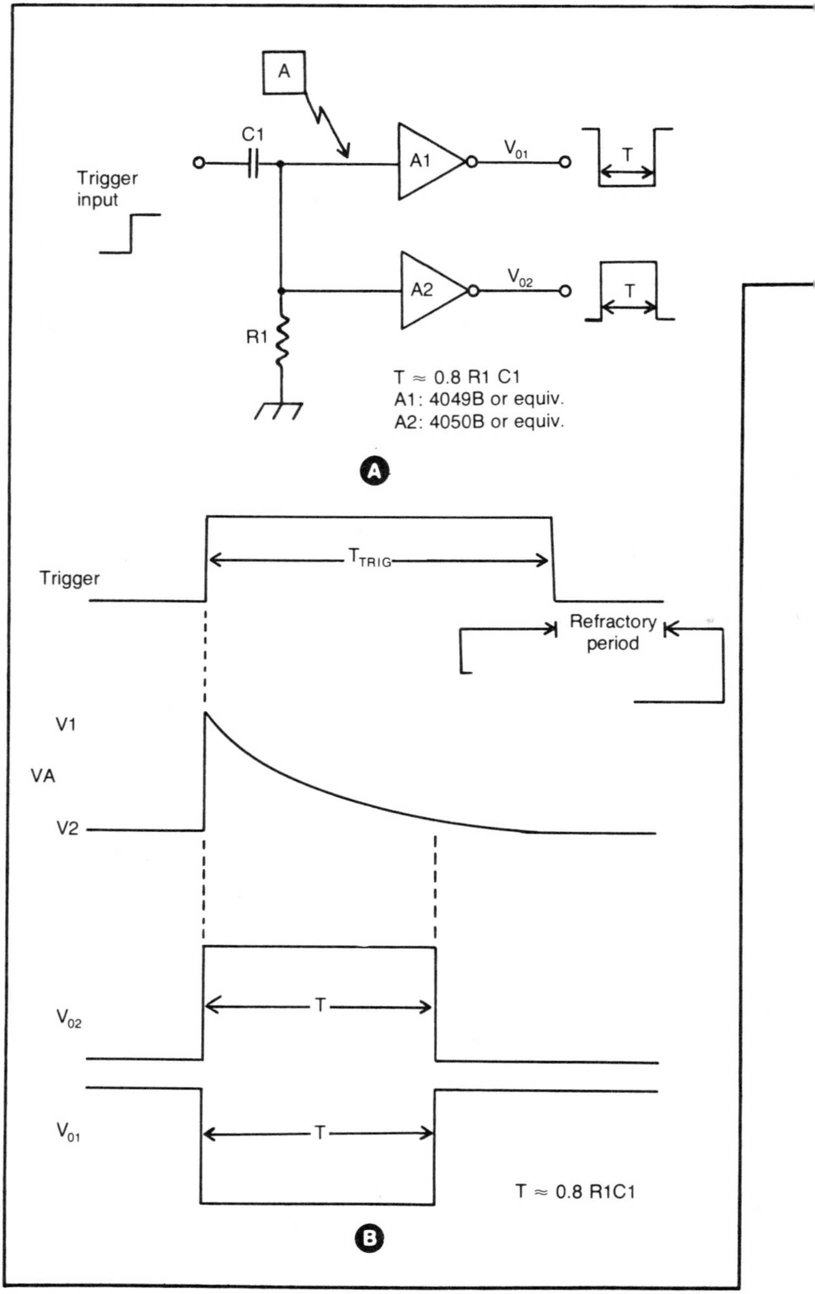

Fig. 13-3. (A) Monostable multivibrator, (B) timing waveforms, (C) alternate triggering.

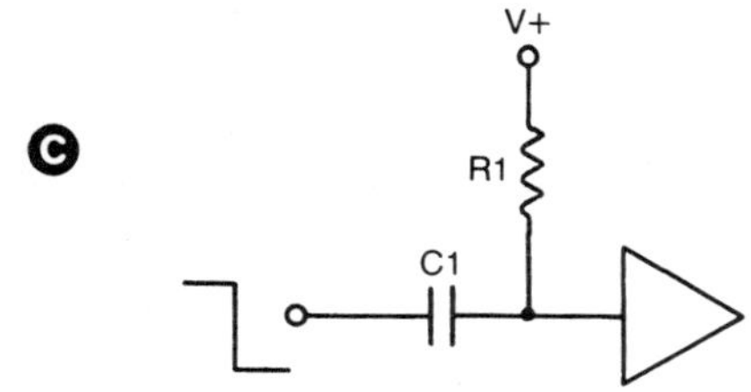

is at the trigger potential for the CMOS device, then the output drops LOW again. This is the reason why the trigger pulse duration must be longer than the output duration; otherwise, the capacitor would never charge.

The circuit as shown in Fig. 13-3A will trigger on positive-going input pulses. If we want to trigger on negative-going pulses, then we must modify the input circuitry according to Fig. 13-3C. In this case, the resistor is returned to V+ rather than ground. Otherwise, the circuit is identical to Fig. 13-3A.

POWER-ON RESET CIRCUIT

A circuit related to the quasi-monostable multivibrator is shown in Fig. 13-4. This circuit is used in computers and in other digital instruments to generate a *power-on reset pulse.* Such a pulse is often applied every time power is applied to the circuit in order to ensure that the computer starts in a correct mode. The pulse will reset counters, flip-flops and microprocessor chips to the state required for initial operation.

Timing of the output pulse is set by resistor R1 and capacitor C1. Initially, as the instant power is applied, the voltage is zero (or anything below ½ V+), the inverter sees a LOW at its input so produces a HIGH output; this HIGH is the reset pulse. After power is applied, however, capacitor C1 will charge from the V+ power supply at a rate determined by resistor R1 and the value of C1. At time t1, when the capacitor potential reaches ½ V+, the output transition occurs on LOW because the inverter now sees a HIGH input. The capacitor will continue charging to a level equal to V+, where it will remain until the power supply is turned off. As a result, the only time the CMOS inverter sees a LOW will be between turn-on at t_0 and t_1; at all other times the input sees a HIGH so the output is LOW. When power is turned off, capacitor C1 will discharge through R1. It is, incidentally, important to make the dura-

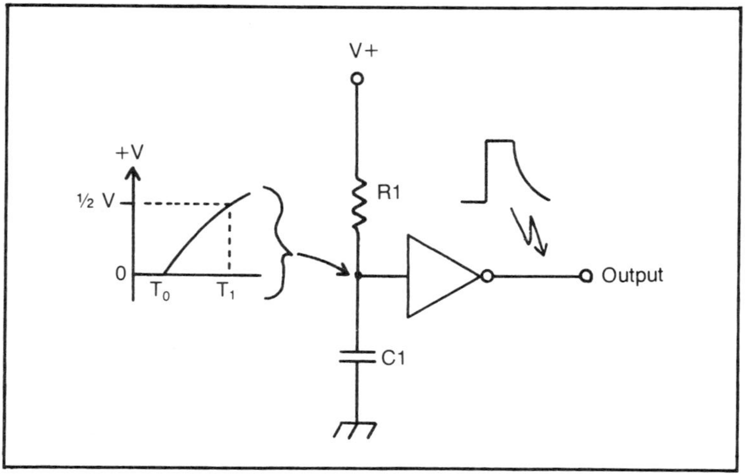

Fig. 13-4. Timing waveforms.

tion of the power-on reset pulse sufficiently long enough to do the job. It should remain HIGH long enough for the dc power supply to come to full operating potential, plus a few milliseconds longer. The reason for this is that not all IC devices can respond at potentials as low as the CMOS inverter can use, so the power-on reset pulse may come and go before some critical chip is ready to be reset or initalized. This may well be the problem with some digital devices that have to be turned on once or twice in order to achieve proper operation.

TTL MONOSTABLES

There are several different TTL monostable multivibrators available, among them the 7412x series (i.e., 74121, 74122, and 74123). These devices can offer nanoseconds to many milliseconds, using capacitors of 10 picofarads, or more, and resistors from 2 to 40 kohms. These devices also offer multiple triggering options including both positive-going and negative-going pulses. Timing of the output duration is by RC network. Although some variation in the timing equation is permitted (see TI data book for TTL devices), it is possible to use the simple approximation of Eq. 13-4:

$$T = 0.69\ RC \qquad \textbf{(13-4)}$$

(all terms are as described for Eq. 13-1)

Figure 13-5A shows the circuit for the 74121 device. The

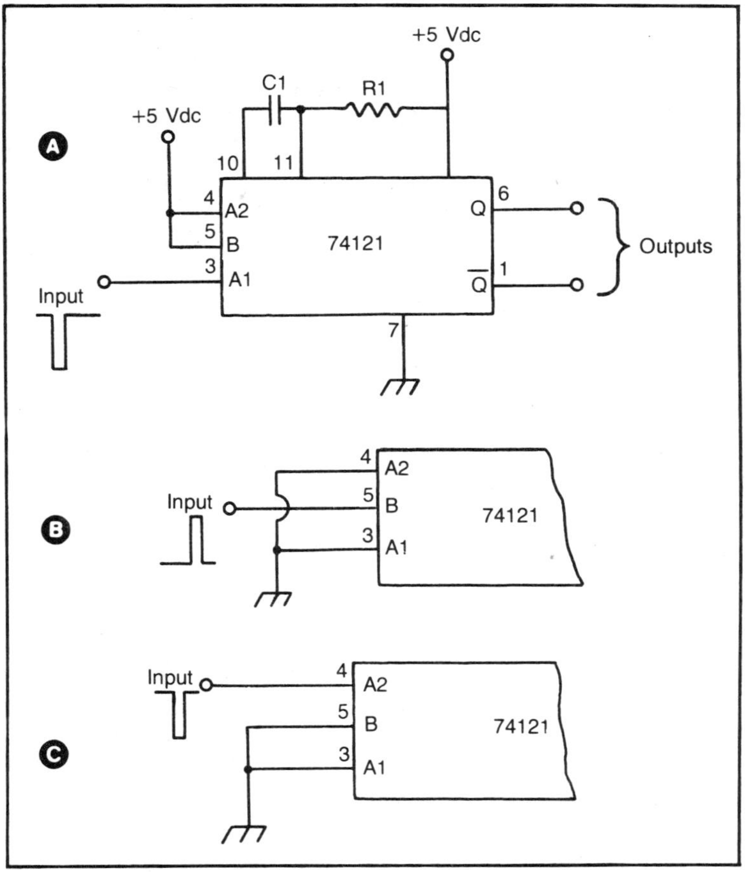

Fig. 13-5. (A) 74121 circuit, (B and C) alternate input circuit.

74121 is a single nonretriggerable monostable multivibrator that offers complementary (Q and not-Q) outputs. Timing is by R1C1. There are three triggering options:

	A1	A2	B
Positive level-triggered at B input, with Schmitt input snap action	LO	LO	T
Negative-going on A2	HI	T	HI
Negative-going on A1	T	HI	HI

In the example of Fig. 13-5A, we see the case of option-3, in which terminals A2 and B are held HIGH (by connection to +5 volts dc), and a negative-going trigger pulse is applied to A1. The remaining two triggering options are shown in Figs. 13-5B and 13-5C.

Figure 13-6A shows the 74122 device, which is described as a single, retriggerable one-shot with an active-LOW *clear* terminal. As in all other digital circuits, the function of clear is to force the chip to the condition of Q = LOW.

Like the 74121 device, the timing of the output pulse is determined by R1 and C1, with value constraints of 5 kohms to 40 kohms for R1, and "greater than 10 pF" for C1.

We can manipulate the *A1, A2, B1, B2* and *clear* inputs to achieve several different triggering options. Put into simplified

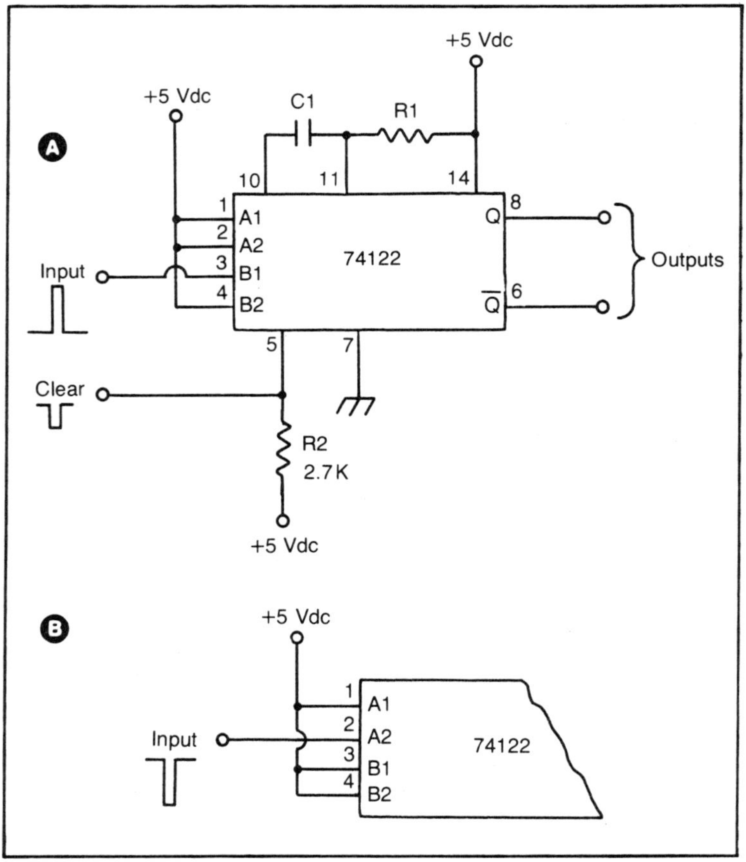

Fig. 13-6. (A) 74122 circuit, (B) alternate input circuit.

form, we can make the following generalizations regarding 74122 triggering:

	A1	A2	B1	B2
Positive-going (LOW-to-HIGH)	H	H	T	H
Negative-going (HIGH-to-LOW)	H	T	H	H

The entire circuit, wired in this case for a positive-going trigger pulse, is the subject of Fig. 13-6A; the modifications of the trigger circuit for a negative-going trigger pulse are shown in Fig. 13-6B, otherwise the circuits are identical.

The *clear* terminal is an active-LOW input, so must be held HIGH all of the time unless it is intended to be active. The HIGH condition is achieved by connecting a 2.7 kohm pull-up resistor between the clear input (pin no. 5) and +5 volts dc. When it is desired to clear the 74122, pin no. 5 is brought momentarily LOW. This action has two effects: (1) it clears the output (i.e., causes Q = LOW and not-Q = HIGH), and (2) inhibits the trigger input from responding to further pulses (this neat trick can be used to disable the one-shot—just keep the clear input permanently LOW unless you want it to respond!).

Figure 13-7 shows the 74123 device, which is a TTL dual retriggerable monostable multivibrator. Both one-shot stages within the 74123 are independent of each other except for power-supply connections. As was true with the two previous one-shot devices, the output time duration is set by an RC network external to the chip; the resistor should be between 5 k and 25 k, while the capacitor must be 10 picofarads or more.

Both positive-going and negative-going trigger pulses can be accommodated. The chip obeys the following protocol:

	A	B
Positive-going (LOW-to-HIGH)	L	T
Negative-going (HIGH-to-LOW)	T	H

As with the 74122 device, the monostable multivibrators in this chip have a *clear* input that functions to drive the Q output LOW, and to inhibit the trigger input. Clear is an active-LOW terminal.

The pinouts shown in Fig. 13-7A show the terminals for both one-shots within the 74123. The pins for one-shot "A" are outside of the parentheses, while those for "B" are within the parentheses. The triggering arrangement in Fig. 13-7A is the positive-going version; negative-going triggering is shown in Fig. 13-7B.

The 7412X series of monostable multivibrators are high-frequency devices that will produce very short pulses, as well as long pulses. The trigger inputs will also respond to short duration pulses, and this makes them subject to noise pulses. Some designers think these chips are a little flaky, so tend to use other circuits when a one-shot is needed, except for the very shortest duration periods.

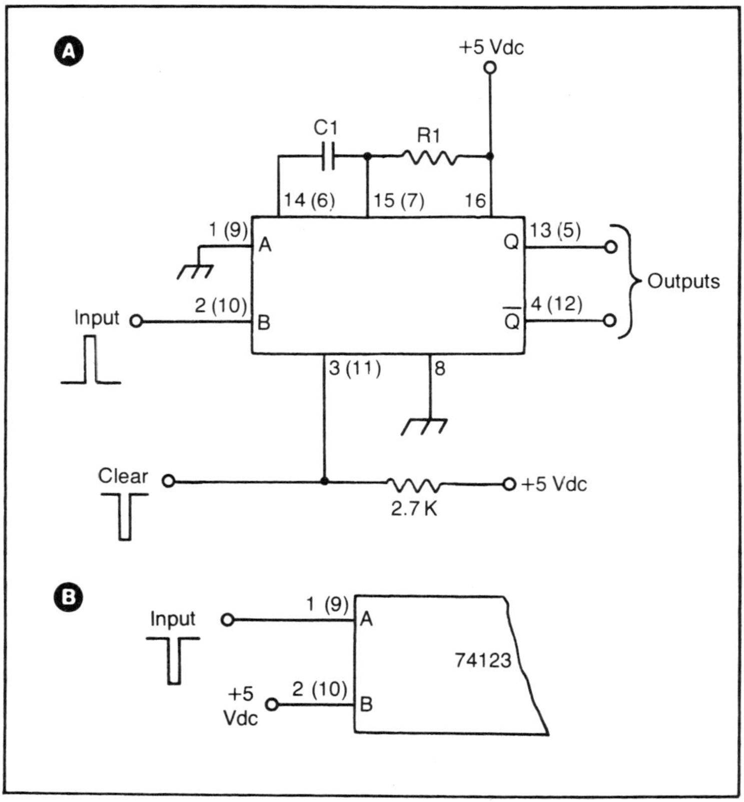

Fig. 13-7. (A) 74123 circuit, (B) alternate input circuit.

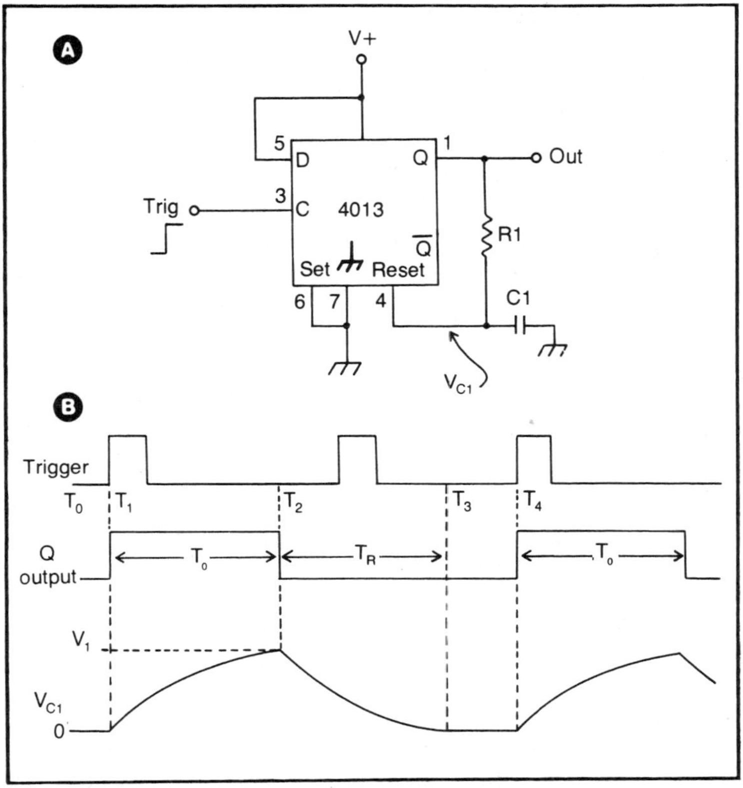

Fig. 13-8. (A) 4013 monostable, (B) timing waveforms.

CMOS ONE-SHOT

CMOS devices can also be used as one-shot stages, even though they are not nearly as fast as the TTL one-shots described previously. In a lot of ways, however, the CMOS devices are more well-behaved than are the TTL devices.

Figure 13-8 shows a one-shot made from a 4013 type-D flip-flop (there are actually two type-D FFs inside of the 4013 device, only one is needed). Before describing the operation of this circuit let's first review the operation of the type-D flip-flop. In the type-D FF, the data applied to the "D" input will be transferred to the "Q" output only when the clock input is HIGH. If, for example, the D-input is LOW, then the Q output will go LOW also when the clock goes HIGH. Similarly, when the D-input is HIGH, the Q output will go HIGH when the clock is HIGH. Of course, if the clock input remains HIGH, then the Q output will follow the data transitions on the D-input. Keep in mind that the Q and not-Q outputs are complementary; i.e., when one is HIGH the other will be LOW. The reset input is active-HIGH, and will force the type-D FF into the condition where the Q output is LOW.

The circuit for the one-shot is shown in Fig. 13-8A, while the timing diagram is shown in Fig. 13-8B. Initially, at time t_0 the one-shot is at rest, so Q = LOW. The trigger input is the clock, so controls the operation of the circuit. The D-input is wired permanently LOW. Timing occurs through the action of R1 and C1. Capacitor C1 charges through R1 when Q = HIGH, and will discharge through R1 also when Q = LOW. When the voltage across C1 reaches a critical threshold, then the reset input thinks it sees a valid HIGH condition, so will force the FF into the Q = LOW condition.

When a trigger pulse is received at time t1, the FF thinks it sees a valid clock pulse, so transfers the HIGH applied to the D-input to the Q output. The output duration commences at this point. Since Q is HIGH, a current will flow through R1 to charge capacitor C1. Voltage V_{C1} rises at a rate that is determined by the values of R1 and C1 until it reaches the threshold voltage (V1) at which the 4013 device thinks it sees a HIGH on the reset terminal. At that instant (t2), the type-D FF is reset, so the Q output goes LOW again.

Note what happens when the pulse following t2 arrives—the one-shot is *not* retriggered despite the fact that a valid trigger pulse is received. The reason for this is that the voltage across C1 has to decay to a point where the 4013 thinks it sees a LOW, otherwise the chip remains reset. The duration of this so-called "refractory period" is exaggerated in Fig. 13-8B for purposes of illustration, but it is nonetheless real and must be considered when designing circuits using this kind of stage.

A "fix" for the refractory period problem is shown in Fig. 13-9A, with appropriate timing diagram in Fig. 13-9B. The solution seems to be placing a diode in parallel with the timing resistor. When Q = HIGH, diode D1 is reverse biased, so is inert. The high leakage resistance of silicon diodes will not materially affect the timing of the circuit, despite the fact that it's in parallel with R1. When Q = LOW, on the other hand, the diode will be forward biased, so will cause the charge in C1 to dump rapidly. This is shown as a greatly reduced refractory period in Fig. 13-9B. There is a problem, however, but it need not be of great concern to most users of this circuit. Note that the V_{C1} voltage never drops all the

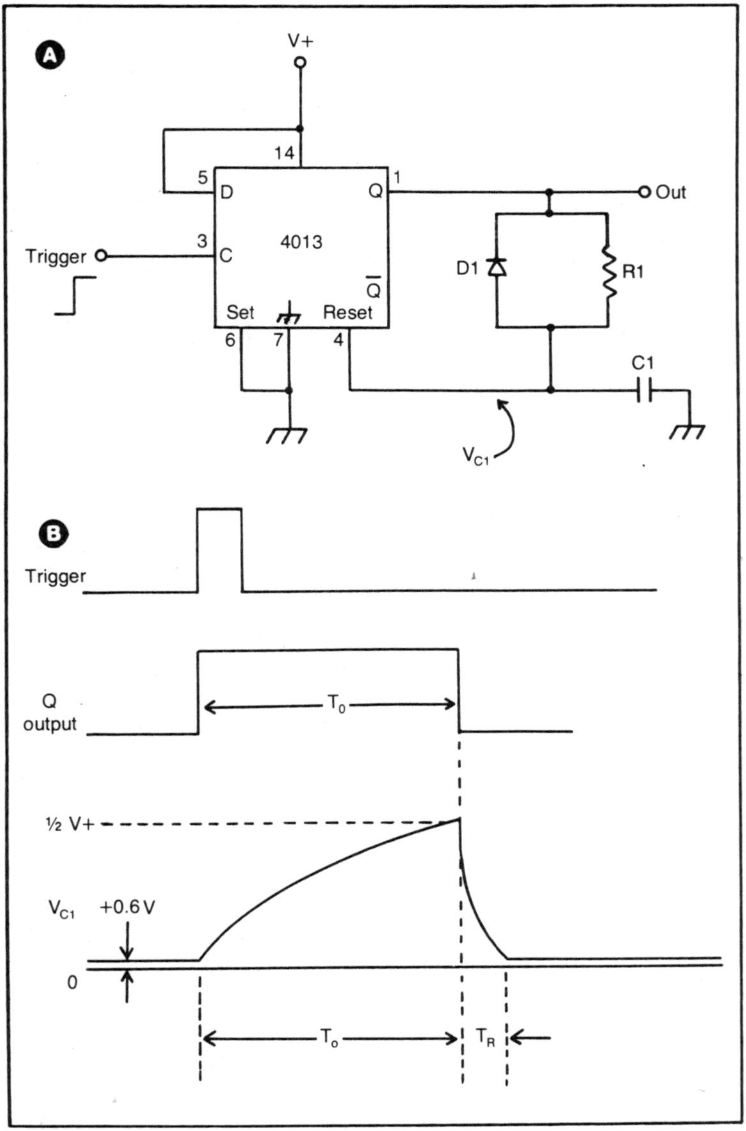

Fig. 13-9. (A) Improved circuit, (B) timing waveforms.

way to zero. It will remain slightly positive because of the junction voltage of D1 (i.e., 0.6 to 0.7 volts).

Figure 13-10 shows a simple method for making the monostable multivibrators shown in Figs. 13-8 and 13-9 *retriggerable*. The idea is to dump the charge in C1 using a new trigger pulse. This is

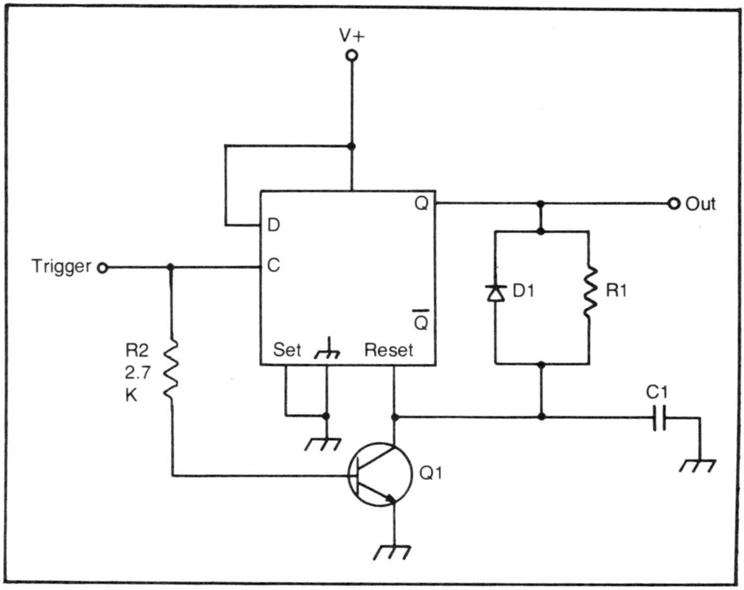

Fig. 13-10. Retriggerable monostable.

accomplished by connecting a transistor switch across capacitor C1. When a trigger pulse is received, it will be a positive level. This positive signal will forward bias transistor Q1, causing it to saturate and thereby short out the timing capacitor. Under this circumstance, the refractory period is reduced almost to zero and the one-shot is set up to continue the output period.

A second CMOS one-shot circuit is shown in Fig. 13-11A. This one is a little more conventional in that the chip was designed to be a one-shot, rather than adapted to that service as was the 4013B device of previous examples. The 4528 device actually contains two independent one-shots. Each section of the 4528 contains two *triggering inputs* (for two options), and a *clear* input. The clear input will return the circuit to the dormant condition of Q = LOW, and will also inhibit further trigger pulses.

Timing in the 4528 device is simpler than in some one-shots because the output time duration is merely the product of the resistance and the capacitance:

$$t = R1C1$$

Where:

t is in seconds.

R1 is the value of R1, in ohms (10 kohm to 10 megohm).

*C*1 is the value of C1 in farads (more than 20 pF, i.e., 0.000000000020 F)

Example:

Find the duration of the output pulse from a 4528 if R1 = 560 kohms and C1 = 470 pF.

Solution:

t = R1C1.

t = $(5.6 \times 10^5$ ohms$) (4.7 \times 10^{-10})$.

t = 0.00026 seconds = 0.26 milliseconds.

Like other one-shot ICs, there are two triggering options used on the 4528 device: positive-going and negative-going. The circuit in Fig. 13-11A shows the positive-going version, while modifications necessary to achieve negative-going triggering are shown in Fig. 13-11B.

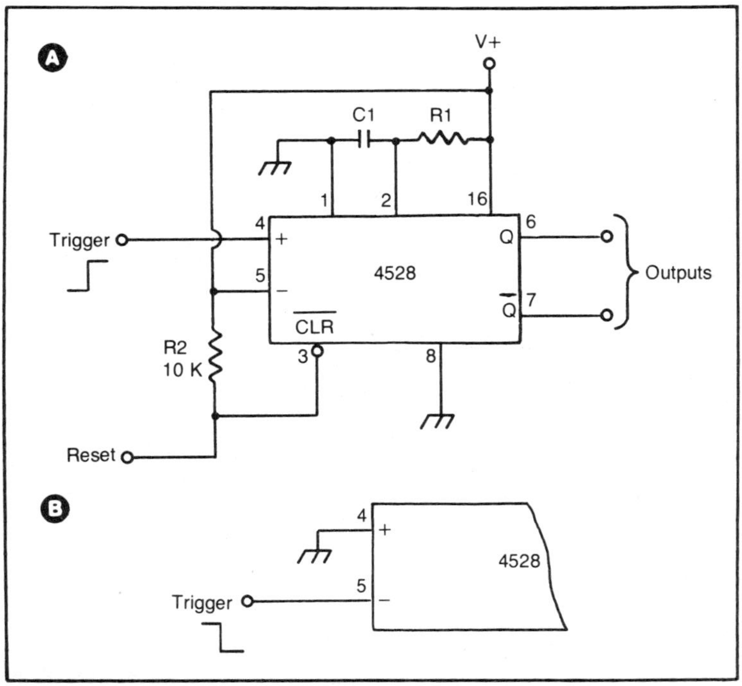

Fig. 13-11. (A) 4528 monostable multivibrator. (B) 4528 monostable multivibrator alternate input circuit.

The reset line shown in Fig. 13-11A is connected to the *clear* input of the 4528. This input is active-LOW, so will be held permanently HIGH except when reset operation is desired. Resistor R2 is used as a pull-up to V+ to ensure $\overline{CLR}$ stays HIGH. Bringing $\overline{CLR}$ momentarily LOW will cause the output to go to the Q = LOW condition.

OP-AMP ONE-SHOTS

Although not strictly a TTL or CMOS circuit, the circuit in Fig. 13-12A can be made compatible with *both* TTL and CMOS devices, yet is based on an operational amplifier. The timing diagram is shown in Fig. 13-12B. When a negative-going trigger pulse is received on the trigger input, the output voltage snaps from $+V_0$ to $-V_0$ where it remains for a period of time *t*. During this period, capacitor C1 will charge at a rate determined by V_0, the value of R1 and the value of C1. When it reaches a potential V_f the voltages on both inverting and noninverting inputs will be equal, so the operational amplifier output will snap back to its original condition. Capacitor C1 will then begin to discharge through *R*1 until it reaches its resting point. This latter potential is clamped to 0.6 to 0.7 volts by diode D4.

Timing is set by V_0, R1, C1, R3, and R4. If R3 = R4, then the output duration is given by:

$$t = R1C1$$

Where:

t is in seconds.
R1 is the value of R1, in ohms.
C1 is the value of C1 in farads.

Example:

Find the output duration when R1 = 100 kohms, C1 = 0.1 μF.

Solution:

$t = 0.8\ R1C1.$
$t = (0.8)\ (10^5\ \text{ohms})\ (10^{-7}\ \text{F}).$
$t = 0.008$ seconds = 8 milliseconds.

Of course, as is too often the case with textbooks, the equation above is not arranged in its most useful form. Usually, we know the desired duration but do not know the capacitor and resistor values

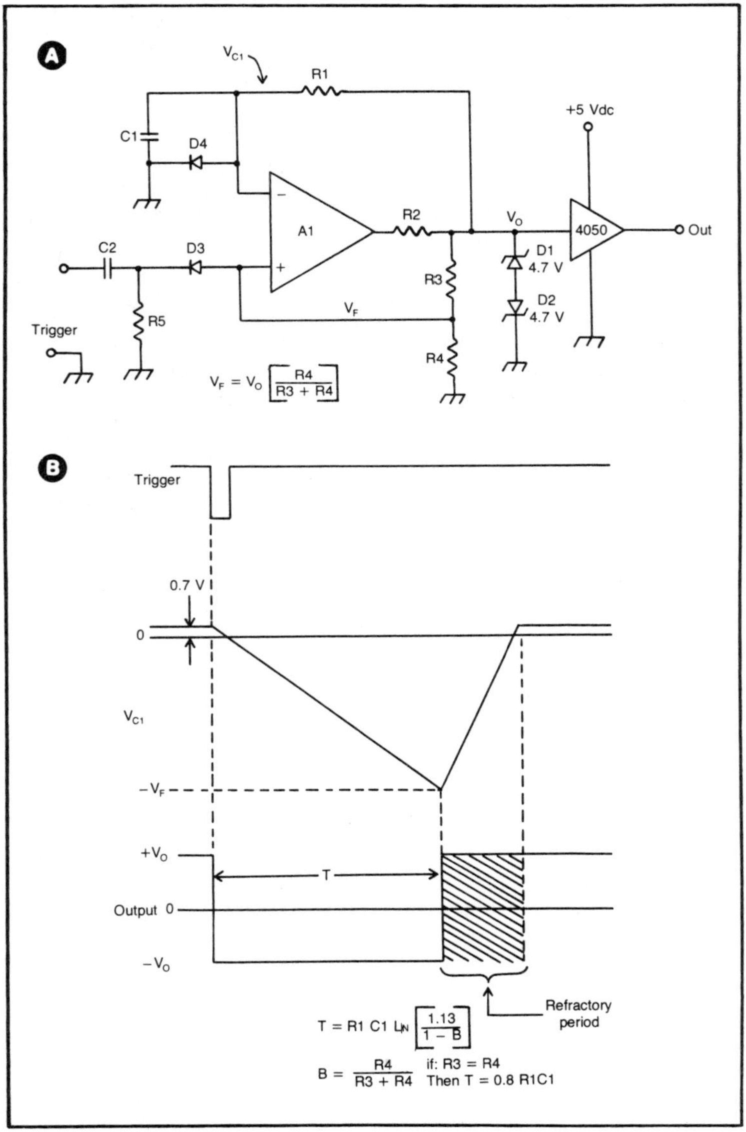

Fig. 13-12. (A) Operational amplifier one-shot with TTL output, (B) timing waveform.

which yield that value. In most cases, the designer will select a value of capacitor from a table of standard values (it's amazing how often 100 pF, 0.001 μF, 0.01 μF and 0.1 μF are selected!) and then calculate a resistor (there are more standard value resistors than capacitors). Therefore, we will rewrite the equation in the following form:

$$R1 = t/0.8\ C1$$

The zener diodes across the output of the operational amplifier are used to clip the output waveform. This accomplishes two jobs. First, it limits the output voltage excursion to some value compatible with stages to follow, and, second, it sharpens the corners of the output waveform. The later action occurs because saturated operational amplifiers do not recover instantaneously.

Ordinarily, we would not need the 4050 CMOS device at the output when CMOS chips follow the operational amplifier one-shot. But when TTL devices are driven by this circuit, we will need the 4050 (or, alternatively, a 4049 if we can tolerate or desire inverted output). The 4049/4050 devices have the interesting property of being TTL compatible *if* the entire package V± voltages are +5 volts dc and ground.

Chapter 14

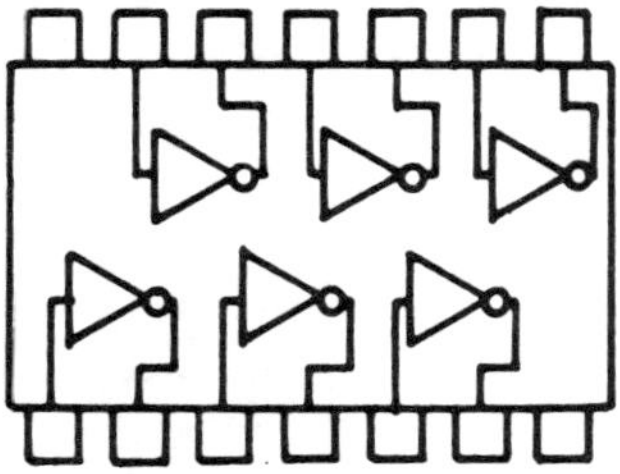

Binary Arithmetic Circuits

Binary Arithmetic operations are performed by the half-adder, and subtractor circuits. These are known as *arithmetic circuits*. Also, in this chapter you will learn about the different types of arithmetic logic chips available to perform these functions.

ADDER CIRCUITS

There are two basic forms of adder circuit. These are called the *half-adder* and *full-adder* (or simply *adder*). Of these, several different types exist. But before examining the various types of circuits, consider the process of binary addition to find out the requirements placed on any circuit that purports to be a binary adder.

The rules for binary addition are as follows:

$$0 + 0 = 0$$
$$0 + 1 = 1$$
$$1 + 0 = 1$$
$$1 + 1 = 0 \text{ plus carry } 1$$

Examine these rules and compare them with the truth table for an Exclusive-OR gate. An XOR gate obeys the following rules:

- ☐ If both inputs are 0, then the output is 0.
- ☐ If both inputs are 1, then the output is 0.
- ☐ A 1 output is created if either, *both not both*, inputs are 1.

The truth table for the XOR gate, then, is

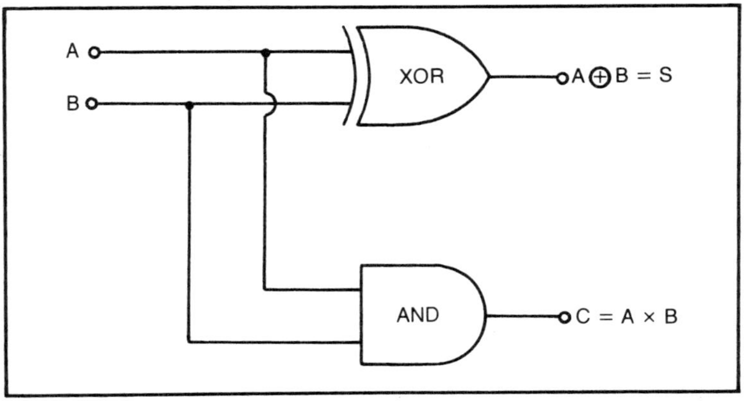

Fig. 14-1. Half adder.

A	B	Output
0	0	0
0	1	1
1	0	1
1	1	0

Note that the sole difference between the truth table for the XOR gate and the rules for binary addition is that, in addition, a carry-one output is created if both A and B are 1. We can therefore create an addition circuit with an XOR gate and a means for generating the carry-one output.

Figure 14-1 shows a circuit for a half-adder. A half-adder is a circuit that generates two outputs: sum ($S = A + B$) and carry ($C = A \times B$) (read: S = A XOR B and C = A AND B). The Exclusive-OR gate generates the sum output, while an AND gate generates the carry output. In many cases, the logic symbol shown in Fig. 14-2 is used to denote the half-adder in logic diagrams, block diagrams, or flowcharts.

A full adder, or simply adder as it is usually called, is a circuit that will accept as a valid input the carry output from a lower order

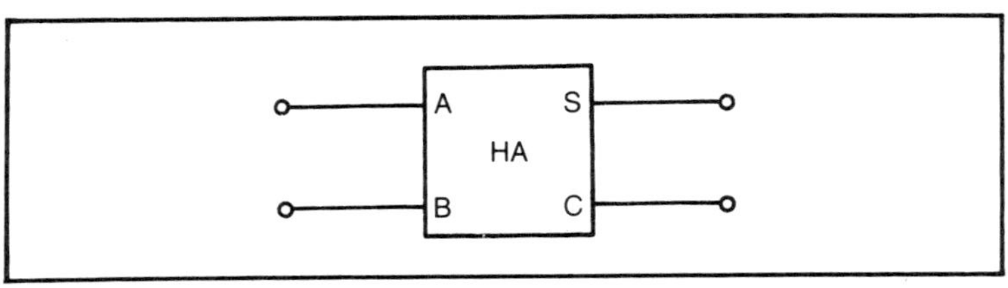

Fig. 14-2. Half-adder symbol.

stage (another adder). Consider again the process of binary addition. Add two binary numbers:

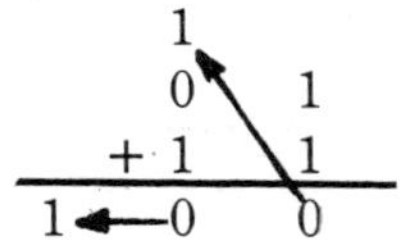

In the first step, we added the two least significant digits. In this case they are both ones, so the result is zero, and a carry output 1 is generated. This carry-one is added to the next most significant column. In the second step, we add a 0 and a 1 to obtain a 1 result. Next we add the carry-one from the previous step, so this is 1 + 1. The result (written down below the line) is a 0, with a carry-1 generated. The result of this operation is a final result of 00 and a carry-1 output. The carry-1 output can be ignored if there are no more significant digits.

The first operation in the above problem could be performed by a half-adder circuit because the least significant digit does not receive any carry bits from a lower order stage (there are no preceding stages to generate any carries). All subsequent stages, however, must be full adder circuits so that they can accommodate carries from previous stages. Otherwise, an erroneous output result could be obtained.

Figure 14-3 shows the logic block diagram for a full-adder, which by definition will:

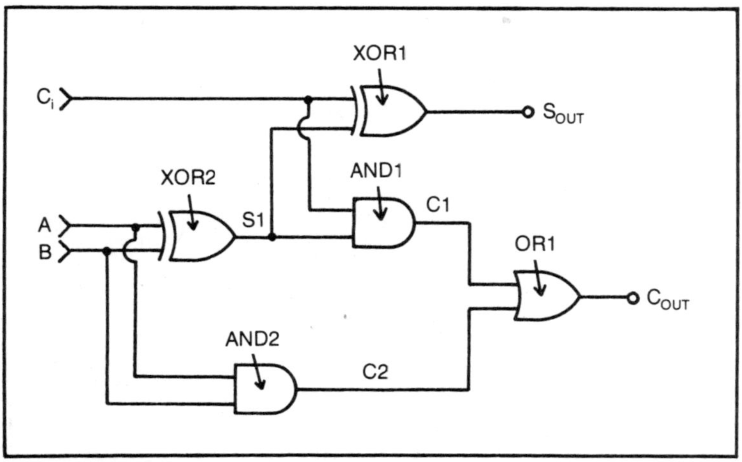

Fig. 14-3. Adder with carry input.

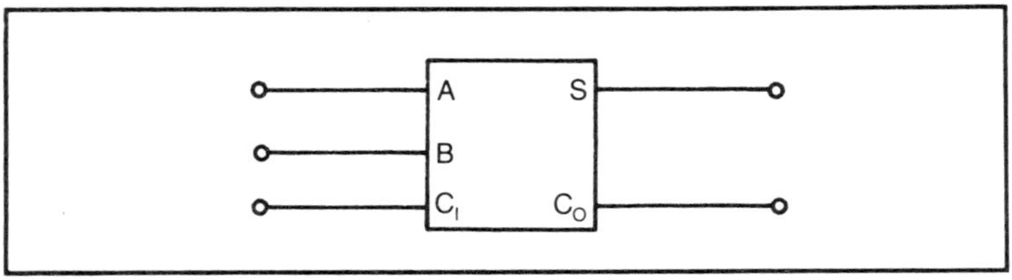

Fig. 14-4. Adder circuit symbol.

- ☐ Add A and B (S = A XOR B).
- ☐ Account for any carry from lower order stages.
- ☐ Produce a carry-out 1, C_o, if needed.

The circuit of Fig. 14-3 consists of two half-adders and an OR gate to generate the carry output. Half-adder HA1 adds together the A and B inputs using the XOR function. This circuit will produce a carry-out 1 if needed. Half-adder HA2 adds together the results of HA1(S') to produce the final output: S= S'XOR C_i. A carry-out may also be generated by this stage. A commonly used logic symbol for the full adder is shown in Fig. 14-4.

SUBTRACTOR CIRCUITS

Addition and subtraction are very similar arithmetic operations. In fact, it is proper to view subtraction as the addition of a positive and a negative number. It is not surprising, then, to find certain similarities between adders and subtractors used in digital circuits; in fact, the principal difference is the capability to account for a borrow-1 from the next higher order digit.

Figure 14-5 shows a simple half-subtractor circuit which pro-

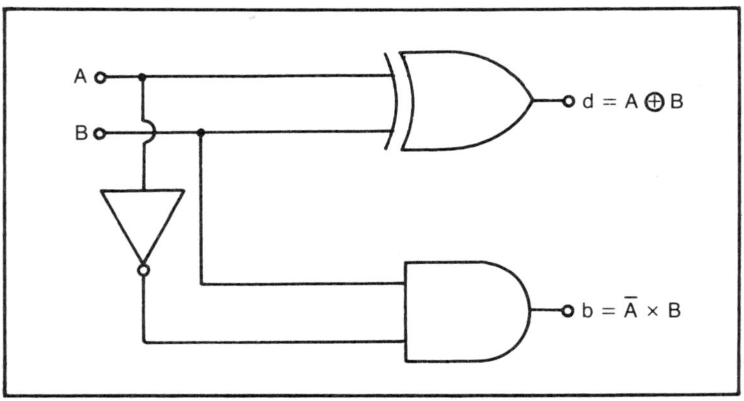

Fig. 14-5. Half subtractor.

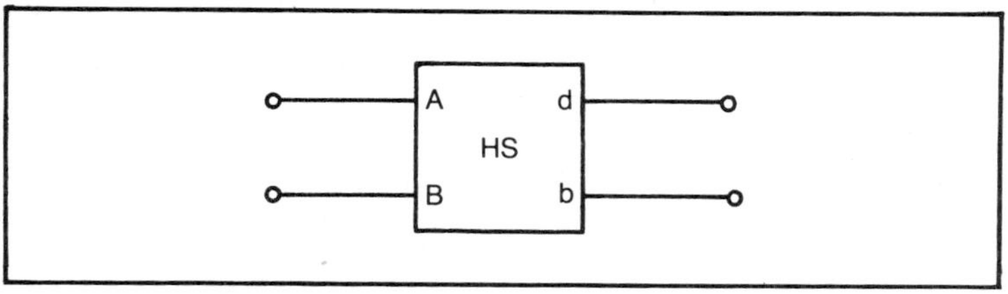

Fig. 14-6. Half subtractor circuit symbol.

duces an output difference *d* of A XOR B, and borrow *b* of $\overline{A}$ AND B. The logic symbol for a half-subtractor is shown in Fig. 14-6.

An example of a half-subtractor circuit made entirely from standard TTL NAND, NOR, and NOT gates is shown in Fig. 14-7.

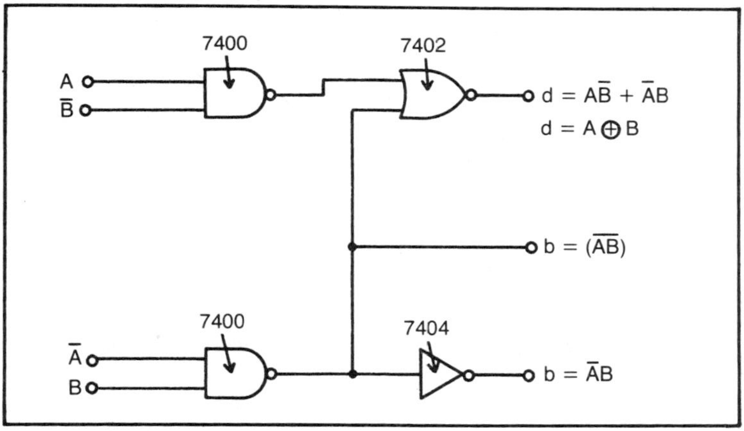

Fig. 14-7. Subtractor with borrow.

This type of circuit is often found in digital equipment because the low cost of TTL IC devices makes it economical.

A full-subtractor must not only find the difference between the two inputs, and create the borrow, but must also account for bor-

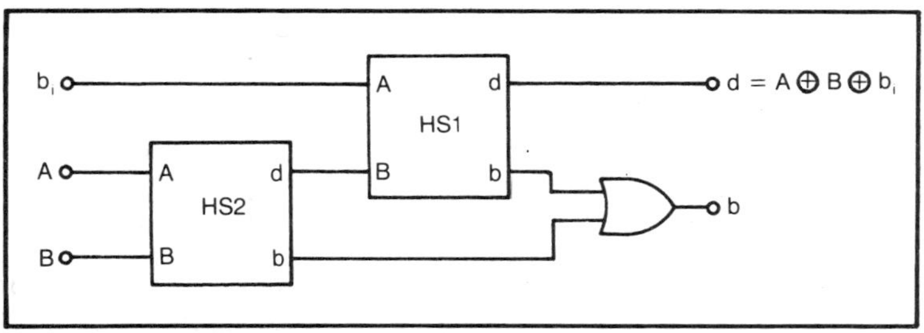

Fig. 14-8. Full subtractor.

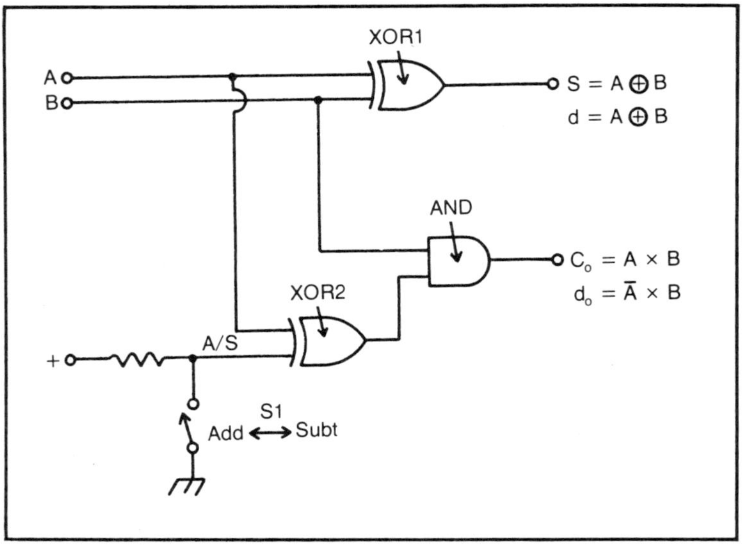

Fig. 14-9. Adder/subtractor.

rows from other stages. Fig. 14-8 shows the full-subtractor circuit, in which two half-subtractors and an OR gate are used.

A universal adder/subtractor circuit is shown in Fig. 14-9. Recall that both adders and subtractors produce an output of A XOR B. The principal difference is that adders produce a *carry* signal (C = A AND B), while subtractors produce a *borrow* signal ($b_o = \bar{A}$ AND B).

In Fig. 14-9, the output of XOR 1 will be A XOR B for both modes; addition or subtraction. The input of XOR 2 marked "A/S" is the mode control terminal and selects whether the addition or subtraction is used. IF A/S is LOW, then the output of XOR 2 will be low if A is HIGH. The output of the AND gate, then, will be HIGH when A is HIGH AND B is HIGH (A × B). This is a carry, so the circuit operates as an adder when A/S is LOW.

When A/S is HIGH, the output of XOR 2 will be HIGH when A is LOW. We can, therefore, state that the output of the NAND gate is A × B, which is the definition of a borrow function. The circuit of Fig. 14-9, then, operates as a subtractor when A/S is HIGH.

Chapter 15

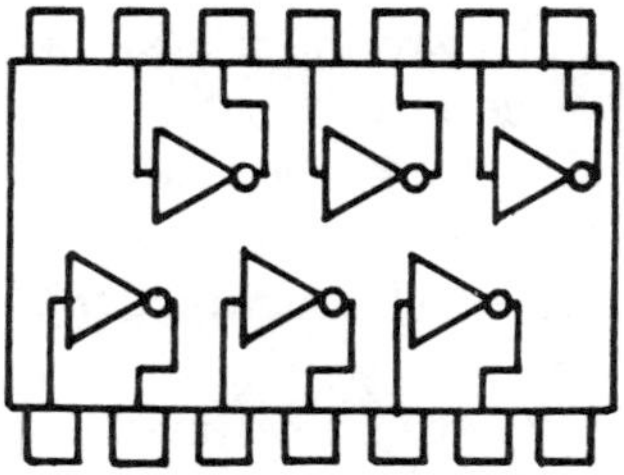

Power Supply Circuits for TTL/CMOS Projects

The dc power supply is at once the most important and the most overlooked or demeaned part of any large-scale electronic project. Any electronic repair technician could tell you that many, perhaps most, fatal faults in electronic equipment involve the dc power supply. I once worked in a hospital electronics laboratory in which we had to maintain and repair a large quantity of medical, scientific, and engineering instruments for the hospital, medical school, and other departments of the university. During that period, records were kept regarding the nature of repairs on all manner of electronic instrument (keep in mind that these were quality instruments, not consumer junk). Those records showed overwhelming predominance of power-supply failures. The second most frequent kind of failure involved power amplifiers such as oscilloscope or strip-chart recorder galvanometer deflection amplifiers (vertical and horizontal stages), audio power amplifiers, servo amplifiers and the like. The lesson to be learned here, I think, is that high power (hence high temperature) circuits fail a lot. As a result, it is incumbent upon the equipment designer to provide adequate power supplies for the project. In this chapter, we will examine the simple power supplies needed to power digital projects, and give some rules regarding power-supply choice and implementation.

There are several features to any dc power supply: transformer, rectifier, filters, load and (in most digital equipments) voltage regulators. Also part of many designs are current limiting circuits and overvoltage protection circuits.

The purpose of the transformer is to change the line voltage to the level of voltage required to power the circuitry. In the United States, the line voltage is 105 to 125 volts ac (nominally 115 Vac) at a frequency of 60 hertz. In other countries, the ac power lines are often 220 Vac, 240 Vac or even 380 Vac, with most being 220 Vac at 50 Hz. For most electronic applications, where timing pulses are *not* derived from the ac power line, the difference between 50 and 60 Hz lines is negligible. In fact, most transformers are labeled "50/60 Hz" so can be used at either frequency (the 400 Hz transformers obtained through military surplus are *not* generally usable at 50/60 Hz!).

When selecting your transformer, keep in mind that most digital circuits require regulated voltage supplies. Most voltage regulators require a dc input potential that is at least 2 volts greater than the output potential, with some requiring 2.5 to 3 volts input-output voltage differential. A typical +5-volt dc power supply, therefore, usually requires a 7.5 to 8.0 volt dc unregulated input supply to the regulator (this is the reason why the S-100 computer bus uses +8 volts dc unregulated for the main bus). Also, the transformer rating will be in terms of rms secondary voltage, and the rectifier operates on peaks. The voltage produced will be approximately 0.9 times the peak voltage, which is (0.9) (1.414) (rms), or 1.27 (rms). A typical 6.3-volt ac filament transformer (terminology left over from vacuum tube days!) will produce an output voltage of (1.27) (6.3 Vac), or 8.001 volts. As a result, we will want to use a 6.3 volt ac filament transformer for the main transformer in a circuit that contains substantial amounts of +5-volt TTL devices. Similarly, for 12-volt power supplies, we will want to use a 12.6-volt filament transformer, and for ±12-volt dc power supplies a 25.2 volt rms ac transformer is indicated.

There are two other ratings of the transformer that must be considered: *volt-ampere rating* and *secondary-current rating*. The VA rating is related to the power that the transformer will deliver (volts times amperes equals watts), but is expressed as "$V \times A$" because reactive loads are not easily expressable in terms of resistive watts. In an ideal transformer, i.e., one with zero losses, the primary and secondary VA product will be identical. In other words:

$$I_{pri} \times V_{pri} = I_{sec} \times V_{sec} \tag{15-1}$$

Where:

I_{pri} is the primary current.

I_{sec} is the secondary current.
V_{pri} is the primary voltage.
V_{sec} is the secondary voltage.

The VA rating, therefore, is the same for both primary and secondary—in *ideal* transformers. But in real transformers, the primary is usually wound closest to the core, so is not as much exposed to air as is the secondary winding. As a result, the VA rating is the ***primary winding VA rating***. This rating expresses the maximum rating of the transformer and should not be exceeded. We sometimes see equipment where the VA rating is exceeded, and the universal result is overheating of the transformer. If you touch the transformer and it blisters your fingers, then it is almost certain that the VA rating is exceeded.

The other major transformer rating is secondary current. This rating is related to the maximum VA rating in that I_{sec} is one term in Equation (15-1). The secondary current rating is the maximum current which can be drawn from the secondary without causing damage. One problem to be on the lookout for is the occasion case where the value of I_{sec} that satisfies the VA rating is less than the rated secondary current. This difference is a contradiction and may point to some creative spec-writing on the part of the manufacturer. It may also, however, point to "different strokes for different folks:" the VA rating might be "free-air" while the I_{sec} rating that is too high might refer to a forced-air rating. In any event, it pays to know the difference and always accept the more conservative of the two! For example, if the transformer is rated at 100 VA for a 6.3 volt secondary, and the I_{sec} rating is allegedly 20 amperes, then we have a problem. For a VA rating of 100, the maximum permissible current is 100/6.3, or 15.9 amperes, not 20 amperes. In that case, derate the transformer to 15.9 amperes.

Another case where the transformer is to be derated is when a bridge rectifier is being used with a transformer that is intended for use with regular two-diode full-wave rectifiers. The bridge rectifier uses the entire transformer secondary winding on each cycle, whereas the regular full-wave rectifier uses half of the winding one on each half of each cycle. As a result, the output potential of the rectifier is twice that of the regular full-wave rectifier. The result is a possible overrating of the transformer. If the transformer has a center-tapped secondary winding, then it is intended to be used in regular full-wave circuits, not bridge rectified circuits. In that case, the secondary current rating will be specified for regular full-wave circuits. If bridge rectifiers are used, then the actual rating for that transformer is *one-half the stated rating*. Thus, if a transformer is rated at 6.3 Vac C.T. (center-tapped) at 20 amperes, then the rating for bridge rectifiers is one-half that amount, or 10 amperes. For small transformers, i.e., those in the 1-ampere range, we can often overlook this limitation. For larger transformers, however, it is important to not overlook this limitation or damage to the transformer may result.

If the transformer was originally intended for full-wave bridge rectifier service, however, the stated rating is selected with the VA rating in mind, so is safe to use. Such transformers do not usually have center taps, but may have several voltage taps. "Control transformers" and "rectifier transformers" often fit this description.

One final problem that causes a need for transformer derating is half-wave rectification. This method is not often used because it is too inefficient. It requires much more filtering action to make it into smooth dc and is wasteful of energy. The transformer VA rating should be 40 percent higher than the wattage of the load. For example, if we design a power supply to deliver 6.3 volts dc at 10 amperes, the VA rating is 63 watts. If the designer was foolish enough to use half-wave rectification, then the transformer VA rating will have to be (1.4) (63), or 88 watts.

Almost all electronic components have at least three different sets of ratings: intermittent commercial/amateur service (ICAS, as it is called in vacuum-tube manuals), continuous commercial service (CCS) and military service. These different means of specifying components reflect the differing intensities of application. Military equipment, for example, is in an intensely harsh environment and *must* work properly. Thus, a half-watt carbon composition resistor might be rated at quarter-watt for military purposes, The typical transformer rating by reputable manufacturers reflects the CCS ratings. If you operate ICAS, then it is possible to safely "push" the rating a little bit. In fact, it is often possible to use 120 percent of the capacity when operating ICAS on CCS rated components. I personally do not recommend that procedure, however, because it is often the case that ICAS becomes CCS, even though the equipment is amateur-operated. A microcomputer, for example, tends to be left on for long periods of time—even in home environments. As a result, the heat build-up will be more similar to CCS situations than to typical ICAS applications. If the transformer is clearly obtained from military surplus, then it is possible to yank a little more current from it than a similarly rated civilian transformer.

As a rule of thumb, select a power transformer that has a Continuous Commercial Service (CCS) rating that is at least 20 percent higher than the need at hand, and preferably 30 percent higher. Thus, if you anticipate a current load of 5 amperes, then order a transformer with 6.5 amperes or more. The idea is to keep the transformer running cool.

Forced-air cooling is also highly useful for power supplies, and is the most overlooked facet of power-supply design. You will greatly reduce the number of power-supply failures, as well as failures in other circuits, if ample forced-air is supplied. I personally prefer to overdesign the air-cooling system than to try getting away with less than needed. In one microcomputer that I put together, there were three fans: one for the transformer, one for the power-supply voltage regulator and rectifier heatsinks, and one for the printed-circuit card cage (a large array of TTL devices will produce immense amounts of heat!). In one computer, the power supply delivered 5-volts at 10 amperes. The main series-pass voltage regulator transistor was mounted on a "standard" hobbyist heatsink as found in blister packs at places like Radio Shack. Before the fan was installed, the TO-3 case of the transistor was so hot that it would take skin off your finger if touched. After the fan, you could easily leave your finger attached to the transistor indefinitely—if you are so inclined.

The main lesson here is to not overlook the forced-air cooling, especially if the current rating of the power supply is more than about 3 amperes. There are a number of different types of fans and blowers, but all are useful. Even a small 40 cubic feet per minute (40 cfm) muffin fan will work wonders.

The rectifier is used to convert bidirectional alternating current into unidirectional pulsating direct current that can be filtered into reasonable pure "smooth dc." The rectifier operates by passing current in only one direction. When the ac polarity is correct then the rectifier will pass current, when it is not correct no current is passed. The semiconductor diode is the most common rectifier element, and for our purposes is the *only* type of rectifier element.

There are basically three forms of rectifier (illustrated in Fig. 15-1): *half-wave, full-wave,* and *full-wave bridge*. The half-wave rectifier (Fig. 15-1A) gets its name from the fact that it uses only half of the ac sine wave to produce the output current and voltage waveforms. The semiconductor diode (D1) used as the rectifier in

Fig. 15-1A passes current only in one direction. When the input sine wave is positive (see inset in Fig. 15-1A), then the diode is forward biased, so current will flow. But, when the input sine wave is in its negative excursion, then the diode is reverse biased so no current will flow. The result is an output waveform that has only the positive humps, with a large space between them. This situation results in inefficiency because of the loss of half of each waveform.

The full-wave rectifier is shown in Fig. 15-1B. Here we use two diodes (D1 and D2), and a transformer that has a center-tapped secondary winding. The center tap is taken as the common reference point in the circuit, so is often found grounded. At any given time, the transformer secondary extremes (A and B) are at opposite polarities. When point "A" is positive with respect to the center-tapped, then point "B" will be negative. Similarly, when point "A" goes negative, then point "B" is positive. This situation is graphically illustrated by the reversed 180-degrees out-of-phase waveforms shown in two of the insets in Fig. 15-1B. Let's consider two scenarios. One half of the ac sine wave applied to the primary, we will find point "A" positive and point "B" negative with respect to the center tap. Considering electron current flow is from negative-to-positive (Ben Franklin fans, *sit down*!), we will find the current flows from the center tap through load resistor R in the direction shown, through diode D1 to point "A;" diode D2 is reverse biased during this period. When the polarity reverses, then point "A" is negative and point "B" is positive. In that case, current flows from the center tap, through the load resistor (R) in the same direction as before, through diode $D2$ to point "B" diode D1 is reverse biased here. It is important to note that the *current in the load flows in the same direction on both halves of the input sine wave*. The result is the "double-humped" waveform (also shown as an inset in Fig. 15-1B) characteristic of full-wave rectified systems. This method of rectification makes more efficient use of the available power, and is a lot easier to filter into smooth dc than is the half-wave rectifier waveform.

The bridge rectifier scheme is also full-wave, and is shown in Fig. 15-1C. Since the same waveforms apply here as in the regular full-wave scheme, we have deleted them in this illustration. Refer to Fig. 15-1B should you want to see the waveforms, points A and B, and the output are as before.

One salient feature of the bridge rectifier is that it uses the entire transformer secondary voltage, rather than the center-tapped voltage. If the transformer is center tapped, then ignore the

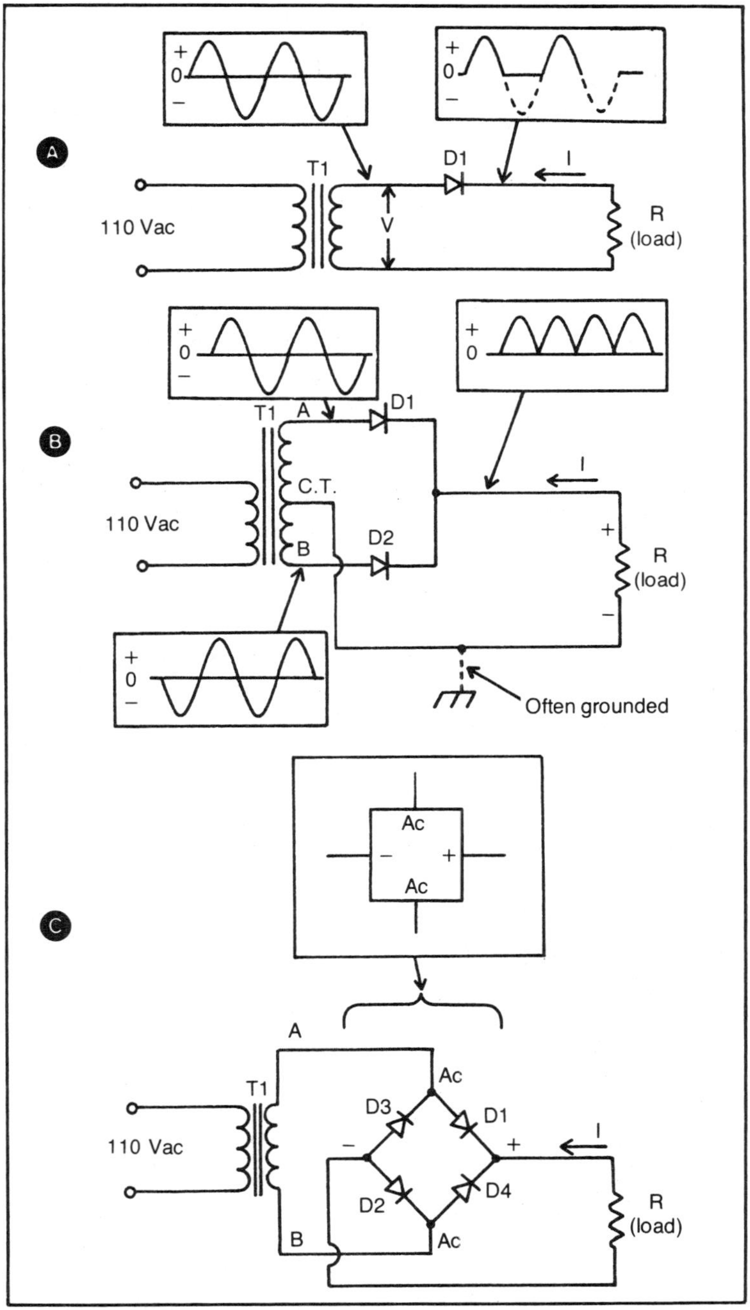

Fig. 15-1. (A) Half-wave rectifier, (B) full-wave rectifier, (C) full-wave bridge rectifier.

center tap and make sure it remains insulated from ground. The voltage rating of the secondary need only be half that of the regular full-wave case. A typical +5 volt TTL power supply will need an 8-volt unregulated dc input, so will require a 6.3 volt ac power transformer (for reasons given earlier). If a regular full-wave circuit is used, then the transformer needs to be 12.6 volts ac center tapped (sometimes expressed as 6.3-0-6.3 Vac), while a full-wave bridge circuit requires only 6.3 Vac noncenter tapped.

Let's consider the full-wave bridge circuit of Fig. 15-1C and see how it works. There are four diodes in the circuit arranged in a bridge ring circuit. The connection of the diodes must be exactly as shown, or damage will result. There are two ac inputs, a positive (+) output and a negative (–) output; the latter two being dc outputs. The positive output is taken from the junction of the cathodes of diodes *D1* and *D4*, while the negative output is from the junction of the anodes of *D*2 and *D*3. The ac points are where the cathode *D*3 joins the anode *D*1, and the cathode *D*2 joins anode *D*4.

Point "A" and point "B" bear the same relationship here as in the previous circuit: when "A" is positive, then "B" is negative, and *vice versa*. Let's walk through circuit operation over one complete sine wave. When point "A" is positive and point "B" is negative, current will flow from point "B," through diode *D*2, load *R* (in the direction shown), diode *D*1 to point "A;" diodes *D*3 and *D*4 are reverse biased. On the second half of the input sine wave, point "B" is positive and point "A" is negative. Under this condition the current will flow from point "A," through diode *D*3, load resistor *R* (in the same direction as before), diode *D*4 to point "B;" diodes *D*1 and *D*2 are reverse biased. Since the current flows through the load in the same direction on both halves of the input ac sine wave, the circuit is a full-wave rectifier.

We rarely have to build a bridge with four discrete diodes, but obtain the bridge ready-built in the form of a block or "stack." The circuit symbol for such a bridge is shown inset in Fig. 15-1C. It is a square with "+," "–," and two "ac" ports indicated. The "ac" ports are often indicated with a small sine wave symbol, but it means the same.

There are two ratings to be considered when selecting diodes: *peak inverse voltage* (PIV) and *forward current* (I). The peak inverse voltage, or PIV, represents the largest reverse bias potential that the rectifier diode will sustain without breaking down with permanent damage. The rule of thumb for filtered power supplies (which includes all power supplies used in electronics) is to make the PIV rating not less than 2.82 times the rms rating of the transformer (some people round off this figure to 3× for safety). Thus, when designing a 5-volt power supply with its 6.3-volt power transformer, we need a PIV of (3)(6.3 Vac), or 18.9 volts. This specification is not inordinately difficult to achieve since the lowest PIV rating normally sold as rectifier diodes is 25 volts PIV! When dealing with higher voltages, however, the PIV rating becomes critical. For example, a 12-volt power supply might use a 12.6-volt transformer in a bridge circuit, or a 25.2-volt center-tapped transformer in a regular full-wave circuit. In either event, the forward voltage applied to the diodes is 12.6 volts, which means a PIV rating of (3)(12.6), or 37.8 volts PIV. In that case, we could not ordinarily use a 25-volt PIV diode, but would have to opt for the 50-volt (or higher devices). Regular diodes in the 1N4000-series are rated at 1-ampere, so are frequently used in small hobbyist power supplies. The 1N4001 will work nicely for +5-volt supplies, but not for 12-volt circuits—it is rated at 25 volts PIV. In that case, the 1N4002 or higher is indicated. I personally prefer the 1N4007 device, which is rated at 1000 volts PIV at 1-ampere. The 1N4007 device is widely available, often at less cost than lesser rated members of the same family, and is thus easily and cheaply procured. The higher PIV rating will give a great deal of protection against such calamities as high-voltage spike transients, either from inductive circuitry being powered or from the ac power mains.

In any event, never use a diode with a PIV rating of less than three times the applied rms. If you have a piece of equipment that too-frequently pops power-supply rectifier diodes, then suspect the original is rated at too low a PIV. This situation occurs somewhat more often than overrating the diode for current, and is the main reason why some popular amateur radio transceivers in the past developed poor reliability reputations!

The forward current rating of a diode is the maximum current that can be passed without causing permanent damage. The normal procedure is to rate this diode according to the free-air current with normal (i.e., 1 centimeter) lead lengths (that's right, folks, the lead length is part of the specification!). If the diode is heatsinked, or, if forced-air cooling is used, then it is possible to "get away" with overrating the diode. Again, I personally prefer the conservative design approach, so will usually go for the design that produces coolest operation. In my own equipment, I prefer to operate the diode at not more than two-thirds of its rated current. If the application calls for 1-ampere, then I will try to obtain a 1.5-ampere diode.

Similarly, the 1-ampere bridge is best operated at currents less than 670 milliamperes. While it should be permissible to go to the limit, I have found from bitter experience that devices operated close to their maximum ratings will not last as long as components more conservatively spaced.

The filter in a power supply is used to smooth the pulsating dc output from the rectifier into nearly smooth dc required by electronic circuits. Figure 15-2 shows the most common form of filter used in power supplies for digital circuits. This "Brute Force" filter consists of a single capacitor.

The circuit in Fig. 15-2A shows a simple filter capacitor connected across the load resistor R, and in parallel with the output of the rectifier. When the ac power is first applied, capacitor C1 will charge by I_1. Part of I_1 also becomes the load current. The capacitor will charge to the same voltage as seen by the load, which is approximately 1.4 times the rms ac voltage (actually a little less). This full-wave action is shown in Fig. 15-2B. When the applied ac passes the peak, however, the voltage in the capacitor exceeds the voltage from the rectifier, so current stored in the capacitor (C1) tends to dump back into the circuit, thereby supplying current to the load during the period when the contribution from the rectifier is diminished. The net effect, as shown in Fig. 15-2B as the shaded area, is to fill in the "spaces" between the humps from the rectifier. The voltage waveform and current flow waveforms (which will have the same shape if the load is resistive) seen by the load is the heavy black "sawtooth" shown in Fig. 15-2B. This waveform is a lot nearer pure dc than is the pulsating dc from the rectifier output, so is a lot more easily handled by the electronic circuits.

If the electronic circuitry does not require regulated voltages, then this circuit is sufficient. Many types of equipment, such as high-fidelity amplifiers, require a nearly pure dc, yet can achieve it by using a large enough filter capacitor at $C1$. The degree of purity is expressed in terms of a so-called "ripple factor," r. For 60 Hz power supplies (with a ripple frequency of 2 × 60 Hz, or 120 Hz), the ripple factor is determined from:

$$r = \frac{1}{416 R_L C_1} \qquad \textbf{(15-2)}$$

Where:

r is the ripple factor, and is nondimensional.
R_L is the load resistance in ohms.
C_1 is the capacitance of capacitor C1 in farads.

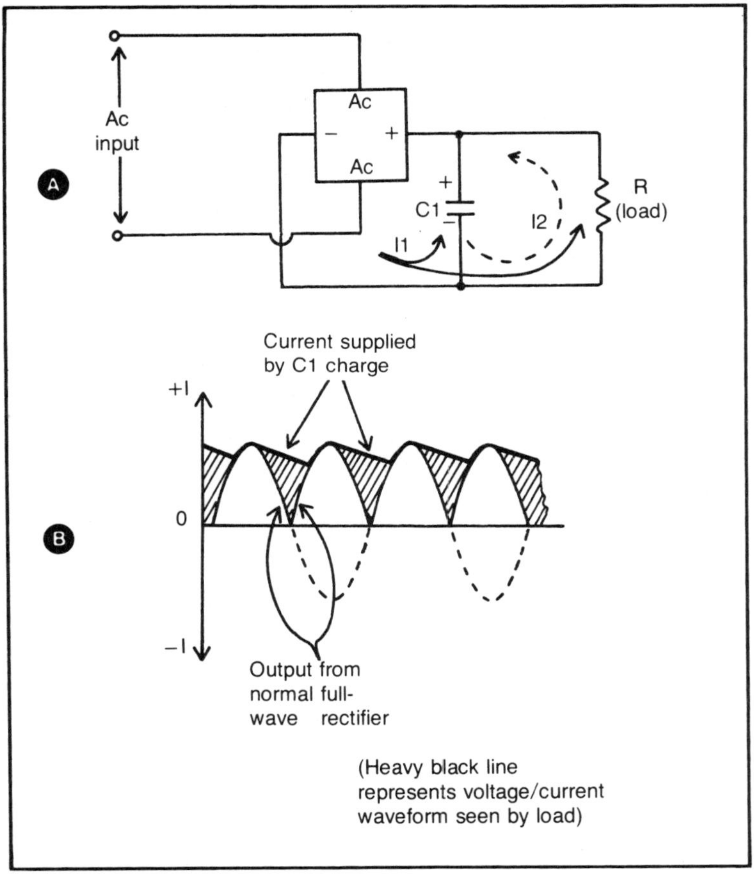

Fig. 15-2. (A) Filter circuit, (B) waveforms.

The ripple factor will depend in large measure on the load resistance. From the fact that the load resistance term is in the denominator of Equation (15-2), we can conclude that the ripple factor is inversely proportional to the load resistance. We will, therefore, design the power supply for the lower ripple factor, hence the large resistance. This value can be obtained from the quotient of the output voltage and output load current. Hence, if the power supply delivers 8-volts at 5-amperes, then the load resistance is (8-volts)/(5-amperes), or, 1.6-ohms.

Practical ripple factors are always under one (1), with 0.1 to 0.5 being common; some applications require ripple factors of 0.01. Let's see how we can calculate the capacitance required for each level. The equation given above is, like many in textbooks, incorrectly stated for the actual practical world. In most cases, we will not want to calculate the ripple factor given by some value of capacitance, but rather, will specify the desired ripple factor and then calculate the minimum capacitance that will achieve that value. In that case, we merely swap the r and C_1 terms of Eq. (15-2):

$$C_1 = \frac{1}{416\, R_L\, r} \qquad \textbf{(15-3)}$$

Where:

All terms are as previously specified.

Similarly, in real practical cases we don't give a hoot about the capacitance being given in *farads* because real capacitors come in *microfarads*. So, by cranking in the correction factor to render the answer in microfarads, we get:

$$C_1 = \frac{10^6}{416\, R_L\, r} \qquad \textbf{(15-4)}$$

Where:

r is the nondimensional ripple factor.
R_L is the load resistance in ohms (V_0/I_L).
C_1 is the capacitance of C1 in *microfarads*.

Equation (15-4), then, is the one which we will use in practical cases. Let's consider two examples to illustrate how Eq. (15-4) might be used. Let's assume that the power supply application requires a ripple factor r of 0.8 for an output load resistance of 1.6 ohms. The capacitor used for C1 should have a minimum value of:

$$C_1 = \frac{10^6}{416\, R_L\, r}$$

$$C_1 = \frac{10^6}{(416)\,(1.6)\,(0.8)}$$

$C_1 = 10^6/532.5 = 1878\ \mu F$ (use 2000 μF standard value)

In our second example, let's assume an output load resistance of 1.5 ohms, and a need for a ripple factor of 0.01 (very low!). The capacitance should be at least:

$$C_1 = \frac{10^6}{416\, R_L\, r}$$

$$C_1 = \frac{10^6}{(416)\,(1.5)\,(0.01)}$$

$$C_1 = (10^6)/(6.24)$$

$$C_1 = 160{,}256\ \mu F!$$

A 160,000 μF capacitor is not overly difficult to obtain, but will occupy a fair amount of space. This value is sometimes found in computer power supplies. Of course, it is possible to obtain the high value of capacitance by connecting several lower value units in parallel.

Later in this chapter, when we discuss voltage regulators, you will see a specification for the filter preceding the regulator as 1000 μF/ampere of load current. For 5-volt regulators, this assumes a pre-regulation ripple factor of approximately 0.5. Of course, the regulator provides a lot of ripple reduction by itself, so this level is not too little for most applications.

That ripple reduction factor caused by regulators is responsible for the claim by some manufacturers of early low-voltage, high-current regulated dc power supplies that their design had the equivalent of one-farad of capacitance in the filter! There wasn't *really* that much capacitance, but it looked that way because of the immense ripple reduction caused by the electronic voltage regulator circuit.

The voltage rating of the capacitor selected for service as a filter must be higher than the rectifier output voltage. I also prefer conservative design here, but that is not always easy to obtain. In some high current supplies (at potentials greater than 5-10 volts) we find that there are fewer capacitors available in the correct capacitance, and they are physically larger. In that case, we would almost certainly have to parallel-connect several of lower capacitance of the correct voltage rating. When parallel-connecting capacitors, keep in mind that the voltage rating of the group of capacitors is no more than the voltage rating of the least among them. For example, parallel connecting a 25-volt unit with a 15-volt unit will yield a 15-volt unit. The voltage rating should be 25 percent larger than the working voltage, or more (I prefer *twice* as high!). Do not confuse the so-called *surge voltage* with the *working voltage dc* (WVdc) rating. It is the *WVdc* rating that is important to us; the surge rating can be 25 percent higher but does not reflect any real capability of the capacitor.

The voltage rating of the capacitor can directly affect reliability, and is not always easily obtained when high capacitances are required—but try we must. I can recall a brand of medical CCU monitoring equipment that used a pair of 270 volt dc power supplies that were filtered with 350 WVdc electrolytic capacitors. We had a large number of service calls in our eight-bed Coronary Care Unit (a place where you definitely don't want unreliable equipment!) that were due to shorted 350 WVdc capacitors in the ±270-volt dc power supplies. A quick analysis showed a disturbing situation in the selection of power-supply filter capacitors. Normally, dc voltages in an unregulated supply can vary ±15 percent, while the voltage rating of electrolytic capacitors can vary ±20 percent, depending upon age, history of use, condition and original manufacturing tolerances. Using the worst case situation where the voltage was over by 15 percent and the capacitor capability was under by 20 percent yielded the startling fact that they were filtering a (270 × 1.15), or 311 volt dc supply with a (350 WVdc × 0.8), or, 280 WVdc filter capacitor! While the real situation was probably not nearly as bad as the worst-case analysis showed, it was nevertheless true that the capacitors were being stressed far too much in the circuit.

In that case, the 350 WVdc, 60 μF, electrolytic capacitors were easily replaced with 450 WVdc 60 μF units. The 450 WVdc and 350 WVdc capacitors were approximately the same physical size, so there was no problem on the printed board. As a result of this change, after all 350 volt units were replaced (two per bedside for eight beds), there were no further capacitor-related failures of that equipment for the next five years we owned the equipment. There were other failures to be sure, but the once a week loss of a capacitor stopped cold. As a result of that experience, I tend to be a little conservative on filter selection. I do not, therefore, recommend (as has one other author) use of 10-WVdc capacitors in 8-volt power supplies used in S-100 microcomputer chassis.

Voltage regulators are used to keep the output voltage steady, despite changes in the load resistance. These circuits are needed because there is a small series resistance associated with the power supply, so as the current demand changes the voltage drop across this internal resistance also changes. The voltage drop is subtracted from the output voltage, so we find the output voltage also changing.

The voltage regulator serves to smooth out that variation.

The simplest form of regulator is the zener diode, but such diodes are not generally suited to the current levels found in most digital projects. As a result, we usually need a better regulator. Fortunately, over the past decade or so manufacturers have offered three-terminal IC voltage regulators that are at least as simple as the zener diode. Most of these are fixed-voltage and are housed in power transistor packages. Three different packages are commonly used:

Package Designation	Current Level (Max)	Equivalent to Transistor Package
H	100 mA 150 mA*	TO.5 (metal can)
K**	1000 mA 1500 mA*	TO-3 (large diamond-shaped metal)
T	750 mA 1000 mA*	TO-220 (also called P66 in some cases—plastic power transistor package)

* When well heatsinked
** Some models available to 10-amperes in same package if heat-sinked

There are several families of devices available as three-terminal IC voltage regulators. All are fixed-voltage, and the voltage rating is part of the part number in all except the oldest forms. The older LM-309H and LM-309K are 100-mA and 1-ampere, 5-volt regulators, respectively. Similarly, the LM-323K is a 5-volt, 3-ampere regulator in the TO-3 power transistor package. The two major positive voltage regulators are the LM-340m-xx and 78xx series. In both cases, the "xx" term is replaced with the voltage rating; in the LM-series the "M" is replaced with the package style. Hence, an LM-340K-05 is a 5-volt regulator in a TO-3 "K" style package; an LM-340T-12 is a 12-volt regulator in a plastic power transistor (T-style) package. The approximate current ratings are given above. The 7805 and 7812 are equivalents in the other series. Some manufacturers tend to use LM-340m-xx and 78xx designations somewhat interchangeable, so one might see a 7805 designation on both K and T packages. Also, some manufacturers tend to list only the "well-heatsinked" current, rather than the free-air current rating; be careful.

The negative voltage versions of the above series are designated LM-320m-xx and 79xx, respectively. Otherwise, the part number rules are the same. There is one other difference, however, and that is in the matter of the pinouts. Figure 15-3 shows the usual pinouts for the three major forms of three-terminal IC voltage regulators. Shown in each section of Fig. 15-3 are the usual IC voltage regulator numberings for the pins (or case) and the equivalents for power transistors in the same package. For example, in the

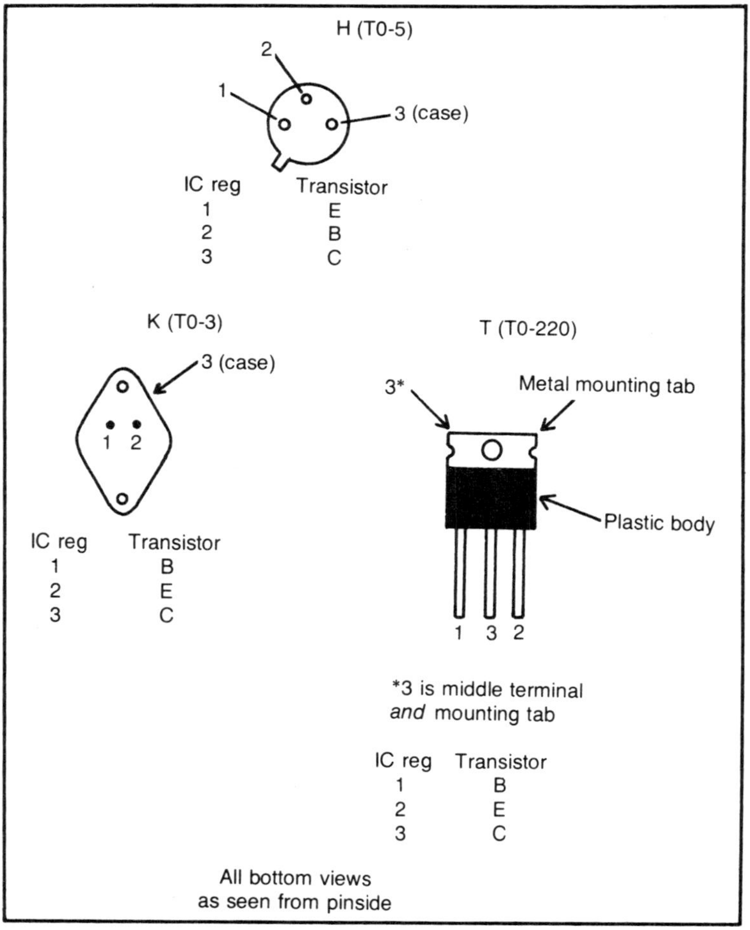

Fig. 15-3. Three-terminal IC regulator packages.

K-style package, which is the same as the TO-3 power transistor package, the transistor emitter terminal is pin no. 2 on a voltage regulator.

In normal voltage regulator service, the input, output, and ground terminals will be as follows:

Function	Positive Voltage Regulator LM-309, LM-340m-xx, 78xx	Negative Voltage Regulator LM-320m-xx, 79xx
Input	1	3 or case
Output	2	2
Ground	3 or case	1

Clearly, from the above it is obvious that the negative voltage regulator will have to be configured so that the case is isolated from ground or problems will result.

The input voltage applied to a three-terminal IC voltage regulator will have to be at least 2-volts greater than the rated output voltage, and in some devices as much as 3-volts greater. This means we have to be prepared to produce at least 8-volts input to a 5-volt regulator, and 15-volts to a 12-volt regulator. Most regulators will permit us to input as much as 35 volts for the rated output voltage, but that is not too smart. The case and junction temperatures depend not on the current drawn (although that is a factor), but on the power dissipation. There are two factors to power dissipation: *output load current* and *input-output voltage differential* ($V_{in} - V_0$). This voltage is shown graphically in Fig. 15-4. If our hypothetical 5-volt power supply has an 8-volt unregulated input potential, then the differential is 8-5, or 3 volts. At 1-ampere, then, the power dissipation is $(I_o)(V_{in} - V_o) = (1\text{ A})(8\text{ V} - 5\text{ V}) = (1\text{ A})(3\text{ V})$, or 3 watts. If we use the maximum rating for V_{in}, then the power jumps to (1 A) (35 V - 5 V) = (1 A) (30 V) = 30 watts. Obviously, if we dissipate ten times as much power, then we will heat up the IC much more. I don't care what the spec sheets say, experience proves that hotter circuits fail earlier.

Figure 15-4 is the normal circuit for three-terminal IC regulated power supplies with positive output voltage. For the negative version of this circuit, merely reverse pins 1 and 3 of the IC regulator and the polarities of capacitors C1 and C4. For +5 volt dc regulated power supplies we may use any of the following devices (arranged by output current):

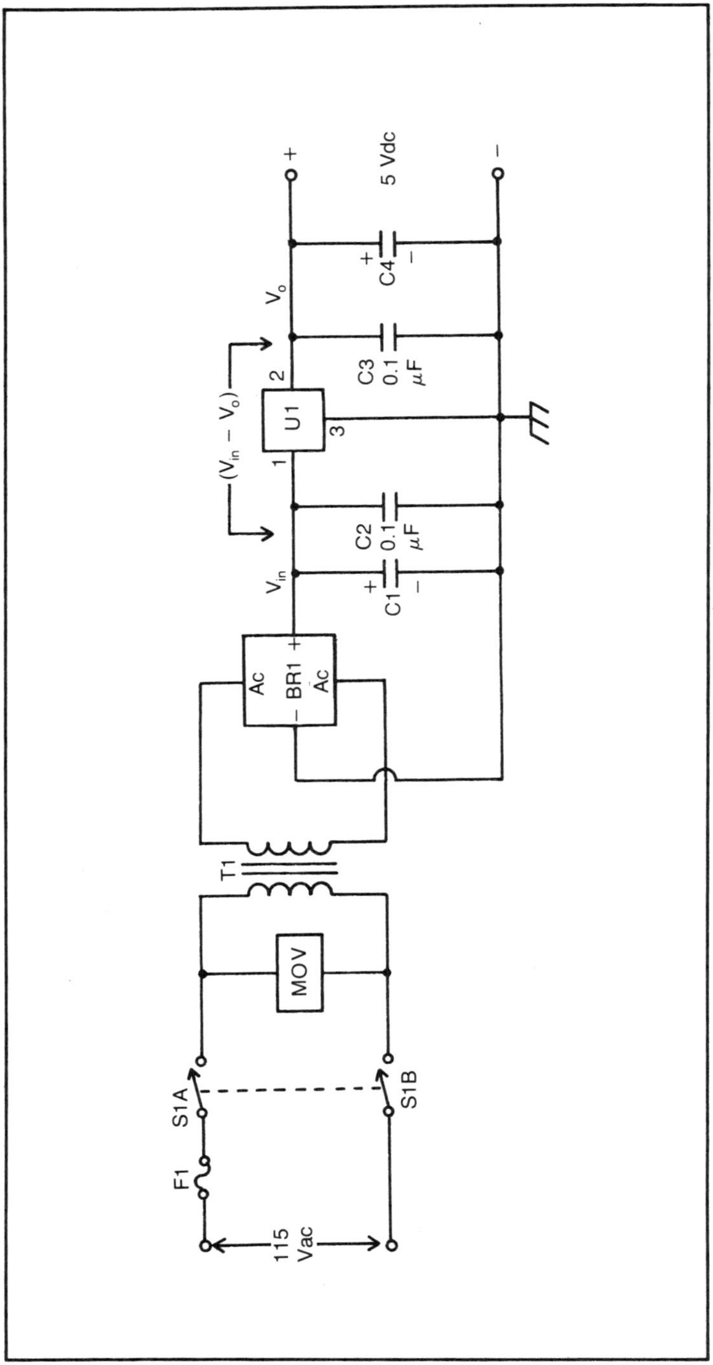

Fig. 15-4. Three-terminal IC regulator circuit.

750 mA	1 Ampere	3 Amperes	5 Amperes	10 Amperes
7805 (T)	7805 (K)	LM-323 LAS-1405[1]	LAS-1905[1]	(none)
LM-340T-05	LM-340K-05			
	LAS-1505[1,2] LAS-1605[1,3]			

Notes:

1. Lambda Electronics, 121 International Drive, Corpus Christi, TX 78410
2. 1.5 amperes rating
3. 2.0 amperes rating

The transformer and filter capacitor (C1) in Fig. 15-4 are selected according to the rules given earlier in this chapter, except that for the capacitor we can assume a ripple factor and modify the rule a bit. For C1, use a capacitance of not less than 1000 μF per ampere of load current (1000 μF/amp). I recommend further a capacitance of 2000 μF/ampere if available. The rule in the rest of this chapter, then, will be 2000 μF/ampere.

Thus, for the more popular values of current capability, we will want to use the following minimum capacitor values at C1:

100 mA	500 μF*
750 mA	1500 μF
1 ampere	2000 μF
2 amperes	4000 μF
3 amperes	6000 μF
5 amperes	10,000 μF

* 100 mA value departs from the rule

The voltage rating of the capacitor must be predicated on the value of V_{in}, *not* on V_0. For example, we might have an output potential of 5 volts dc, and an input voltage of 35 volts. If we used the output figure, we might conservatively select a 15-WVdc capacitor—which will promptly go up in smoke upon application of power to the circuit! For the 35-volt case, I would use not less than 50 WVdc as the proper rating.

Capacitors C2 and C3 are used to improve the noise immunity

of the regulator. These 0.1 to 0.5 μF capacitors must be mounted physically as close as possible to the body of the regulator. Most builders mount them actually to the body, using the pins as support.

Capacitor C4 is used to improve the transient response of the circuit. This capacitor is a recommended option, and should have a value of approximately 100 μF/ampere. The purpose of this capacitor is to provide a small reservoir of charge that can be dumped into the circuit between the time when a larger current demand presents itself and when the regulator can supply the need. The circuit will still work without this capacitor, but it is desirable in digital circuits because of the frequent large shifts in load experienced in those circuits.

The regulator in this circuit can be any of those listed above. Keep in mind, however, that all voltage regulators last longer when properly heatsinked, and that becomes especially true for current levels above 1 ampere. It is not recommended that 3 to 5 ampere voltage regulators be operated without a heatsink. I recommend that "standard" large-size finned heatsinks normally used for power amplifier service. Also, if the application will allow a blower or muffin fan for the regulator and its associated heatsink, then by all means provide one for this circuit.

The MOV device in the primary circuit of transformer *T*1 is a General Electric *Metal-Oxide Varistor* (MOV), and is used to kill power line glitches. These high-voltage transients can not only sometimes cause damage to the circuitry, but have the unfortunate habit of temporarily disrupting digital circuits. This becomes especially true when clocked or resettable circuits are used. In some microcomputers, we find the program takes off and "bombs" for no apparent reason. Checking the hardware, the program, and the input data will reveal no fault—yet, that one time, the program bombed. One probable cause for such a happening is power line high-voltage transients. These spikes can top 2000 volts and last for 30-50 microseconds. The MOV operates much like a pair of back-to-back zener diodes and will clip off any transients greater than 200 volts (which is higher than normally expected peak voltages on a 110 volt line). The MOV device is a low-cost insurance policy that you may well have to provide later anyway, so why not incorporate it now and never know whether it would have been needed or not?

Assuming that the transformer ratings, the capacitor working voltages and the regulator types are adjusted accordingly, this same circuit will work for any voltage that three-terminal devices will supply: 5, 6, 6.2, 8, 10, 12, 15, 18, and 24 volts dc.

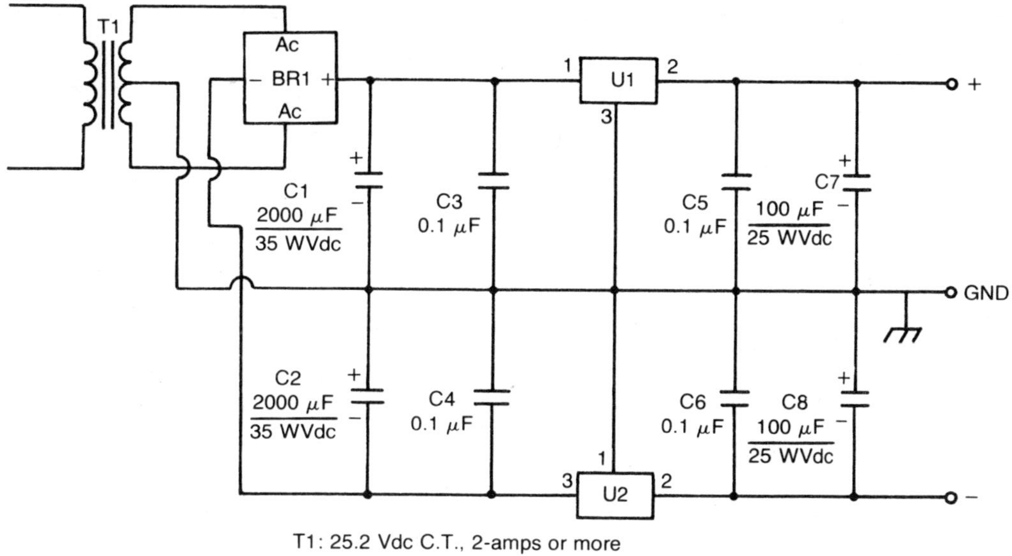

Fig. 15-5. Dual-polarity regulated supply.

±12-VOLT DC, 1-AMPERE SUPPLY

The power-supply circuit shown in Fig. 15-5 provides bipolar 12-volt (i.e., ±12-volt dc) power at currents to 1-ampere per side. This bipolar power supply can be used for those CMOS circuits that operate from ±12 volts, or, for operational amplifier and other analog circuits that also operate from those potentials. Such power supplies might have to be provided if there are analog components in the circuit.

Basically, the circuit is the same as Fig. 15-4, with due allowances for the negative power supply. Note that the pinouts are different on the negative side. Failure to observe this little difference will cause immediate loss of the power-supply regulator, and possibly the rectifiers (*sigh*).

The only feature of this circuit which might provoke a little wonderment is the transformer. Note that we are simultaneously using a bridge rectifier *and* center-tapped transformer. In this case, the center tap is common (hence it is grounded). One half of the secondary provides power to the positive regulator while the other half provides power to the negative regulator. In this case, the bridge is used as a pair of half-bridges in parallel, so we are able to get away with using a single prepackaged bridge stack; the negative output of the bridge becomes the negative output of the power supply, and the positive output of the bridge is the positive output of the power supply.

Regulators that can be used for this circuit include any of the LM-340/78xx series for the positive side, and LM-320/79xx series for the negative side. Others are also available, but those are common. Remember the current ratings given above for T and K packages.

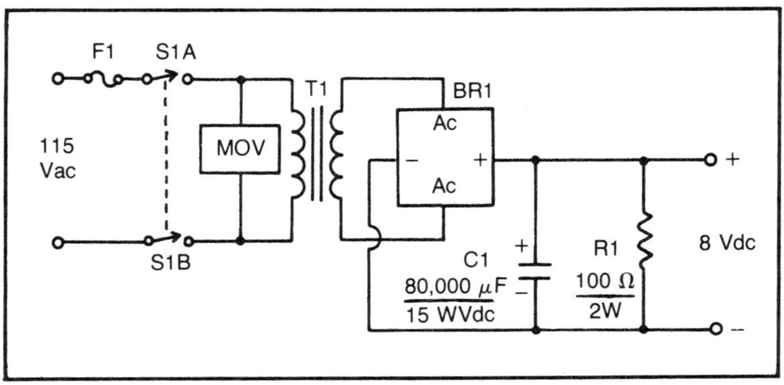

Fig. 15-6. S-100 power supply.

The rating of the transformer will be at least 25.2 volts rms center-tapped (some transformers are available to 28 volts rms, and are equally useful). The current rating will be not less than 2 amperes if the rating is 740 mA per side. If either side is to be heavily loaded, i.e., operated extensively close to its full-rated capacity, then be a lot more conservative and select a transformer with not less than 3.5 to 4 amperes secondary current (again, center-tapped). Keep in mind that the method of operation here is half-wave bridge, so we can expect the current rating to be 2.4 times the normal current drain. For a total of 1 ampere, then, the minimum rating will be 2.4 amperes. I recommend a 5-ampere transformer for this service.

S-100 POWER SUPPLY

Figure 15-6 shows the circuit for a simple power supply that can be used in S-100 microcomputers. Those computers use distributed voltage regulation, which means that each plug-in printed circuit board has its own +5 volt regulator. Some cards will have multiple regulators (I have seen as many as five!) each supplying 1 to 3 amperes of current at +5 volts dc. The main power bus is specified at +8 volts unregulated, and must be capable of providing a large current. Even small S-100 computers require 10 to 15 amperes at +8 volts, while some need to 30 amperes at the same potential.

The transformer in Fig. 15-6 must have a current rating of twice the load current, unless a large amount of forced-air cooling is provided. This requirement means a 10-ampere power supply will require a 20-amperes transformer. In one case of what seems like overly conservative design, a pair of 6.3-volt ac, 25-ampere transformers were connected so that their primaries were in parallel, and the secondaries were in series-aiding. This means that we could treat the pair of transformers as a single 12.6-volt, 25-ampere center-tapped transformer. The rectifier stack could then be replaced with a pair of 50-ampere stud-mounted diodes connected as a regular full-wave rectifier.

In the case shown in Fig. 15-6, however, a bridge rectifier is used. This rectifier can be made from ordinary stud-mounted diodes, or brought as a stack. Such bridges usually have four spade-lug terminals, and a mounting hole in the middle of the block, or, tabs out to the sides. It is always preferable to heatsink these stacks because of the high currents used. Be aware that many such stacks have a little guide pin on the base. The usual procedure is to drill two holes, a large one for the main mounting lug and a smaller one for the guide pin. If you prefer to not use the guide pin, then file it off flush with the case. Otherwise, when you torque down the nut holding the mounting screw, then it is possible that the rectifier will be cracked—too bad.

The filter capacitor values is selected for most applications, and is based on a ripple factor of approximately 0.5. This means a current rating of 25 amperes or less. I recommend not less than 80,000 μF, and as much as 120,000 μF. Keep in mind that higher values of capacitance may cause excessive charging currents at initial turn-on each time the equipment is used, so some means of softening the blow may be indicated here.

The resistor is used to provide a certain minimum load to the rectifier. It is found that the voltage across the capacitor tends to go quite high when there is no load, so one must beware of popping capacitor C1 (which is usually a 15 WVdc unit in S-100 systems). The 100-ohm, 2-watt resistor is used for this purpose. In most cases, the capacitor used at C1 will have screw-type terminal lugs, so the resistor can be mounted directly to the filter capacitor.

6-AMPERES FROM 78XX REGULATORS

There are several options available that will yield higher current operation. One is to buy a larger current regulator. This is a reasonable solution unless (a) the cost is too high, or, (b) the higher current regulators are not easily obtained for whatever reason (it happens, especially in hobbyist circles). The circuit in Fig. 15-7 shows a means of overcoming that problem. Here we use a 78xx or

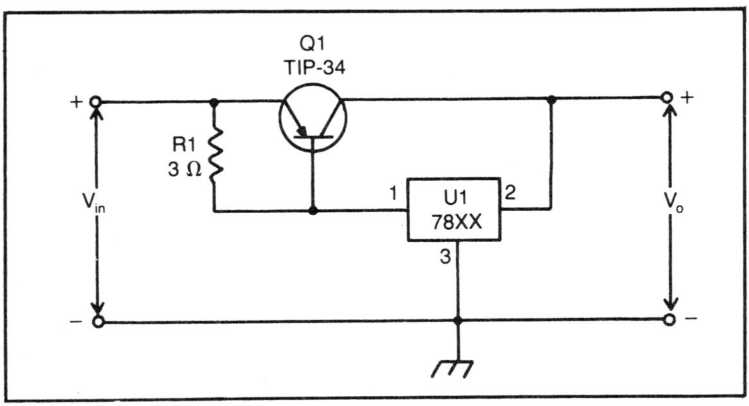

Fig. 15-7. Series-pass transistor increases current capability.

LM-340m-xx regulator as a reference to the base of a series-pass transistor that actually passes the current. The output voltage will be slightly less (by 0.6 to 0.7 volts) than the rated regulator potential, but that is not a real problem. The main bulk of the output current is carried by the *TIP-34* (or equivalent, there are several) power transistor.

ADJUSTABLE 5-AMPERE VOLTAGE REGULATOR

There are times when we need an adjustable output voltage. For some people, it is necessary to have a wide range of voltages for different occasions. Such an application might be a bench power supply. In other cases, we might want a smaller range of potentials in order to make the actual voltage some precise value. This situation is sometimes seen in TTL digital instruments that use centralized voltage regulation. The higher current +5 volt regulated supplies will experience a relatively large voltage drop between the output of the supply and the point where the voltage is needed. This situation is often seen where the supply is outboard to the digital PCB, or, even on another chassis or another cabinet altogether. In that case, the +5 volts might be as low as 3 or 4 volts when it is actually applied to the digital PC board, the remainder being dropped in the connecting conductors. If we have an adjustable voltage regulator, we can adjust the voltage to be +5 volts at the point where it is needed, rather than at the output of the regulator. Figure 15-8 shows an adjustable-regulator power supply based on the easily obtained LM-338 device.

The circuit in Fig. 15-8 is of the regulator and filter capacitors only. The rectifier transformer is missing, but is designed along lines described previously. Since the LM-338 device is a 5-ampere regulator, we will use a filter capacitor (C1) of 10,000 μF, based on the 2000 μF/ampere rule given earlier.

The circuit becomes adjustable because of resistor network R1/R2. The adjustment potentiometer (R1) is set to the desired output potential. This potentiometer can be either a panel-mounted type (if front-panel control is desired), or, a printed-circuit "trimmer" potentiometer (i.e., multi-turn) that is preset once, and then forgotten. The output voltage will be:

$$V_o = (1.25\ \text{V}) \left[\frac{R1}{R2} + 1\right] \qquad \textbf{(15-5)}$$

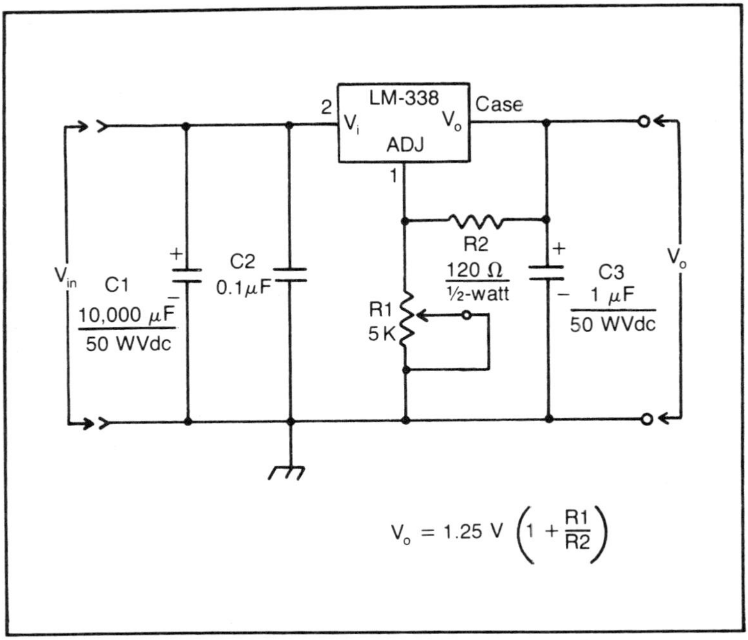

Fig. 15-8. LM-338 provides 5-ampere and variable output voltage.

An implication of Eq. (15-5) is that the output voltage will not drop exactly to zero, even when potentiometer R1 is at zero ohms. In that case, the output voltage will be (1.25 V) (0 + 1) = 1.25 volts.

If we want the power supply to hover around +5 volts, or some other specific potential, with high resolution (i.e., volts per turn of R1 being low) traded off for narrower range, we can make the potentiometer a combination of a fixed resistor and a potentiometer. For example, if we replace R1 with a series combination of a 330-ohm resistor and a 100-ohm 20-turn potentiometer, we will have (according to Eq. 15-5), an output voltage range of 4.7 to 5.7 volts dc, with a "tuning ratio" of 50 millivolts per turn of the adjustment. Such a situation would make it easy to home in on exactly five volts dc.

The pinouts for the LM-338 device are a little different than other three-terminal IC voltage regulators, so some caution is in order. The input is pin no. 2, while the adjustment network connects to pin no. 1; the output on this device is the case ("pin no. 3," even though no pin is used!). Since it is advisable to heatsink the TO-3 case of the LM-338, we must make provisions for either use of mica insulators and heat transfer grease, or, insulating the entire heatsink from ground.

+5 VOLT, 20-AMPERE AND 30-AMPERE REGULATED POWER SUPPLIES

Certain large digital projects require tremendous amounts of dc power at a potential of +5 volts. Some microcomputers, for example, require 20 to 30 amperes at the magic +5 volt potential. None of the simple-minded regulator circuits normally published for low current applications are directly applicable to very high current supplies. The series-pass transistor circuit given earlier, for example, works nicely to about 10-amperes. But above 10-amperes we find that transistor *beta* figures, drive requirements, and the gains needed cause problems to look very large that could be ignored at lower levels. The matter of thermal drift, for example, is exacerbated by two factors: the larger amount of heat generated and the larger current gains needed in the regulator. As a result, the design of high-current regulators is a bit more complex than might otherwise be true. There is, however, a solution that should appeal to most readers. *Lambda Electronics, Inc.* makes several hybrid block regulators at current levels as high as 30 amperes. I have personally used the 20-ampere version (LAS-5005) in my *Digital Group, Inc.* Z80 computer, and it worked well. The basic circuit, using the LAS-70xx device that permits up to 30 amperes is shown in Fig. 15-9A. For the user, this regulator is simplicity itself. There are two external connections for the heavy input current (pins 1 and 3), a V+ terminal for powering the internal circuitry (pin no. 20, which is normally strapped to pin no. 1), two heavy current regulated output terminals (5 and 7), a pair of *sense* terminals (9 and 12, of which more in a moment), and a pair of terminals that are normally strapped together (15 and 16).

The input capacitor (C1) is the regular filter capacitor, and is selected according to the 2000 μF/ampere rule. Obviously, this capacitor will be a large beastie! The output capacitor (C2) is used to improve transient response of the circuit, and should be selected according to the 100 μF/ampere rule.

The value of resistor R1, which sets the output potential, is determined from:

$$R1 = \frac{(0.25\ V_o)\ (1000\ \text{ohms})}{(\text{volt})} \qquad \mathbf{(15\text{-}6)}$$

Example

Assume a potential of +5 volts dc to power a TTL compatible microcomputer. Find the minimum value for R1:

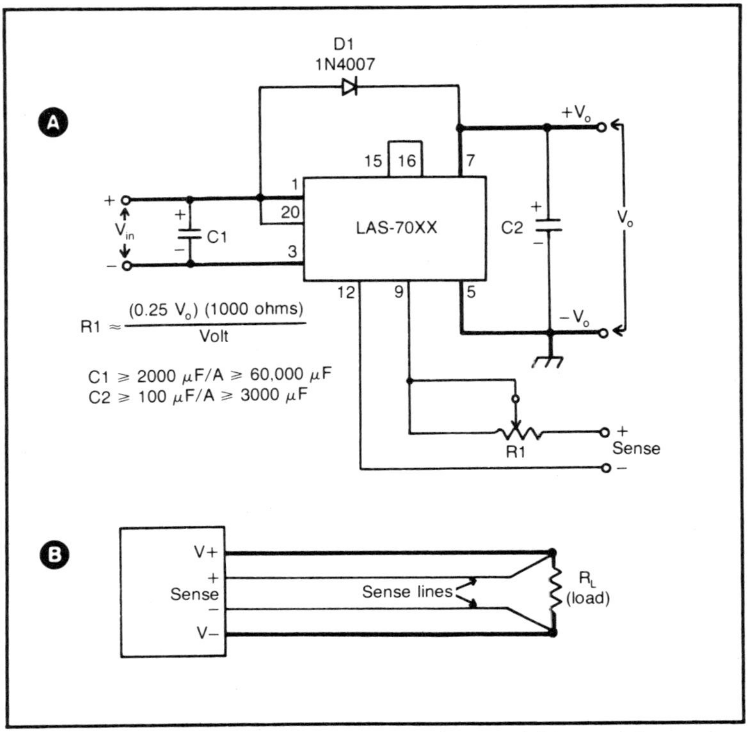

Fig. 15-9. (A) LAMBDA LAS-70xx regulator, (B) use of the *sense* lines.

$$R1 = \frac{(0.25)\ (5\ \text{V})\ (1000\ \text{ohms})}{(\text{volt})}$$

$$R1 = 1250\ \text{ohms}$$

This resistor should have a minimum value of 1250 ohms, so I recommend placing a potentiometer in series with a fixed resistor to allow the total resistance to swing from slightly below this value to two or three times the value.

The LAS-5000 and LAS-7000 series devices are equipped with *sense* inputs. These are lines that will tell the internal regulator circuitry exactly the potential that is being output. If the load is located very near the regulator, then it might be wise to simply strap the +*sense* line to the $+V_o$ line, and the −*sense* line to the $-V_o$ line. If there is a problem of voltage drop in the long lines between the output terminals and the load, then we will connect the sense lines to the power lines at the load (Fig. 15-9B). The regulator will then see the actual load voltage, rather than the voltage at the output of the regulator, thereby eliminating the voltage-drop problem. This feature is especially useful when the power supply is remotely located from its load; when 20- to 30-ampere currents are used, even a few inches of #12 wire constitutes "remote" location.

POWER SUPPLY PROTECTION

There are still a couple of matters which we will have to consider before closing this chapter on power supplies: protection circuits. There are two different types of protection circuit needed, especially in high-current power supplies often found in digital equipment: *overvoltage protection* and *output current limiting.*

Overvoltage protection is needed because the chips used in digital electronic circuits will not easily tolerate too great a voltage. TTL chips, for example, must operate at +5 volts dc, which must be regulated. If we attempt to operate TTL devices at, say, +6 volts, then they will fail prematurely. The need for protection is created by the fact that the input voltage to the regulator is almost always two or three volts higher than the nominal output voltage. For a +5 volt TTL power supply, therefore, the input voltage to the regulator will typically be +8 volts dc—far too much for TTL circuits. If some defect, such as a shorted series-pass transistor or open zener diode, causes the output voltage to rise to the level of the input voltage, then the TTL circuitry receiving power will get a sunburn. Your multi-hundred buck computer will fry in quick order, damage will be extensive.

The circuit in Fig. 15-10 is a type of overvoltage protector called an *SCR crowbar.* The active protection element is a high-current silicon-controlled rectifier (SCR), diode D2 in Fig. 15-10. This type of diode will remain turned off, i.e., it will not pass current in either direction, until a voltage is applied to the gate (G). When the gate signal is applied, however, the diode will break down and conduct in one direction from cathode (K) to anode (A), just like any other diode. The PIV rating of this diode should be two or three times the value of the input voltage V_{in}, and the current rating at least twice (preferably several times) the normal rated output current of the regulator.

The gate terminal is controlled by the network containing diode D1 and resistors R1 and R2. The zener potential of D1 is selected to be slightly above the nominal output potential of the regulator. For a +5 volt power supply, therefore, the diode potential should be 5.6 to 6.2 volts. If the output voltage goes too high,

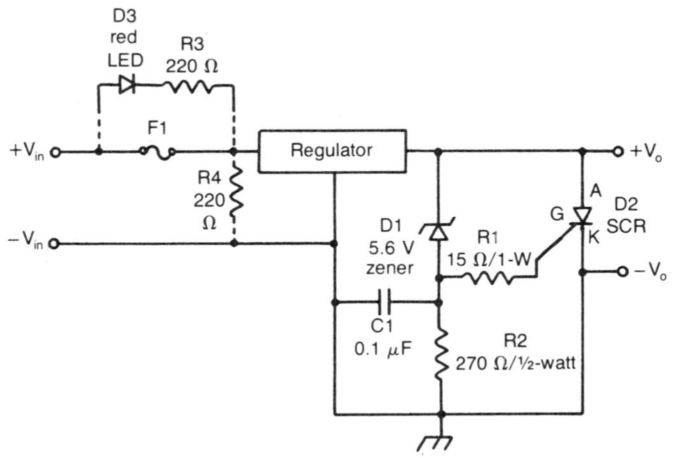

Fig. 15-10. SCR crowbar circuit.

then diode D1 will break over, i.e., conduct current, creating a voltage drop across resistor R2. This potential is applied to the gate through resistor R1, so will cause the SCR (D2) to trigger on. The SCR will remain turned on until the anode-cathode current is reduced below a critical hold current close to zero. For all practical purposes, this current is too low to permit erroneous, and harmful, turn-off prior to removing the harmful power.

The shorting of SCR D2 will cause fuse F1 to blow, thereby protecting the circuit. In most cases, both the fuse and the SCR will be destroyed but the expensive digital equipment being powered is saved.

The red LED (D3) and the associated resistors (R3/R4) constitute an optional *blown fuse indicator*. As long as fuse F1 is intact, the diode is shorted out, so cannot turn on. If, however, the fuse opens for any reason, current will be conducted by diode D3, resistor R3 and resistor R4 thereby causing the LED to light up. Some manufacturers build this circuit into the fuse holder to permit the user to identify when (or which) fuse is blown.

A circuit showing output current limiting is shown in Fig. 15-11. This same type of circuit is often already built into the IC voltage regulator, so might not be necessary in all cases. Transistor Q1 is the normal high-current series-pass transistor, which is controlled by zener potential V_z. The output potential is approximately equal to V_z less the base-emitter voltage V_{be}.

Transistor Q2 is the control element. The base-emitter voltage for this transistor is developed across resistor R2. Since this resistor is in series with the high current line, the Q2 base-emitter voltage is proportional to the output current, I_o. Since a silicon transistor requires 0.6 volts before it is turned on fully, we must pass a large enough current to produce more than 0.6 volts before Q2 turns on. The value of R2, therefore:

$$R2 = \frac{0.6}{I_{o(max)}} \qquad \textbf{(15-7)}$$

Where:

$R2$ is in ohms.

$I_{o(max)}$ is the maximum rated output current.

Example:

Find the appropriate value for resistor R2 in Fig. 15-11 if the rated output current of the power supply is 15 amperes.

$$R2 = 0.6/I_{o(max)}$$
$$R2 = 0.6/15$$
$$R2 = 0.04 \text{ ohms (i.e., 40 milliohms)}$$

The value 40 milliohms is not exactly a standard value, but can be created. For example, a standard "fuse resistor" value is 0.12 ohms. These under-1-ohm resistors are used extensively to protect power amplifier transistors as well as providing a limited amount of

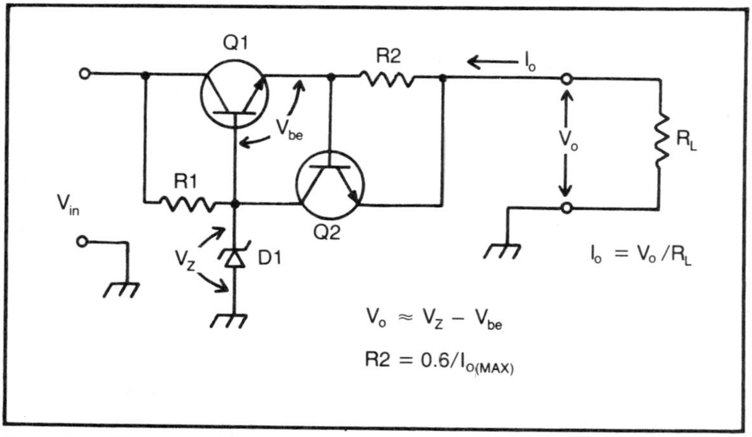

Fig. 15-11. Overcurrent protector.

negative feedback to such amplifiers. We can press them into service for current-limiting as well. Standard values include 0.1 ohms, 0.12 ohms, 0.15 ohms, 0.2 ohms, 0.25 ohms, 0.33 ohms and 0.47 ohms. Since 0.12 ohms is exactly three times our required 40 milliohms, we can parallel three 0.12 ohm resistors to make a single 0.04 ohm resistor. Most of these fuse resistors are 3- to 5-watt wirebound resistors, so will work nicely.

When the voltage across R2 rises above a certain figure determined by I_o R2, then transistor Q2 turns on thereby shorting the zener potential to ground. Resistor R1, incidentally, should have a power rating sufficient to handle $(V_{in})^2/R_1$. The power rating of R2, need only be $(0.6)^2$/R2, which for the case in which R2 = 0.040 ohms, would be (0.6) (0.6) (0.04) = 14 milliwatts for normal operation, and $(V_o)^2$/R1 for shorted-output operation.

Chapter 16

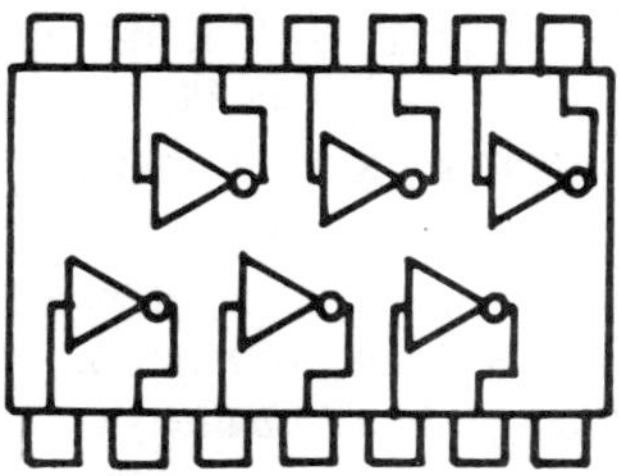

Using TTL/CMOS Devices in Microprocessor Interfacing

The low-cost TTL integrated circuit revolutionized digital electronics by increasing its range of application. Once IC logic blocks became available at low cost the chore of designing digital circuits was easier and more products could afford to be made digital.

CMOS extended the usefulness of digital electronics by reducing power consumption. The TTL device requires considerable chunks of current (25 milliamperes per package is typical), while equivalent CMOS devices operate with microamperes. The CMOS device draws appreciable current only during high/low or low/high state transitions, so the *average* current is quite low. In a typical digital circuit with, for example, 30 chips, the TTL version will require (30 × 25 μA), or 750 μA. That's ¾ ampere! The equivalent CMOS circuit would require (30 × 10 μA), or 300 μA. Translated into graphical terms the TTL circuit requires 2500 times as much current. Consider the implications of that difference in heat generation and operating time for battery-powered equipment.

One product which became possible with the advent of CMOS is the IC microprocessor chip. Previously (with only bipolar TTL technology available) the central-processing unit (CPU) of a programmable digital computer required many chips and generated lots of current. In the minicomputer world, 50 to 100 ampere, +5-volt, dc power supplies were often needed. CMOS technology allowed construction of low current, low heat dissipation IC CPU circuits, hence the microprocessor was born.

Even though MOS processes (i.e., CMOS, PMOS, NMOS) are used to make microprocessor and related support chips, most microcomputers are designed with TTL-compatible pinouts. This design permits use of TTL devices when building microcomputers from the μP chip. Therefore, both CMOS and TTL integrated circuits are used in microcomputer design.

Partially because of the heat dissipation problem, microprocessor outputs usually sink only 3.2 milliamperes, i.e., they have a TTL fan-out of two. Such an output will, therefore, drive only two standard TTL inputs without external buffering.

Because of microprocessor drive limitations most microcomputers require buffer devices between the microprocessor chip and both address and data busses. There are TTL "bus drivers" or "buffers" available with fan-outs of 30, 100, or more. Data passes over the *data bus* (eight bits in most microcomputers) to and from either of two places: *memory* or *input/output* (I/O) ports. The *memory* section of a computer may be likened to a large cabinet full of cubby holes. An analog is the rack used by postal workers to sort mail. Each cubby hole has a specific location—i.e., an "address." In the postal analogy, the address can be designated by an X-Y coordinate system: so many rows over by so many columns down. In a computer, the address of the electronic "cubby hole" is defined by a binary (or hexadecimal equivalent) number. For most 8-bit machines (which typically use 16-bit address busses) there may be up to 65,536 individual memory locations (65,536 is called "64K" in computer jargon because "K," kilo; means the nearest power of two to 100, i.e., 1024).

Memory operations require at least two signals from the microprocessor chip: one is the 16-bit address of the designated location, while the other is some indication that a *memory read or memory write* operation is to take place.

The address, of course is passed over the *address bus*. The signal denoting the type of operation is derived from the microprocessor *control signals*. All four are *active-low* outputs. The MREQ signal tells the world that a memory operation is taking place. The RD (read) and $\overline{WR}$ (write) tell whether the operation is a read or write procedure. For a memory read, therefore, MREQ and RD will be low simultaneously. Similarly, a *memory write operation* will see both MREQ and WR low. External logic is required to tell the memory section what to do. Three inputs are required: Address, $\overline{MREQ}$, and either $\overline{WR}$ or $\overline{RD}$.

The Z-80 microprocessor has unique input/output instructions. When an I/O instruction is executed the CPU will either read or write data from one of 256 discreet I/O ports. During these operations, the $\overline{\text{IORQ}}$ signal will be LOW, along with either $\overline{\text{WR}}$ or $\overline{\text{RD}}$ (Fig. 16-1).

The port address in Z-80 machines is passed down the lower order eight bits of the address bus (i.e., A0-A7). If it is an output operation, then at the same time the contents of the accumulator are passed over both the data bus (D0-D7) *and* the upper eight bits of the address bus (A8-A15). Four control signals, then, define all Z-80 read/write operations: $\overline{\text{MREQ}}$, $\overline{\text{IORQ}}$, $\overline{\text{WR}}$ and $\overline{\text{RD}}$. These must be used in conjunction with the address code.

Some microprocessors (e.g., 6502) do not have discreet I/O instructions so must dedicate memory locations for I/O data. Such a system is called *memory-mapped I/O*. Suppose we have a computer in which location 65,536 (i.e., $FFFF_{16}$ in hexadecimal) is designated as the output port to a mechanical printer. When we want to write a character to the printer we merely execute a *memory write* to address $FFFF_{16}$.

In the rest of this chapter we will discuss the use of popular TTL and CMOS devices to interface to microprocessors. Considered will be address decoders/operation decoders and the design of eight-bit I/O ports. The following examples are for Z-80 circuitry but the methods will be applicable to other devices as well. When trying to apply to other μP chips, study the control signals, timing diagrams and adapt or adopt as needed.

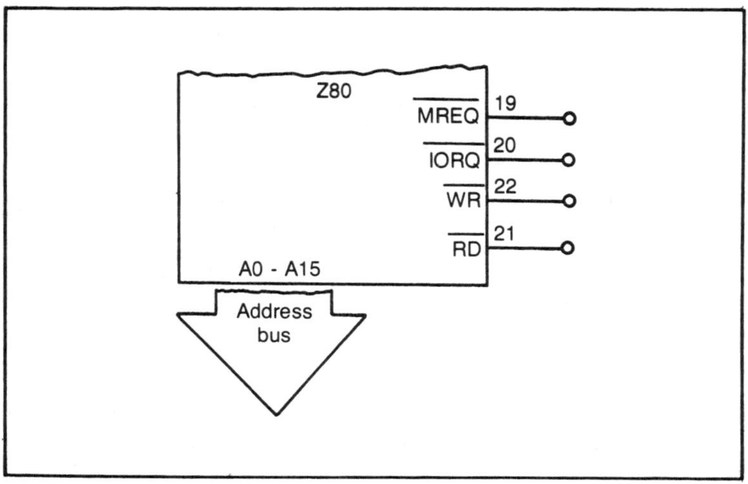

Fig. 16-1. Z-80 microprocessor control signals.

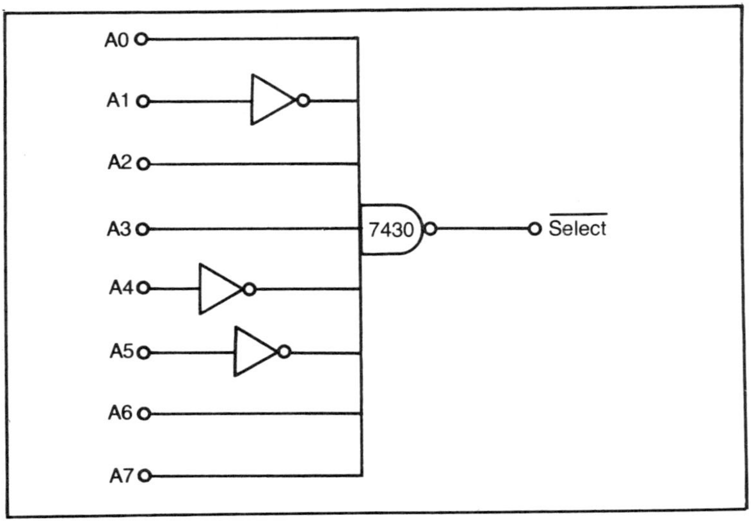

Fig. 16-2. Simple 8-bit address decoder.

ADDRESS DECODERS

The address bus of a typical microcomputer uses sixteen address bits to define 65,536 discrete memory locations. In order to identify either a memory location, or, an I/O port, we must be able to decode these addresses. We will begin out discussion with a look at eight-bit decoder circuits. These circuits will decode the Z-80 I/O port addresses, and will also be used in BANK select applications for memory.

EIGHT-BIT DECODERS

The eight-bit decoders to follow are designed to examine bits A0-A7 of the address bus and issue an active LOW SELECT pulse when the correct port address is present on the bus. The first decoder circuit is shown in Fig. 16-2 and is based on the 7430 TTL device. The 7430 is an eight input NAND gate. Reviewing the rules for NAND gates will show us the way to using this chip as an address decoder:

1. If *any* input of the 7430 is LOW, then the output will be HIGH.
2. All eight inputs must be HIGH for the output to be LOW.

According to the rules, therefore, we must conspire to somehow force all inputs HIGH *only* when the correct address is present on the address bus. Since the possible codes for 256 ports ranges from 00000000_2 (0H) to 11111111_2 (FFH) most of our ports will have one or more *LOW* lines in the correct address.

One solution, viable when there is only one port, is to designate the port as No. 255, the code for which is FFH. Under this condition, the correct code on the address bus will produce all ones (i.e., HIGHs) at the inputs of the 7430 device.

In most cases, however, we will not be able to use such a simple scheme because more than one I/O port will be designated. In those situations we will have to use one or more inverters in the lines between the address bus and the 7430 input terminals.

The decoder in Fig. 16-2 is set up with inverters on three input lines so will decode binary address 11001101_2 (CDH). If you examine the binary representation (i.e., 11001101) you will note that bits A0, A2, A3, A6, and A7 will be HIGH when the correct address is on the bus. For these lines no additional "processing" is needed. Bits A1, A4, and A5, however, will be LOW for a correct address. These lines must be inverted prior to being applied to the 7430 inputs. TTL inverters, sections of 7404 hex inverter IC, may be used for this purpose. With the inverters in place, the only code on the address bus that will produce 11111111_2 at the 7430 inputs is 11001101_2, the correct port address.

The output of the 7430 goes LOW when the correct address appears on the bus, so it can be used as a NOT-select (SELECT) signal to tell the whole wide world that the correct address has been received. In a later section of this chapter we will see how SELECT or SELECT signals are used in conjunction with microprocessor control signals to turn on the specific I/O port or memory bank being addressed.

Figure 16-3 shows another circuit for decoding addresses. In this case, a TTL 7442 device is used as a four-bit decoder. The 7442 was originally designed as a *BCD-to-1-of-10 decoder* and was used to provide discreet active-LOW outputs that could turn on archaic numerical indicators (e.g., lamp columns) or *Nixie®* tubes.

The 7442 device sees a binary coded decimal (BCD) four-bit input word representing numbers 0000_2 (0_{10}) to 1001_2 (9_{10}). If, for example, the input word is 0110_2 (6_{10}), then the 7442 output corresponding to "6" (i.e., pin no. 7) will go LOW; all other outputs remain HIGH. We can, therefore, use the 7442 alone as a four-bit address decoder.

So why would we want a four-bit decoder in an eight-bit system? Fair question! The answer is why use a 12 gauge shotgun to do the job of a peashooter! The 7442 will accommodate up to ten I/O ports; few microcomputers will have more than ten I/O ports! Using the 7442 gives us a single-chip decoder that will accommodate the needs of most users. All we have to do is connect the four input lines of the 7442 to bits A0 through A3 of the address bus.

Figure 16-4 shows the use of two 7442 devices to achieve eight-bit operation. Here we use one 7442 to examine the lower order half-byte (i.e., "nybble"), A0 through A3; the other 7442 examines the higher order nybble (A4 through A7). The appropriate outputs from the 7442s are connected to inputs of the NOR gate.

So what are "appropriate" 7442 outputs? Let's consider an example. Suppose we want to select I/O port 23_{10}. The code would be 001000112. When the correct code is present on the address bus, the "2" output of the high-order 7442 (pin no. 3) and the "3" output of the low-order 7442 (pin no. 4) will *both* be LOW.

The rules governing the operation of the NOR gate are as follows:

1. A HIGH on *either* input causes the output to be LOW.
2. *Both* inputs must be LOW for the output to be HIGH.

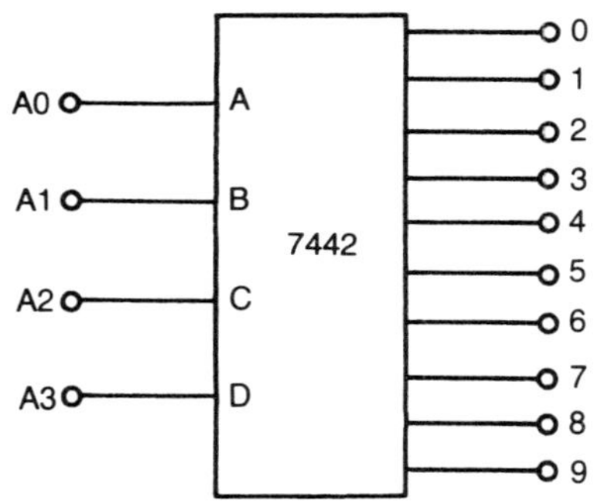

Port No.	A3	A2	A1	A0
0	0	0	0	0
1	0	0	0	1
2	0	0	1	0
3	0	0	1	1
4	0	1	0	0
5	0	1	0	1
6	0	1	1	0
7	0	1	1	1
8	1	0	0	0
9	1	0	0	1

Fig. 16-3. I/O port select circuit (four-bit-to-1-of-10).

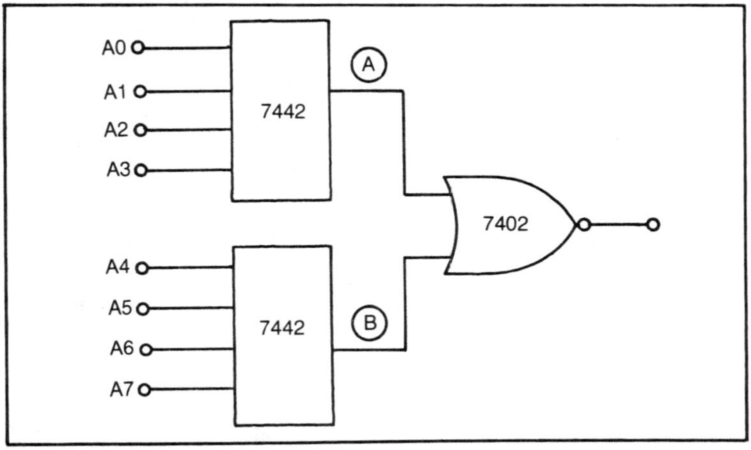

Fig. 16-4. 8-bit decoder.

From the foregoing we may conclude that the output of the 7402 NOR gate in Fig. 16-4 will be HIGH if and only if the correct code is on the address bus.

Another four-bit decoder is shown in Fig. 16-5. This circuit uses the TTL 7485 device, but there is an equivalent CMOS device (but not pin-for-pin compatible) that will do the same job. The 7485 is a four-bit *magnitude comparator* that examines two four-bit binary words (designed "A" and "B") and issues three outputs. They denote whether A equals B, A is greater than B, or A is less than B; we use the A = B output here.

The four inputs for the "A" word are connected to bits A0 through A3 of the address bus. The "B" inputs are connected to a DIP switch. Such a switch contains four SPST switches that are "toggled" by a pencil-point or similar tool. The reason it is called a "DIP" switch is that it mounts on a printed wiring board in the same space as a DIP integrated circuit—it has two rows of 0.1 inch center pins, with the rows spaced 0.3 inch apart. One side of each switch in the DIP package is grounded. When the switch is closed, therefore, the corresponding 7485 input is grounded; when the switch is open the 7485 input is HIGH. A pull-up resistor between each input and +5 volt dc power supply ensures that the input remains HIGH when it is supposed to be, and does not go inadvertently LOW under noisy conditions.

The "A=B" input of the 7485 remains LOW unless the correct address is present on the address bus. When the correct address (i.e., it matches the code set by the DIP switch) is present, however, the "A=B" output goes HIGH. It will remain HIGH only so long as the correct code is present on the address bus.

There are two common variations on the circuit of Fig. 16-5. One is to replace the DIP switch with wire jumpers on the PC board. This design reduces the cost slightly, but at the expense of the flexibility provided by the switch.

A second variation is to replace the DIP switch with a rotary switch (either "thumbwheel" or screwdriver PC type) that provides four-bit BCD or hex outputs (providing either 10 or 16 combinations, respectively). This circuit modification will retain the flexibility, improve human engineering by eliminating the need to convert port numbers or address locations into binary number format, at a small increase in switch cost.

The 7485 device, and its CMOS equivalent, can be cascaded to provide larger word lengths (in four-bit increments). Figure 16-6 shows the method for cascading 7485s. Each device is equipped with three outputs (described above), and three matching cascading inputs. When the cascading inputs are used, the "A=B" output of the last 7485 in the chain will go HIGH only when the "A=B"

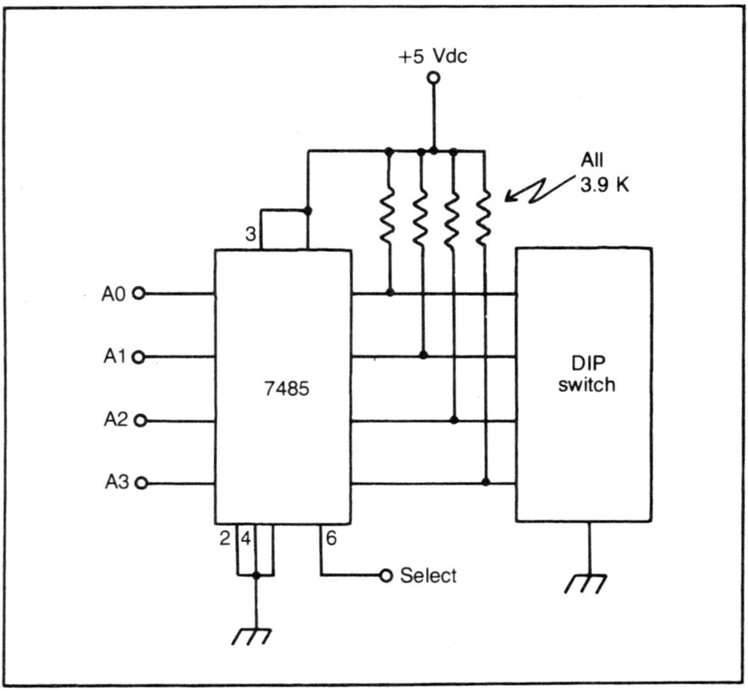

Fig. 16-5. Programmable 4-bit decoder.

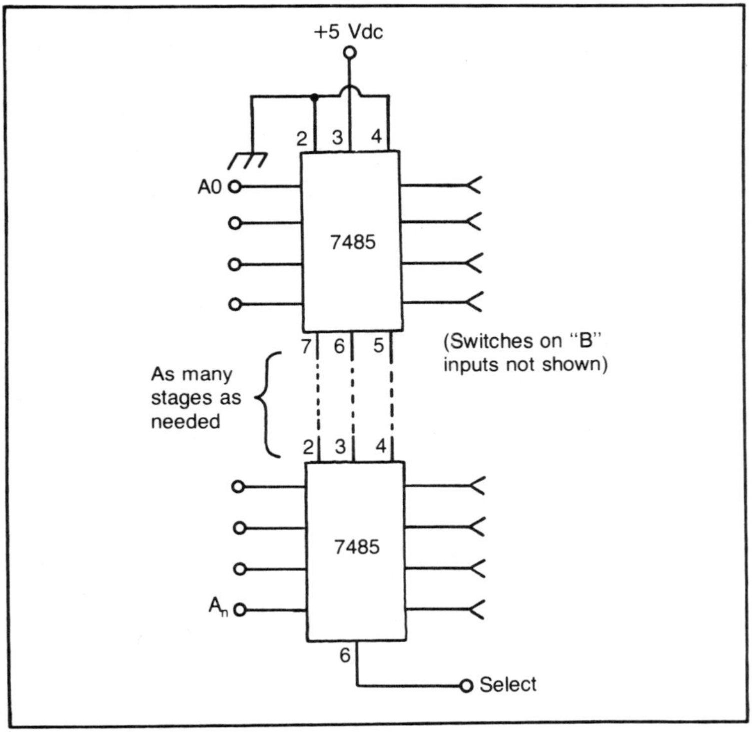

Fig. 16-6. Cascade connection of Fig. 16-5.

outputs of all lower order 7485s are also HIGH, *and* its own A/B are equal. Under that condition, a SELECT signal is generated.

We can accommodate word lengths that are not multiples of four bits by grounding the unneeded terminals on the highest order 7485. It is *essential* that *both* equivalent bits on word-A *and* word-B are grounded.

MEMORY ADDRESS DECODING

Computer solid-state memory chips have the number of pins required to form the address for all locations within. Thus, a 1K (i.e., 1024 byte) will have 10 pins ($2^{10} = 1,024$) for addressing. Memory devices, therefore, have built-in decoding.

Because memory chips have built-in decoding in discrete sizes (e.g., 1K, 2K, . . . 3K), we use an array, or bank, type of organization in designing computers. The size of the bank is the size of the memory chips used. For purposes of discussion we will use a 1K bank size, even though that size is now made obsolete by larger size memory chips. Recall from above that a 1K memory chip requires ten memory address lines. Therefore, we will use bits A0 through A9 of the address bus.

Figure 16-7 shows a bank organized system using the 7442 device to decode address bits A10 through A13. We could just as easily use other techniques (see before), but each of the others would require each bank to have a unique decoder of its own. The circuit of Fig. 16-7 requires but one device, a single 7442, for all banks (the 7442 will accommodate up to ten banks).

Recall, please, that the 7442 is a BCD-to-one-of-ten-decoder. It will examine a four-bit binary-coded decimal word (coded in 8-4-2-1 system) and generates a unique active-LOW output that indicates the decimal value of the BCD input word. The ten address inputs of all 1 K memory chips are bussed together with the address bus (A0 of each chip connected to A0 of the address bus, and so forth).

The chip enable ($\overline{CE}$) lines of each memory IC are connected to the active-LOW outputs of the 7442. (The $\overline{CE}$ line being LOW turns on the chip, HIGH turns it off.) The code on A10 through A13 determines which 7442 output goes LOW. We can, therefore, turn on the various banks from the code that would normally appear on

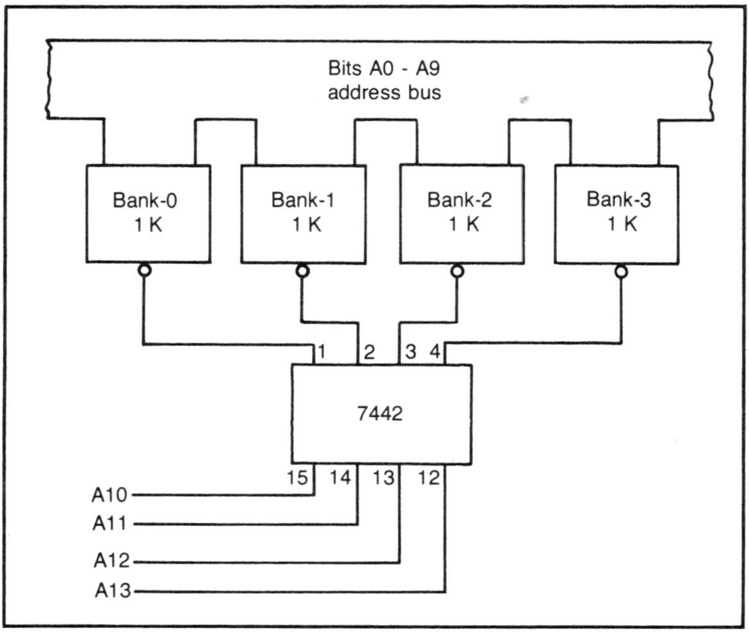

Fig. 16-7. Decoding for banked memory.

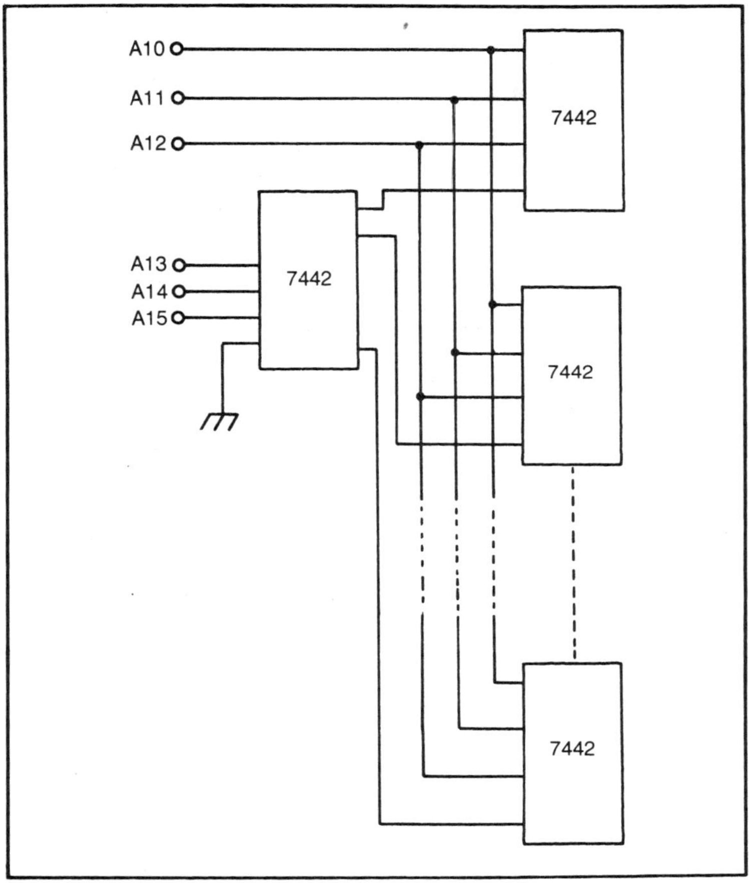

Fig. 16-8. Decoding for banked memory.

the address bus when a higher location than 1 K is called for. With the scheme shown, we can address memory to 10 K in 1 K increments. Table 16-1 shows the 7442 pinouts and A10-A13 code for up to 8 K of memory.

We can extend memory all the way to the 64 K limit of most 8-bit microcomputers (which have 16-bit address busses) by forming 1 K banks into 8 K banks, and then using a similar circuit to select which 8 K bank is in operation (see Fig. 16-8). Coding in Table 16-2 shows how A13 through A15 will appear for locations 8 K through 64 K.

Each 8 K bank will have its own 7442 (to select 1 K sub-banks), with the circuit being as in Fig. 16-6. We use the "8" input of each

Table 16-1. Memory Coding.

Memory Locations	Block Number	7442 Output	7442 Pin Number	A13	A12	A11	A10
0-1 K	0	0	1	0	0	0	0
1 K-2 K	1	1	2	0	0	0	1
2 K-3 K	2	2	3	0	0	1	0
3 K-4 K	3	3	4	0	0	1	1
4 K-5 K	4	4	5	0	1	0	0
5 K-6 K	5	5	6	0	1	0	1
6 K-7 K	6	6	7	0	1	1	0
7 K-8 K	7	7	9	0	1	1	1

7443 (pin 12) to turn the stage on and off. This pin is used in BCD systems to denote the decimal numbers "8" and "9," which don't exist in our 64 K system (examine *bank no*. column in Table 16-2). If we force the "8" input HIGH, then there will be no valid 0 through 7 outputs so the chip is turned off for all practical purposes.

The outputs of the main bank select 7442 are used to turn on the sub-bank select 7442 devices. When an appropriate main bank 7442 output is LOW it will turn on the associated sub-bank 7442. The address called for on the lower order lines (A0-A9) will then be selected from the specific 1 K sub-bank determined by the high order bits (A10-A15).

DESIGN OF I/O PORTS (PARALLEL)

Input/output ports are the means by which the computer communicates with the outside world. In a parallel port circuit, all data bits are transmitted simultaneously. This job requires eight or sixteen (depending on data bus size) independent lines for each port. Since the techniques for 8-bit and 16-bit systems are essentially identical, we will consider only the 8-bit system.

Table 16-2. Memory Coding.

Memory Location	Bank No.	7442 Output	7442 Pin	A15	A14	A13
0-8 K	0	0	1	0	0	0
8 K-16 K	1	1	2	0	0	1
16 K-24 K	2	2	3	0	1	0
24 K-32 K	3	3	4	0	1	1
32 K-40 K	4	4	5	1	0	0
40 K-48 K	5	5	6	1	0	1
48 K-56 K	6	6	7	1	1	0
56 K-64 K	7	7	9	1	1	1

Figure 16-9 shows a complete I/O port. Please note two things. First, the particular IC devices selected are workable examples only and are not intended as be-all-and-end-all. The idea is to show *technique*, not specific circuits. Second, the input and output sides of the circuit are independent of each other. If you only need one of the two functions, then it is not *necessary* to include the unneeded function. For example, if you need only an input port, then only IC2 is needed, delete IC1.

An output port is used to capture and remember *data* on the *bus* when an output instruction is being executed by the CPU. The period required to execute such an instruction is only a few microseconds, so the output port must be capable of *latching*, i.e., remembering, the data after it disappears from the data bus. Very few peripheral devices that would normally be connected to the port can react that rapidly; latching must be provided either by the computer output (most sensible), or, the peripheral input circuit (not the best method).

The 74100 device used in this output is a dual *quad-latch*. In other words, it contains two banks of four latches each. A latch, of

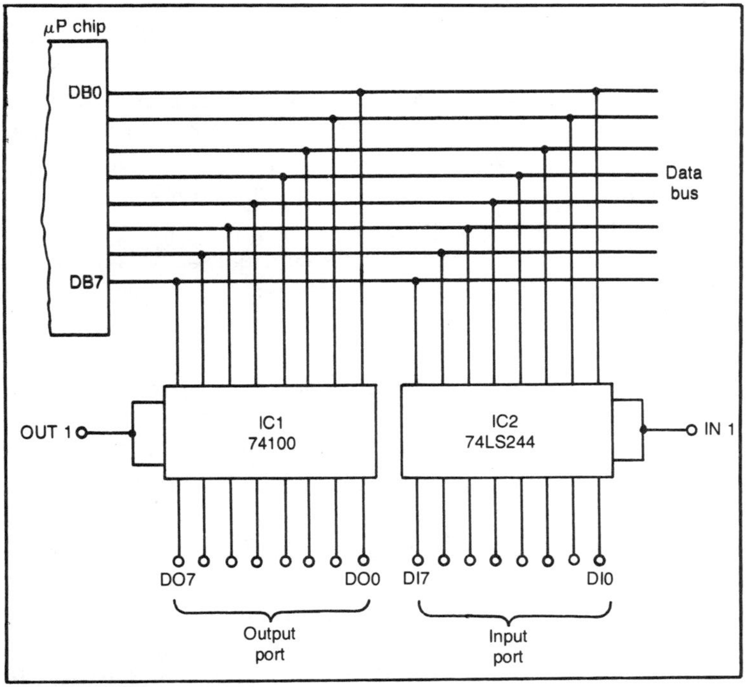

Fig. 16-9. Typical I/O port.

course, is an R-S flip-flop with the *clock* line used as a *strobe*. Let's review the operation of the R-S flip-flop (alias *data latch*).

1. When the *clock* (i.e., *strobe* line is HIGH, the Q-output follows the HIGH/LOW data on the D-input; and
2. When the *clock* is LOW, then the Q-output retains the last valid data that existed before the clock dropped LOW.

By connecting the two strobe terminals of the 74100 together, therefore, we can cause eight bits of data to be stored. The eight R-S flip-flop inputs are connected to the eight lines of the data bus, and the eight Q-outputs become the output port parallel lines.

An OUT 1 signal is used to strobe the data signal to the outputs. This signal is an example of a *device select* pulse; of which more later. The input port device is *IC2*. This chip contains eight *tri-state noninverting buffers* that are turned on or off by an active-LOW. We tie these CE lines together to form an IN 1 device select line. We will discuss generation of IN and OUT signals shortly.

The reason for requiring a tri-state device for the input port is that we do not want the port loading the data bus *except* when the CPU commands it into operation by sending IN 1. At all other times we want the input port to float at high impedance across the bus.

Figure 16-10 shows several methods of interfacing other devices to a parallel output port. Not shown, incidentally, is the essentially trivial case of connecting one parallel port to another. For example, connecting the parallel output port of an 8-bit computer to the 8-bit input ports of either another computer, or, a peripheral device such as a printer. In that case, merely wire the respective ports together bit-for-bit.

Returning to Fig. 16-10, let's consider some interfacing methods. First, let's see how we might interface a light-emitting diode (LED), or, an incandescent panel lamp to an output port. We select (arbitrarily, by the way) bit DO0 to drive the LED circuit. The programmer must take whatever action is needed to set SO0 HIGH whenever the panel LED/lamp is to turn on.

The actual LED driver is an open collector TTL inverter device. The LED is connected to the dc power supply through a current-limiting resistor (R1). If Vt is a +5 volt dc supply, then a 340-ohm resistor is usually indicated for R1. The lamp can be connected directly between the inverter output and Vt, provided that the normal lamp current does not exceed the current rating of the open-collector output.

When the driving bit of the output port (DO0 in this case) is LOW, then the inverter output is HIGH so the LED/lamp is off.

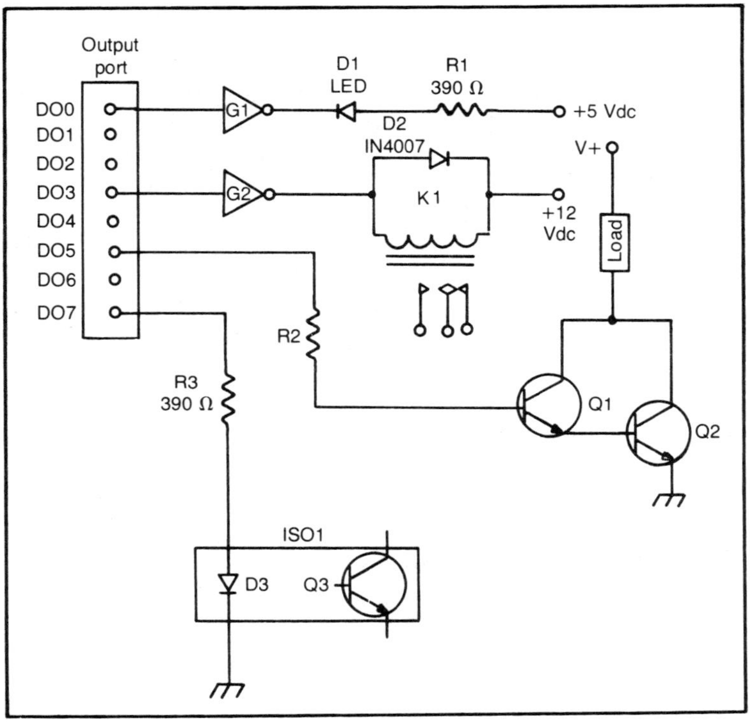

Fig. 16-10. Interfacing with the external world other than other computers.

When the driving bit is HIGH, however, the inverter output is LOW so the LED/lamp has a ground path and will thereby turn on.

The diode across the relay coil (1N4007) is used to suppress the reverse-polarity high-voltage spike that is generated by the process of "inductive kick" when the relay is de-energized. This spike can reach hundreds of volts in amplitude, so it may damage components. The reason why an ancient device like an electromechanical relay is sometimes used with a computer is *isolation*. Some loads, e.g., 110 volt ac circuits are dangerous to connect to the output port—isolation is critical.

Another isolation technique is shown at bit DO7 of the output port. Here we use an *optoisolator* (ISO1) to achieve the needed isolation. An optoisolator is an IC device that contains an LED and a phototransistor positioned so that the light from the LED impinges on the base of the phototransistor.

The case of Fig. 16-10, the LED of ISO1 is driven directly by the output port bit (DO7); this assumes that the port can drive the

LED. If the port cannot directly drive an LED, then use an open-collector TTL inverter LED driver as at DO0.

A driver for heavy loads, such as large lamps or solenoids, is shown connected to bit DOS of the output port. Here we see a Darlington amplifier consisting of Q1 and Q2. In a Darlington circuit, both collector terminals are tied together. Furthermore, the emitter of the input transistor drives the base of the output transistor. In essence, the Darlington pair forms a transistor in which the "base" is the base of Q1, the "emitter" is the emitter of Q2 and "collector" is the common collector connection Q1/Q2. The main reason for using the Darlington pair is *beta* ratings of Q1 and Q2. If Q1 and Q2 are the same, then the total *beta* is the square of the transistor *beta*. For example, if two beta = 100 transistors are used, then the total beta is (100) (100), or 10,000.

The beta amplification of the Darlington pair allows us to drive large loads with low currents. Recall that *beta* is the quotient I_C/I_B. Since I_B is limited to the driving current available from the output port, the *beta* limits the maximum current that can be accommodated. For example, many microcomputer output ports are limited to 3.6 mA of drive current. If the driver transistor has a *beta* of 100, then the maximum load current will be (3.6 mA) (100), or 360 mA. If the Darlington is used, then the *beta* is 10,000 so the device will handle up to (3.6 mA) (10,000), or 36 amperes!

Of course, the output transistor (Q2) has to be capable of sustaining the load current and collector dissipation. You can buy Darlington power drivers that have both transistors inside a single TO-5 package. Alternatively, one can also make the Darlington pair from discrete transistors; a 2N2053 for Q1 and 2N3055 for Q2 is a popular choice.

If the load connected to the Darlington collector is a solenoid, then a 1N4007 diode must be connected in parallel with the coil. Otherwise, we will face the same high-voltage spike problem as in the relay circuit. In fact, the problem will generally be worse because most solenoids carry higher current than relay coils.

Several input port scenarios are shown in Fig. 16-11. As before, the trivial case of connecting the parallel input port to other parallel output ports is not considered.

At positions DI0 and DI2 we see the method for interfacing switches to the computer. These switches may be front-panel switches for user interface, or, may be internal for use in setting options the customer pays for or certain operating conditions. With respect to the latter, some companies who offer microprocessor

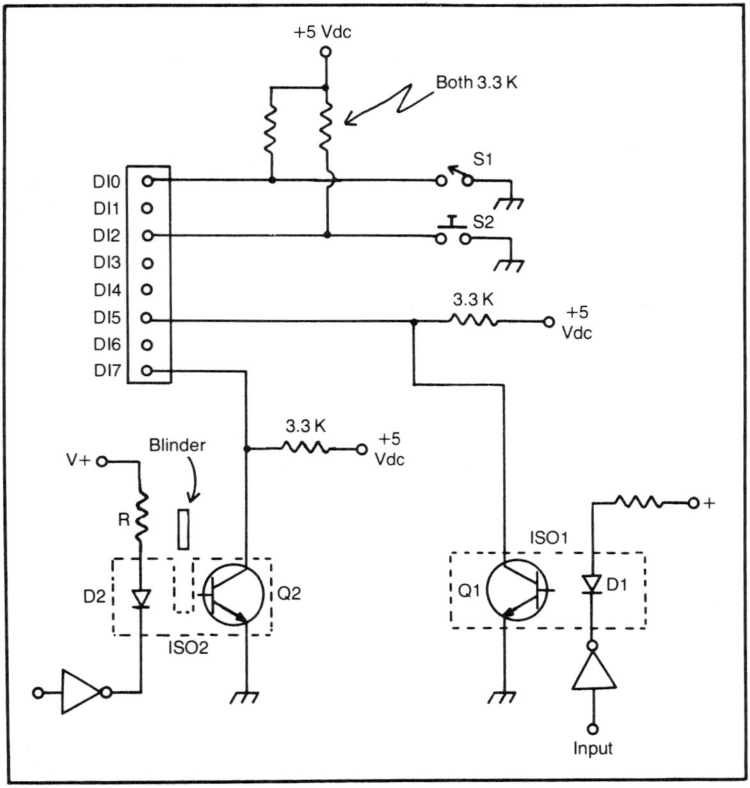

Fig. 16-11. Interfacing switches.

based instruments have cost-options that are purely a function of software. They will sometimes go ahead and program the read-only memory (ROM) with the needed software and include an option select switch at the input port. If the customer buys the option, then the switch is set to the active position. We could, for example, make a bit HIGH to turn off the option, and LOW to turn it on. An eight-bit port will accommodate light option-select switches such as that in position DI0. A program must be written that will scan the port when the computer is turned on to determine which options are authorized by the HIGH/LOW conditions on the port.

Although S1 is a toggle switch and S2 is a pushbutton type, both are treated the same way in the circuit: one side is grounded and the other is connected to a bit at the input port. A pull-up resistor is used between the "hot" side of the switch and +5 volts in order to ensure the input remains HIGH when the switch is open. Some special-purpose input ports have built-in pull-up resistors so will not require resistors as shown. The circuit arrangement shown in Fig. 16-11 keeps the bit HIGH when the switch is open, and LOW when the switch is closed.

An optoisolator (see above), ISO1, can be used to isolate the computer from dangerous loads. The LED in ISO1 is driven in the usual manner by an open-collector inverter stage. The phototransistor is connected to the input port as if it were a grounding SPST switch. The other optoisolator used in Fig. 16-11 is not for isolation but serves as an example of a photodetector circuit to register events.

Assemblies are available that provide the LED and phototransistor in the same housing. There is a cut-out in the housing that permits a flat vane to interrupt the light path between D2 and Q2. An example of this system is found in the microprocessor-based control circuits of the Heathkit model H14 printer. The optoisolator is used to indicate the print-hand side of the paper. An opaque metal vane mounted to the print-head assembly blinds the light path, thereby turning off the phototransistor. The computer periodically scans the optoisolator bit to see if it is HIGH or LOW. In the circuit of Fig. 16-11, the bit is HIGH when the head carriage is in position and LOW when it is elsewhere.

SERIAL I/O PORTS

A serial I/O port transmits data one bit at a time over a single wire or communications channel. Although much slower than parallel communication, serial transmission can be less costly because of fewer common links needed.

Although we will not consider the various serial methods (RS-232, etc.), we can discuss the *universal asynchronous receiver transmitter* (UART) that makes it easy to convert back and forth between serial and parallel formats. The universal asynchronous receiver/transmitter, or UART, is a special digital IC that contains independent data-transmitter and receiver sections. The transmitter accepts *n*-bit parallel-format data and transmits it in serial form. The receiver, on the other hand, accepts serial-format input-data and reassembles it into parallel format. The UART makes the task of designing serial I/O ports in parallel-format computers easier.

Asynchronous transmission is preferred over synchronous transmission because it is not necessary to precisely track the clocks at each end. The clocks must be operating at very nearly the same frequency, but they need not be locked together. This eliminates the added circuit or extra communications channel needed to synchronize the two clocks. The tolerance of the clock rates in asynchronous transmission is 0.02 percent, but this is easily obtained if crystal-control is used.

The UART is a single large-scale integration chip that performs all of the data transmission functions on the digital side. Bit length, parity, and the overall length of the stop bits can be programmed into the UART.

The block diagram of a common UART IC is shown in Fig. 16-12. This particular device is the TR1602A/B by Western Digital. Note that the pinouts for the UART are almost universally standardized and are based on the now obsolete AY-1013. Most UARTs are capable of all three communications modes: *simplex, half-duplex,* and *full-duplex*. This feature is due to the fact that the transmitter and receiver control pins are independent of each other.

The UART is capable of being user-programmed to determine transmitted word-length, baud rate, parity type (odd/even, receiver-verification/transmitter-generation), parity-inhibit and stop-bit length (1, 1.5, or 2 clock periods). The UART also provides seven different status flags: *transmission completed, buffer-register, transfer completed, received data-available, parity error, framing error*, and *overrun error.*

The maximum clock speed is between 320 kHz and 800 kHz, depending upon the particular type selected. Note that the clock rate actually used in any given application is dependent upon the baud rate. The clock frequency is always 16 times the data baud-rate.

The receiver output lines are three-state, which means there is a high impedance to both ground and positive voltage when the outputs are inactive. This allows the outputs of the receiver section to be connected directly to a data bus without extra circuitry.

The transmitter section uses an eight-bit input register. This feature makes it capable of accepting an eight-bit parallel word from a source such as a keyboard, computer output port, data bus, etc. It assembles these bits and then transmits them at the designated time, adding any demanded parity or stop bits.

The receiver data-format is a logical mirror image of the transmitter section. It inputs serial data-bits; strips off the start, stop, and parity bits (if used); and then assembles the binary word in parallel form. In addition, it tests the data for validity by comparison of the parity bits and stop bits.

The standard UART data format is shown in Fig. 16-13. The data line (transmitter serial-output or receiver serial-input nor-

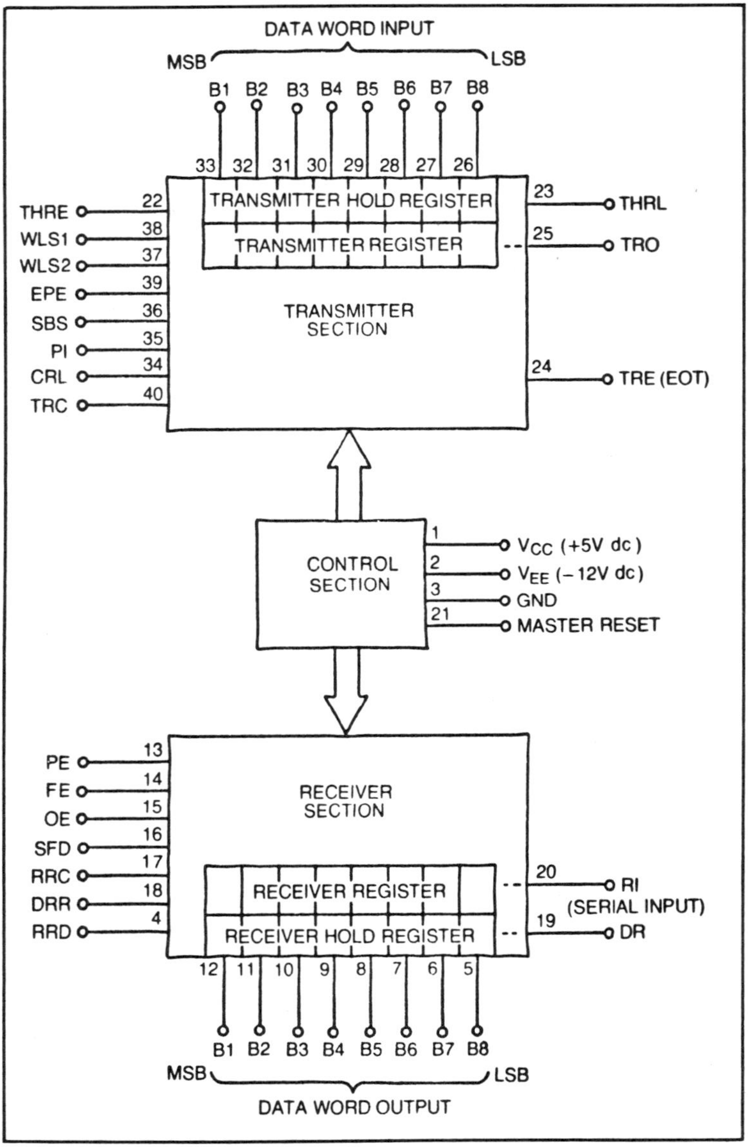

Fig. 16-12. UART block diagram.

mally sits at a logical high level unless data is being transmitted or received. The start bit (B0) is always LOW, so the HIGH-to-LOW transition is what the UART senses for the starting of a word or transmission. Bits B1-B8 are the data bits loaded into the transmitter register from the outside world. Although 16-13 shows all

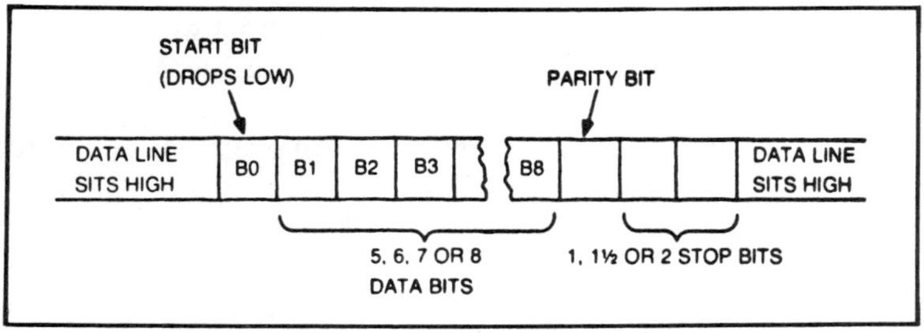

Fig. 16-13. Serial data format.

eight bits, you can program the device for fewer if needed. Lengths of 5,6,7, or 8 bits are allowed. Bits are dropped in shorter formats from the B1 side of the chain.

Figure 16-14 shows typical receiver and transmitter configurations for the UART. The transmitter section is shown in Fig. 16-14A while Fig. 16-14B shows the circuitry for a receiver section. If the serial output of the transmitter is connected to the serial input

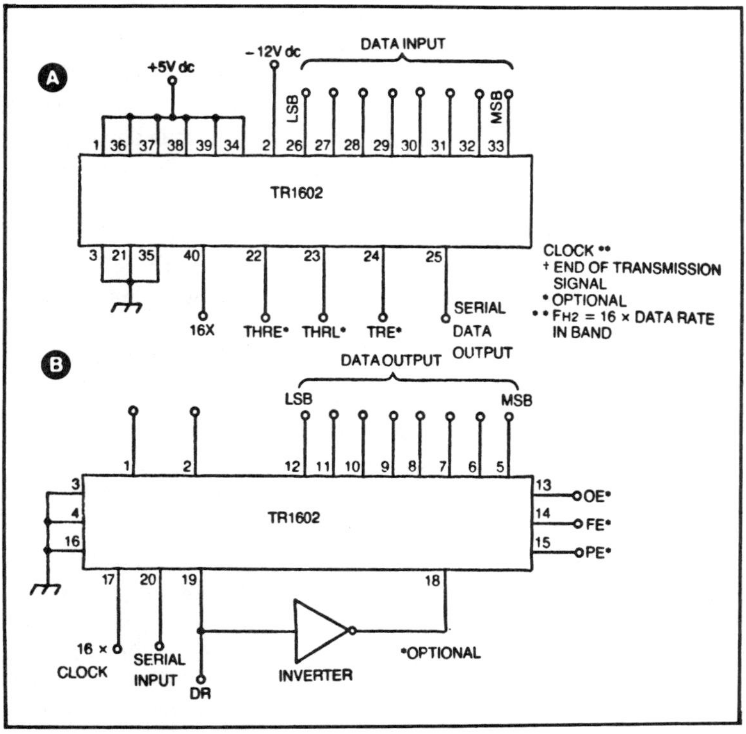

Fig. 16-14. (A) Receiver UART connections, (B) transmitter UART connections.

of the receiver, a closed loop exists; and the output word from the receiver matches the transmitted word. In most cases the UART is used to drive some external communications channel, such as an audio-frequency shift-keyer, for transmission over some standard communications media.

The UART can be connected as in Fig. 16-14 with separate transmit and receive lines, but that method is not the most optimal and requires separate I/O-ports for the connection. You can also connect the UART directly to the data bus of a microcomputer!

Figure 16-15 shows the basic connections for the standard UART to an eight-bit data-bus. The receiver-output-register lines are tri-state; so they can be connected directly to the lines of the bus. The transmitter-register lines also have a HIGH impedance; so they also can be connected to the data bus. You can enable the transmit-hold register by bringing the line LOW momentarily when output data is present on the line. This is done by connecting the transmit-hold register-line to an output-device select-line. Circuits to generate in and out pulses are discussed later.

The receiver register is dumped to the data bus when the IN 2 line is dropped low. The IN 2 line is connected to the RRD and DRR lines. The IN 1 line is an active-LOW control-line that connects the status flags to the data bus when LOW. Keeping this line prevents the status flags from entering the data bus. The status flag lines are set to HIGH impedance (i.e., three-state) when the first input line is HIGH.

In the circuit of Fig. 16-15 the programming lines (CRL, PI, SBS, WLSI, W12, and EPE) are set permanently HIGH through a pull-up resistor. You can program the UART any way that you wish by setting specific lines HIGH or LOW (see Table 16-3). Sometimes, however, you might not wish to have a preprogrammed UART; you might want to operate UART programming under software control. You can adapt the UART to this mode of operation, which is common among UARTs that are designed for use with specific microprocessor chips by using a 74100 data-latch as an output port-register (Fig. 16-16). Six control lines of the UART (pins 34-39) are connected to the outputs of the 74100. The inputs of the 74100 are connected to lines B0-B5 of the data bus. To set the UART load the CPU accumulator with a binary word that contains the correct bit pattern for the programming desired. For example, if you want to program the UART with the hardwires in Fig. 16-15 you need to load the accumulator with the binary word xx111111 (x indicates "don't care"). You can load the accumulator with either FF

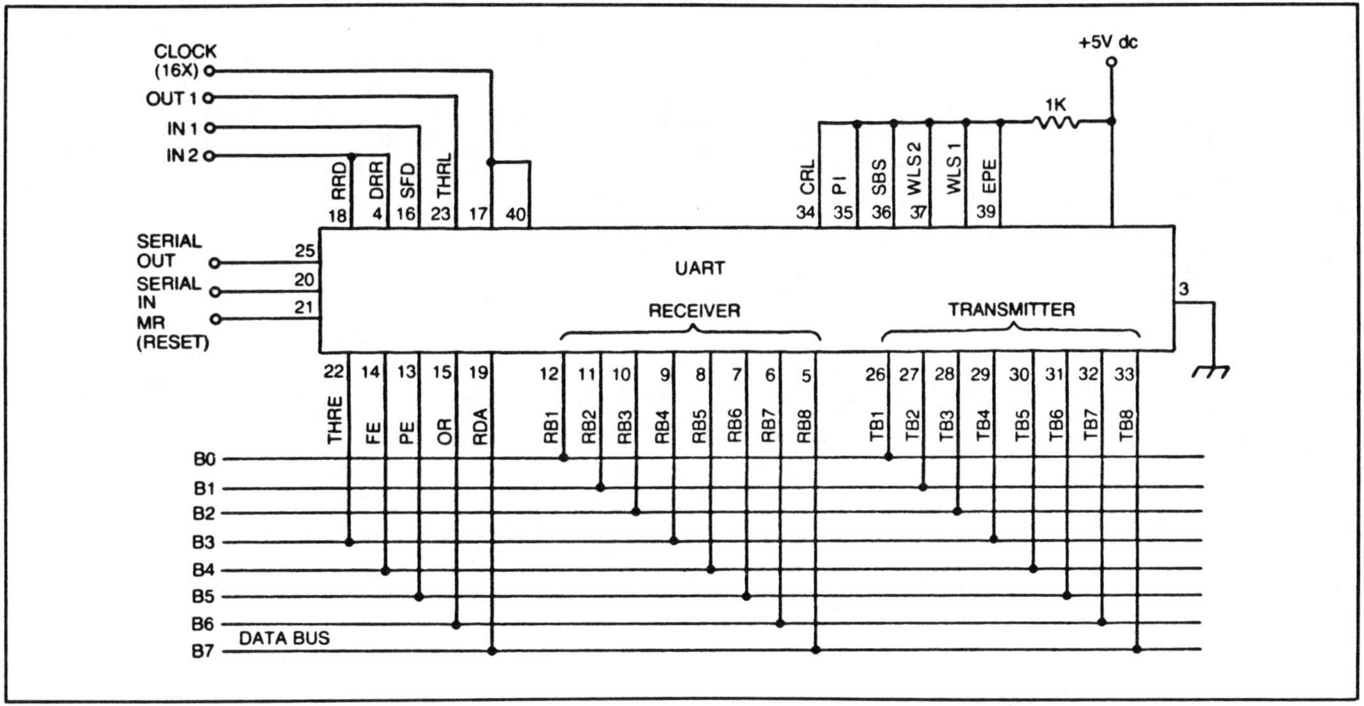

Fig. 16-15. Data bus connection of the UART.

Table 16-3. UART Pinout Functions.

Pin No.	Mnemonic	Function
1	V_{CC}	+5 volts DC power supply.
2	V_{EE}	−12 volts DC power supply.
3	GND	Ground.
4	RRD	Receiver Register Disconnect. A high on this pin disconnects (i.e., places at high impedance) the receiver data output pins (5 through 12). A low on this pin connects the receiver data output lines to output pins 5 through 12.
5	RB_8	LSB
6	RB_7	
7	RB_6	
8	RB_5	Receiver data output lines
9	RB_4	
10	RB_3	
11	RB_2	
12	RB_1	MSB
13	PE	Parity error. A high on this pin indicates that the parity of the received data does not match the parity programmed at pin 39.
14	FE	Framing Error. A high on this line indicates that no valid stop bits were received
15	OE	Overrun Error. A high on this pin indicates that an overrun condition has occurred, which is defined as not having the DR flag (pin 19) reset before the next character is received by the internal receiver holding register.
16	SFD	Status Flag Disconnect. A high on this pin will disconnect (i.e., set to high impedance) the PE, FE, OE, DR, and THRE status flags. This feature allows the status flags from several UARTs to be bus-connected together.
17	RRC	16 × Receiver Clock. A clock signal is applied to this pin, and should have a frequency that is 16 times the desired baud rate (i.e., for 110 baud standard it is 16 × 110 baud, or 1760 hertz).
18	DRR	Data Receive Reset. Bringing this line low resets the data received (DR, pin 19) flag.
19	DR	Data Received. A high on this pin indicates that the entire character is received, and is in the receiver holding register.
20	RI	Receiver Serial Input. All serial input data bits are applied to this pin. Pin 20 must be forced high when no data is being received.
21	MR	Master Reset. A short pulse (i.e., a strobe pulse) applied to this pin will reset (i.e., force low) both receiver and transmitter registers, as well as the FE, OE, PE, and DRR flags. It also sets the TRO, THRE, and TRE flags (i.e, makes them high).
22	THRE	Transmitter Holding Register Empty. A high on this pin means that the data in the transmitter input buffer has been transferred to the transmitter register, and allows a new character to be loaded.
23	THRL	Transmitter Holding Register Load. A low applied to this pin enters the word applied to TB1 through TB8 (pins 26 through 33, respectively) into the transmitter holding register (THR). A positive-going level applied to this pin transfers the contents of the THR into the transmit register (TR), unless the TR is currently sending the previous word. When the transmission is finished the THR→TR transfer will take place automatically even if the pin 25 level transition is completed.
24	TRE	Transmit Register Empty. Remains high unless a transmission is taking place, in which case the TRE pin drops low.
25	TRO	Transmitter (Serial) Output. All data and control bits in the transmit register are output on this line. The TRO terminal stays high when no transmission is taking place, so the beginning of a transmission is always indicated by the first negative-going transition of the TRO terminal.
26	TB_8	LSB
27	TB_7	
28	TB_6	
29	TB_5	Transmitter input word.
30	TB_4	
31	TB_3	
32	TB_2	
33	TB_1	MSB

(Continued on page 351.)

34	CRL	Control Register Load. Can be either wired permanently high, or be strobed with a positive-going pulse. It loads the programmed instructions (i.e., WLS1, WLS2, EPE, PI, and SBS) into the internal control register. Hard wiring of this terminal is preferred if these parameter never change, while switch or pro-gram control is preferred if the parameters do occassionally change.
35	PI	Parity inhibit. A high on this pin disables parity generation/verification functions, and forces PE (pin 13) to a low logic condition.
36	SBS	Stop Bit(s) Select. Programs the number of stop bits that are added to the data word output. A high on SBS causes the UART to send two stop bits if the word length format is 6, 7, or 8 bits, and 1.5 stop bits if the 5-bit teletypewriter format is selected (on pins 37-38). A low on SBS causes the UART to generate only one stop bit.
37 38	WLS_1 WLS_2	Word Length Select. Selects character length, exclusive of parity bits, according to the rules given in the chart below:

Word Length	WLS1	WLS2
5 bits	low	low
6 bits	high	low
7 bits	low	high
8 bits	high	high

39	EPE	Even Parity Enable. A high applied to this line selects even parity, while a low applied to this line selects odd parity.
40	TRC	16 × Transmit Clock. Apply a clock signal with a frequency that is equal to 16 times the desired baud rate. If the transmitter and receiver sections operate at the same speed (usually the case), then strap together TRC and RRC terminals so that the same clock serves both sections.

(hex) or 3F (hex) to accomplish the job! Then output the word stored in the accumulator to output-port 2 (i.e., make the OUT 2 line high). Of course, any combination can be programmed into the 74100, and the UART responds accordingly. The output lines of the 74100 are latched; they remain at the word programmed.

GENERATING IN/OUT SELECT SIGNALS

Several circuits appearing in this chapter have called for IN or OUT *device select* signals. Such signals are used to turn on/off input and output ports, respectively.

The device SELECT signals are created by the confluence of the correct control signals that indicate the CPU wants that particular port on or off at that instant. In systems using the Z-80 microprocessor chip we will need *input/output request* ($\overline{IORQ}$), either *read* ($\overline{RD}$) or *write* ($\overline{WR}$) and a SELECT/$\overline{SELECT}$ pulse from an address decoder.

Figure 16-17 shows several alternate circuits for creating IN and OUT signals. One method uses our old friend the 7442 *BCD-to-1-of-10 decoder.* We can connect the A, D, and C inputs to the $\overline{IORQ}$, $\overline{RD}$ and $\overline{WR}$ Z-80 signals, respectively. The "D" input of the 7442 is connected to a $\overline{SELECT}$ signal from an address decoder. With this arrangement, we find the following codes appearing at the 7442 inputs when $\overline{IN}$ or $\overline{OUT}$ is called for:

Function	$\overline{\text{SELECT}}$ (D)	$\overline{\text{WR}}$ (C)	$\overline{\text{RD}}$ (B)	$\overline{\text{IORQ}}$ (A)
IN	0	1	0	0
OUT	0	0	1	0

Hence, an input operation, requiring an IN signal, will cause $0100/_2$ (i.e., 4_{10}) to appear on the 7442 input lines. Therefore, we may use the "4" output of the 7442 as an active-LOW IN signal. Similarly, during an output to that same port, the binary word 0010_2 (2_{10}) appears on the 7442 inputs. We may, therefore, use the "2" output (pin no. 3) for an active-LOW $\overline{\text{OUT}}$ signal.

Another method shown in Fig. 16-17 uses the 7485 four-bit magnitude comparator. Recall from earlier that the 7485 compares two four-bit words designated "A" and "B," and issues a HIGH output through pin no. 6 when A=B.

We can apply $\overline{\text{IORQ}}$, $\overline{\text{WR}}$, $\overline{\text{RD}}$, and $\overline{\text{SELECT}}$ to the "A" word inputs of the 7485, and then program the "B" inputs according to what needs to be accomplished. In Fig. 16-17 we use switches

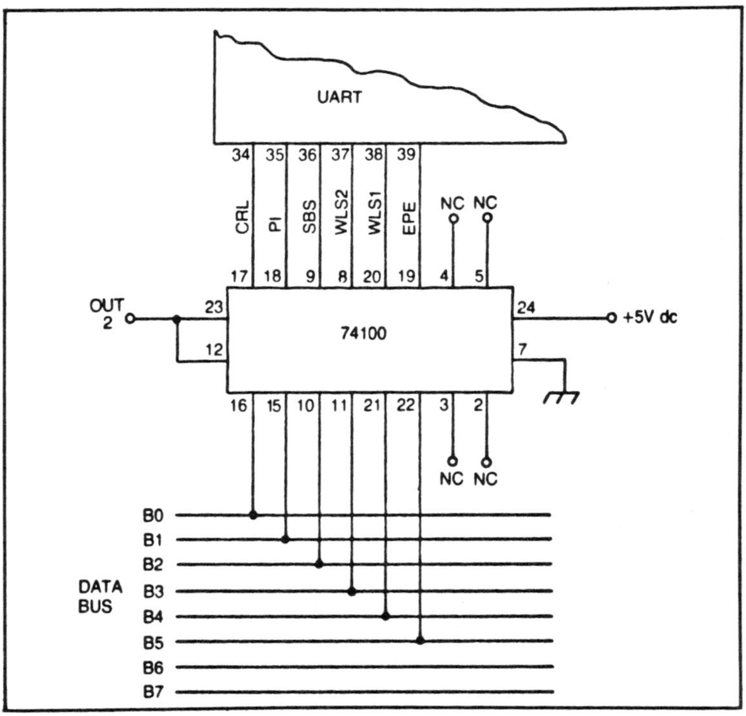

Fig. 16-16. Data latch used for programmable UART.

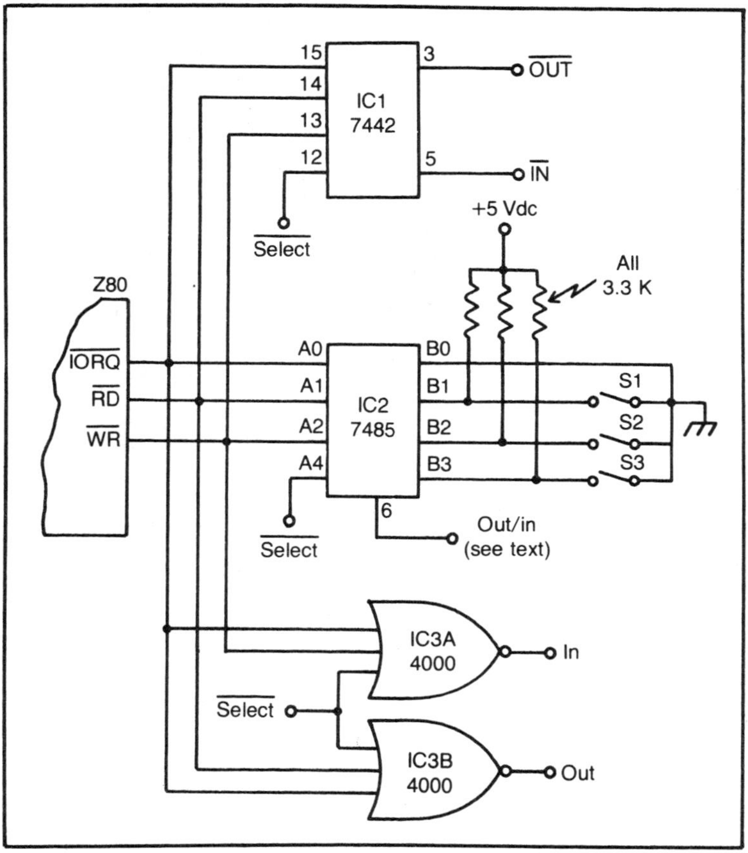

Fig. 16-17. IN/OUT device select signals.

S1-S3 for this purpose. Of course, IORQ will always be LOW during I/O operations, so B0 is wired permanently LOW (or MREQ if a memory operation).

The B1 and B2 inputs of the 7485 determine whether IN or OUT is intended; only one of these bits can be LOW (switch closed) at a time. Close S1 for a Read (i.e., input) operation, or, S2 for a *Write* (i.e., output operation). When S1 is closed, the signal at pin 6 will be IN, but if S2 is closed the pin 6 signal is OUT. Switch S3 determines whether the circuit responds to an active-LOW or active-HIGH address select decoder output. For SELECT LOW leave S3 open, for SELECT HIGH close S3.

Our last method uses a pair of NOR gates to generate IN and OUT. This circuit uses CMOS type 400 IC gates, and is almost self-explanatory. In some cases, a computer will have more than one I/O port, so it will save parts count by generating *System IN* and *OUT* signals, but not with address decoding. Each I/O port will then have its own address decoder and either NAND or NOR gates to turn on the port when the correct address is present *and* either *IN* or *OUT*, as appropriate, is present.

Index

D

M

N

O

P

U

V

W

X